Medical Geography

Medical Geography

Melinda S. Meade
John W. Florin
Wilbert M. Gesler
University of North Carolina at Chapel Hill

THE GUILFORD PRESS
New York   London

© 1988 The Guilford Press
A Division of Guilford Publications, Inc.
72 Spring Street, New York, NY 10012

Printed in the United States of America

Last digit is print number: 9 8 7 6 5 4 3

Library of Congress Cataloging-in-Publication Data

Meade, Melinda S.
  Medical geography.

  Bibliography: p.
  Includes index.
  1. Medical geography. I. Florin, John William
II. Gesler, Wilbert M., 1941–   III. Title.
RA792.M42   1988   614.4′2   87-19673
ISBN 0-89862-781-8

Preface

In many countries throughout the world, medical geography has become a well-established specialization within the field of geography. A course in medical geography is now taught in at least 50 colleges here in the United States and in Canada. Unfortunately, the literature on medical geography is scattered throughout many international journals (although the journal *Social Science and Medicine* has been something of a vehicle for new ideas in the area). The few published collections of medical geography articles and symposia are now quite obsolete. We receive frequent requests for a course curriculum and up-to-date lists of readings. Putting together such reading lists and preparing lectures for unfamiliar aspects of the field is demanding for a teacher and often limiting for students. There has long been a need for a medical geography textbook.

This text has two purposes: (1) to provide a broad-based, comprehensive survey of the rich diversity of medical geography for upper-division undergraduate and graduate students; and (2) to provide a sound reference for the complexities of such matters as biological classifications, chemical pathways, radiation breakdown sequences, and health care systems. Our perspective is holistic and international. We hope it will provide the necessary biological background for geographers to understand disease processes and the necessary geographical background for health researchers to understand spatial processes. We have tried to meet the needs of different levels of scholarly inquiry and variety of technical backgrounds by separating the more technical explanatory material from the text in the form of vignettes (set at the end of each chapter), which instructors can use as they feel appropriate for different classes.

We hope that this text will be a sound foundation for the future development and practice of medical geography and that it will inspire geographers and others to bring their own special subdisciplinary knowledge to enrich and advance this growing course of study.

Finally, we wish to acknowledge the special contribution to this text by three of our former students who labored hard and skillfully for little recompense. The figures were all put in final form and prepared for publication by Margaret B. Pierce, Geraldine A. Roberts, and Judith Waechter Lobe.

Melinda S. Meade
John W. Florin
Wilbert M. Gesler

Contents

Vignettes

ON AIRS, WATERS, AND PLACES

Whoever wishes to investigate medicine properly, should proceed thus: in the first place to consider the seasons of the year, and what effects each of them produces (for they are not at all alike, but differ much from themselves in regard to their changes). Then the winds, the hot and the cold, especially such as are common to all countries, and then such as are peculiar to each locality. We must also consider the qualities of the waters, for as they differ from one another in taste and weight, so also do they differ much in their qualities. In the same manner, when one comes into a city to which he is a stranger, he ought to consider its situation, how it lies as to the winds and the rising of the sun; for its influence is not the same whether it lies to the north or the south, to the rising or setting sun. These things one ought to consider most attentively, and concerning the waters which the inhabitants use, whether they be marshy and soft, or hard and running from elevated and rocky situations, and then if saltish and unfit for cooking; and the ground, whether it be naked and deficient in water, or wooded and well watered, and whether it lies in a hollow, confined situation, or is elevated and cold; and the mode in which the inhabitants live, and what are their pursuits, whether they are fond of drinking and eating to excess and given to indolence, or are fond of exercise and labor, and not given to excess in eating and drinking.

From these things he must proceed to investigate everything else.

—Hippocrates (c. 400 B.C.)

1

1

Sources and Questions of Medical Geography

Medical geography uses the concepts and techniques of the discipline of geography to investigate health-related topics. Subjects are viewed in holistic terms within a variety of cultural systems and a diverse biosphere. Drawing freely from the facts, concepts, and techniques of other social, physical, and biological sciences, medical geography is an integrative, multistranded subdiscipline that has room within its broad scope for a wide range of specialist contributions. Medical geography is both an ancient perspective and a new specialization. As illustrated by the quote from Hippocrates, he (460?–377? B.C.) was familiar with the importance of cultural–environmental interactions more than 2,000 years ago. This ecological perspective on disease and health continued to be philosophically important, even dominant, until the emergence of germ theory in the second half of the 19th century. Thus, the 18th- and 19th-century physicians who first used the term "medical geography" and who struggled in dozens of works to describe and organize the avalanche of new information about human diseases, cultures, and environments, were continuing the holistic Hippocratic tradition (Finke, 1792–1795; Fuchs, 1853; Hirsch, 1883–1886). Their descriptions are being rediscovered and reevaluated by geographers and other scientists once again concerned with disease ecology (Barrett, 1980).

Geographic variation in health has long been studied under such interdisciplinary rubrics as geographic pathology, medical ecology, medical topography, geographical epidemiology, geomedicine, and so forth. Only within the last 30 years has the perspective and methodology of geography been applied to the study of health, disease, and care. The emergence of a systematic interest in medical geography can be dated from the first report of the Commission on Medical Geography (Ecology) of Health and Disease to the International Geographic Union in 1952. Another 15 years passed before the work of pioneering researchers and teachers in a dozen countries resulted in a substantive focus on medical geography in the international community (for the English-speaking world, see especially Stamp [1964] in the British Commonwealth and May [1950, 1958] in the United States).

This chapter describes the evolution of medical geography and its relationship to geographic questions. Definitions of health are considered, and commonly used terms are defined. The sources and nature of the data that most medical geographers use are examined. The chapter finishes with a discussion of the interdisciplinary response to inadequacies in the current medical paradigm and with a description of the challenge to geography posed by the study and improvement of health.

TRADITIONS AND EVOLUTION

Geographers argue endlessly about the nature of their discipline. Some dichotomize geography into the study of physical or human phenomena. They separate, for example, the study of geomorphology or climatology from the study of economic processes or politics. This separation often provides a useful framework for structuring programs of study, but it is simplistic for research purposes, for many geographical subdisciplines pose questions that link physical and cultural dimensions. A more useful organization is defined in terms of the nature of questions posed and the approach to answering them.

In general, most research in geography tries to account for *spatial variation*. The focus is on understanding *why* phenomena are distributed *where* they are across the surface of the earth. This often involves analysis of phenomena (independent variables) thought to be associated with the study (dependent) variable, in order to determine whether their distribution varies spatially in the same way. This tradition emphasizes spatial organization. It uses analysis of pattern to get at the process that generated it.

Given some understanding of the processes that generate locational distributions, some geographers address more applied questions: What is the optimum location for a specific purpose? How can the greatest efficiency in spatial structure (as transportation systems, urban places, communications, or administration) be advanced? In other words, the conceptual and methodological understanding is used and extended in *locational decision making*.

The "man–land" research tradition is concerned with the impact of human activity on the physical environment and what influences variations in the physical environment have on human activity, culture, or biology. In the beginning of the 20th century, many geographers followed a basic premise called *environmental determinism*. The assumption was that the physical environment caused, or at least had a dominating effect on, human environments, activities, and biology. However, the simplistic methodologies of the time could not handle the complexities to which they were applied. There followed a philosophical progression to the idea that a wide variety of human activity developed *within* the constraints of the environment. This was called *possibilism*. Ultimately, environmental determinism and possibilism were replaced by a *behavioral* approach that treated the environment as totally subservient to the will of humans, or ignored it completely. Today, the approach of *cultural ecology* considers humans to be cultural beings whose existence is inextricably interwoven with the environment.

Geographers have more recently researched how we actually interpret, evaluate, understand, and interact with our environment. Studies in *cognition and perception* have shown that what we actually learn of our environment and how we interpret that knowledge are strongly influenced by our particular economic, social, and cultural background: the perception of hazard posed by earthquake, hurricane, and flood varies; the perception of distance varies with consequence for spatial activity. People living in downtown Singapore, for example, may consider a bus trip to the other side of the island to be a very long journey, whereas people in the western United States may routinely drive a hundred miles to see a show. How people organize their space and relate to their environment depends in large part on how they perceive it.

Regional geography has a long tradition. It focuses on integrating all the variable phenomena in order to characterize the special identity of particular places and areas. True integration of the great complexities of culture and environment is the highest expression of regional geography but is seldom achieved. More often geographers work with limited regions created for partic- ular research purposes. Some regions, called *uniform* or *formal* regions, de- limit areas that are homogeneous for certain variables. In this way, the Islamic world or Anglo-America is a region, as is the Cajun dialect area or the grassland biome (vegetational region). Other regions, called *functional* regions, delimit areas within which interaction occurs. Areas in which people patronize a certain shopping mall, read a certain newspaper, or root for a certain ball team are functional regions. The epistemological purpose of regionalizing is the reduction of variance through classification. In other words, grouping all of an area's "like" things together helps organize the cacophony of variation into patterns that are useful for recognizing the underlying processes.

Cartography, the construction and interpretation of maps, holds a central place in geography. Most geographers have a profound love affair with maps. An old saying in the field is, "If it can't be mapped, it's not geography." Although a wide variety of statistical techniques are used in geography, as in other social sciences, maps are a unique, powerful, and flexible tool for analysis of geographical phenomena (see Vignette 1-1). A map is a model of the world that, through the use of point, line, and area symbols, can integrate many dimensions of reality. The development of computer-automated cartog- raphy is revolutionizing the field (Marble, Calkins, & Peuquet, 1984). The geographic information systems that are being created hold profound signifi- cance for the work of medical geography (see Vignette 9-1).

The strands of medical geography have evolved from all these traditions in the larger discipline. They have been variously emphasized in Soviet, German, British, Australian, Indian, French, and North American schools. The many strands are crossing, twining, and bifurcating. The following six questions, however, presently include all areas of research in medical geography (largely paraphrased from Amedeo & Golledge, 1975).

1. *Why is a phenomenon distributed in a particular way?* There is enormous variation in the incidence of diseases on the surface of the earth. Rates for cancer of the esophagus, nasopharynx, stomach, colon, and other sites may vary 100-fold between countries; 10 babies in a thousand die in one country, 300 in another; measles, leprosy, tuberculosis, and malaria are mild in one place and deadly in another. There is also enormous variation in the access of people to physicians, hospital beds, medicines, and other health services.

The first step in addressing why phenomena are distributed in certain ways is to accurately describe where they are located. Traditionally, various forms of mapping have fulfilled this purpose. Today, maps are sometimes abstracted into graphs, or their x and y coordinate systems are used to delimit classes for statistical frequency distributions. Regardless of the means of construction or of use, accurate maps are a valuable first tool of analysis.

National atlases of mortality have been produced in Great Britain, Japan, and other countries. Atlases of mortality from cancer have had a strong impact in the United States and in China. Indeed, atlases of disease occurrence and diffusion such as the *Welt-Seuchen-Atlas* (Rodenwaldt & Jusatz, 1952–1961) have been the most recognized contribution of geography to the health professions. British geographers have been especially vigorous in mapping disease distribution with accuracy and sophistication. The role of statistical mapping and the analytical use of mapping in general are not, however, widely recognized outside of the field.

Some geographers have used aggregated, census-derived demographic and socioeconomic data in multivariate statistical analyses to explain the distribution of such phenomena as physician specialists, bronchitis, suicide, or teenage pregnancy. Some geographers have used microscale, interview, and field-mapped data to illuminate connecting paths between environmental, cultural, and demographic patterns and occurrence of disease or of health care practices. Some have generalized distributional patterns of vegetation and topography to delimit regions of landscape within which certain kinds of disease transmission occur. Whatever the scale or approach, accounting for the spatial distribution of health-related phenomena is the dominant purpose of medical geographic inquiry.

2. *Why are facilities and businesses located where they are?* Why are the offices of physicians, public clinics, or research hospitals located in certain places and not in others? How do the locations of different levels of specialization relate to each other? Why, for example, do facilities to handle heart attacks and those to give cancer radiation therapy have different locational distributions? Are health facilities located in the most efficient places, and can knowledge of the processes behind their locational needs be used to optimize location?

These types of questions were first emphasized in geography in the United States but today are attracting international attention. Whatever the political economy of a country's health service system, there is a need to optimize the location of emergency service facilities and to build expensive facilities like those for dialysis in places that will be accessible to the *future* distribution of people in need.

3. *Why do people move in certain directions for certain distances?* The movement of people over space has inevitably attracted a lot of geographic attention. Population geography studies human mobility at many scales. Economic geography investigates consumer behavior. Medical geography is concerned with how far people will travel to get different health services, and why they go to one place and not another. It is concerned with patterns of human mobility and frequency of contact as these affect the diffusion of contagious disease or the exposure of people to places of disease transmission. Medical geographers have studied the transfer of diseases from Europe and Africa to the New World, the exposure of people to the hazard of schistosomiasis in certain bodies of water at certain times of day, and how far a teenager will travel to get contraceptive information. Why people move as they do is a question basic to understanding health service utilization and the transmission of disease.

4. *Why do innovations spread as they do?* Not only people, but ideas and material goods diffuse across space through time. A few medical geographers have addressed the diffusion of medical technology, such as changes in diagnosis, procedure, nomenclature, or concepts of disease causation. Most have considered the spread of infectious agents analogous to the spread of other innovations and have studied the relationship to settlement systems and activity patterns. How do changes in the size of cities, the density of populations, or transportation links affect the diffusion of disease? Can one use the spatial pattern that results from the spread of a disease in a neighborhood, within an urban system, or across a region, to understand the process? Can one learn enough of barriers to and corridors for diffusion that one can learn to control an epidemic?

5. *Why do people vary in perception of the environment?* If distance means different things to different people, then they will use health care services differently. Indeed, their perception of what causes illness will result in different preventive and curative behavior. When several medical care systems are available, people will choose among them according to their perceptions of efficacy for particular health problems. The occurrence of illness itself will vary, often by ethnic group, according to how the sick role or pain thresholds are defined. Thus, one office worker will be absent to go to the doctor with a sore throat, while another will work with terrible congestion and a mild fever, which he or she does not report.

Just as people vary in their differing perceptions of the hazards of flood, so do they vary in perceptions of the hazards of unboiled water, unbelted automobile accidents, or malaria transmission. Such perceptions affect the material environment as well as human activities and influence the planting of trees, drainage of water, and other alterations of the earth.

6. *How do objects, ideas, processes, and living beings interact to characterize and constitute places?* Some of the medical topographies of the 17th and 19th centuries attempted to answer this question and explain why places were or were not healthy (Chalmers, 1776; Dickson, 1860; Ramsey, 1796). May (1958, pp. 30–32) addressed the problem well in his classic description of how the rice-farming peasants of Vietnam contracted diseases from the way they lived on their land. The regional (country) monographs in medical geography currently being produced by Jusatz and others in Heidelberg systematically consider all the dimensions of environment, population, and health care that determine the health status of populations and their subgroups. In general, however, there have been few attempts to understand how the health status of a certain population in a certain place has resulted from the interaction of the people, their environment, and their culture, or with how it might change in the future.

DEFINITIONS AND DATA

Definitions of Health and Disease

Everyone knows what health is, and yet, a precise definition of it is difficult to come by. This problem is shared by researchers who, in studying health, ironically tend to measure disease. Health, however, is more than the absence of disease. We know that greater health is usually equated with lower mortality and morbidity rates and that health, of course, is a good thing in itself. The problem remains of how to define health without reference to disease.

The first major definition to present health as a positive entity, a presence to be promoted and not merely an absence to be regretted, occurs in the 1946 charter (preamble to the constitution) of the World Health Organization: "Health is a state of complete physical, mental, and social well-being and not merely the absence of disease or infirmity." This influential statement was important for the philosophical position it stated and for the goals it set for government programs and research funding. It has not proved very useful, however, for implementing any standards or research designs that require criteria. It is utopian.

May's definition of disease was for many years referred to by geographers. He stated that disease is "that alteration of living cells or tissues, that jeopardizes survival in their environment" (May, 1961, p. XV). There are several important points in this definition of disease. The organism has an environment to which it relates. The idea that disease jeopardizes *survival* implies that there may be different levels of health without there being disease. An office worker, for example, need not have the physique or eyesight of a hunter. One may be shy or born with a physical handicap and lead a productive life into old age, depending on the society and technology one lives within.

An influential definition comes from Dubos (1965, p. XVII): "States of health or disease are the expressions of the success or failure experienced by the organism in its efforts to respond adaptively to environmental challenges." This definition implies a system whose parts can exist in different states of interaction. Health is not necessarily a condition of physical vigor but is suited to reaching goals defined by the individual. The most important word is *adaptively*. Dinosaurs were highly *adapted* to their environment but could not cope with environmental changes. There is a dynamic quality to health. Dubos's definition, however, defines what health results from, not what it is.

In this book, we adhere mainly to Audy's definition: "Health is a continuing property that can be measured by the individual's ability to rally from a wide range and considerable amplitude of insults, the insults being chemical, physical, infectious, psychological, and social" (1971, p. 142). This definition will be elaborated upon in Chapter 2, but here it should be noted that health is present until death and is a dynamic quality continually engaged in coping with a changing social and physical environment. Audy says that health can be measured, but this has proved very difficult. Reseachers have used many indices and surrogate measures over recent years, and they have filled government reports and suggested criteria, but each definition is limited to its narrow purpose and has never become widely accepted. In the end we are still measuring the absence of disease.

Terminology

A familiarity with some terminology is necessary before the availability and limitations of data on health and disease can be appreciated. Some of the most commonly used terms are presented in this section.

Diseases are referred to as *congenital* when they are present at birth. These may be of genetic origin, as in certain heart defects; they may be acquired in the womb, as with drug addiction or chemical-induced deformity; or they may be acquired during the process of birth itself, as when severe inflammation of the eyes results from passage through a birth canal infected with the bacteria of gonorrhea. Diseases are referred to as *chronic* when they are present or recur over a long period of time, and as *acute* when their symptoms are severe and their course is short. *Degenerative* diseases are characterized by the deterioration or impairment of an organ or the structure of cells and the tissues of which they are a part. *Infectious* diseases result from the activities of living creatures, usually microorganisms, that invade the body. *Contagion*, transmission of infectious disease agents between people, may be direct through person-to-person contact or indirect through the bites of insect *vectors* or via *fomites* or *vehicles* such as contaminated blankets, money, or water.

One way of looking at the continuum of health and disease is illustrated in Figure 1-1. The term *clinical* refers to conditions that have symptoms that can be presented to a physician for observation and treatment. In a *subclinical* condition, an infectious agent may enter the body, multiply, stimulate the production of antibodies, and be eliminated from the body without the person being consciously aware of any illness. Usually the only way that subclinical infections can be detected is through *serology*, or the identification of antibodies and other immune reactions in the blood. Quite a few diseases, such as forms of encephalitis or Rocky Mountain spotted fever, produce acute reactions in only a small proportion of those infected. Other common diseases are very mild infections when acquired in childhood and often pass unnoticed. In either case, public health officials are sometimes startled to find from serology that a "rare" disease has in fact infected the majority of the population.

Sometimes an infectious disease has a period of *latency* or *incubation* between the time when infection occurs and the appearance of clinical symptoms. Although the disease is not manifest, people are sometimes infectious. The common cold, for example, usually has an incubation period from 1 to 3 days, but people may be infectious for 24 hours before their own symptoms appear. Measles has an incubation period of about 10 days until onset of fever; people are then infectious for the 3 or 4 days until the rash appears, as well as for several days afterward. Since disease data usually are not produced until clinical symptoms are diagnosed, statistics at any given time usually underestimate the amount of disease in a population. Problems with time lag, inapparent disease, and time of diagnosis are accentuated when degenerative diseases are studied. The latency period between initial stimulus and the diagnosis of presented symptoms for cancer, for example, is commonly more than 20 years and may vary by several years for individuals, depending at what stage of the disease they are diagnosed.

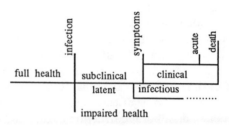

Figure 1-1. The health continuum, illustrating terminology and stages of ill health between full health and death.

A disease is *endemic* when it is constantly present in an area. It may occur at low levels, occasionally popping up here or there, as typhoid does in the United States (hypoendemic), or it may occur with intense transmission, as malaria does in parts of Africa (*hyper*endemic). Sometimes a hypoendemic disease can flare into rapid spread, perhaps in response to the dislocations of warfare or malnutrition from crop failure. Sometimes diseases that are not endemic are introduced and spread rapidly. A disease is said to be *epidemic* when it occurs at levels clearly beyond normal expectation and is derived from a common or propagated source. Epidemic disease may inlcude an outbreak spreading through a population or widespread degenerative diseases, such as lung cancer.

The terms "incidence" and "prevalence" are often confused or used loosely and wrongly. *Incidence* refers to the number of cases of a disease being diagnosed or reported for a population *during a defined period of time*. It refers to new cases. *Prevalence* refers to the number of people in a population sick with a disease *at a particular time*, regardless of when the illness began. Thus, the incidence of tuberculosis in Texas in 1985 refers to the number of new cases diagnosed there that year, whereas the prevalence of tuberculosis in Texas in 1985 includes the total number of Texans suffering from the disease that year. The incidence and prevalence rates of chronic diseases, especially, can be very different.

Vital Statistics and Other Sources

Research findings cannot be discussed or questions be clearly posed without familiarity with the availability and limitations of medical data and statistics. This section describes the main sources of medical information and discusses some of the problems with using them.

Historical information on specific diseases is difficult to obtain. Records that are otherwise useful for demographic statistics are seldom helpful for disease-specific information because there have been major changes in beliefs about disease etiology and classification. Until a century ago illness was often ascribed to imbalance in body humors. Fever or diarrhea, which today are considered merely symptoms of entities usually identified by the microorganism that causes them, were discussed as though they were the disease. Even the best physicians, for example, might have written that malaria "became typhous in its course" or might have failed to distinguish yellow fever and cholera. One can search the medical descriptions to see if particular diseases might have been present in certain times and places, and then say that syphilis was not known in medieval Europe or that smallpox was introduced to the Aztecs by the Spaniards. Total deaths can sometimes be collected from church records or private journals, and these sources are also used to estimate mortality from malaria, scarlet fever, and other diseases that can be identified. For most diseases, disease-specific mortality rates cannot be estimated, however, except possibly for the catastrophic epidemics; and even then several diseases may have been occurring concurrently.

Vital registration provides the most important single source of data. In 1842 Massachusetts became the first state to enact legislation requiring the systematic registration of vital events. In 1880 the United States Census Bureau established a national registration area for deaths. Initially only Massachusetts, New Jersey, the District of Columbia, and a few large cities were included. It was not complete in coverage until the last southern state achieved standards in 1932. During the latter part of the 19th century, as the New York City Board of Health codification of cause of death spread and was more widely used, it became possible to study patterns of mortality. The 1890 *Census of Population* included survey questions on cause of death that, although subject to much error, provide landmark coverage for the entire population.

There is great variation internationally in the reliability of cause-of-death data. In many countries the World Health Organization maintains special registration areas in which cause of death is better diagnosed and reported than elsewhere in the country. The organization's statistics are often based on these areas, which may not be typical of rural areas or remote regions with different ethnic groups. Nevertheless, there is extensive reporting of cause of death in most countries according to the categories of the International Classification of Disease (World Health Organization, 1978).

Even in economically developed countries where mortality is well reported, the only accurate data are those derived from autopsies. Autopsy surveys have demonstrated that cause of death is inaccurately diagnosed much of the time—as much as 20% or 30% for some causes. It is very difficult, for example, to determine cause of death from cancer of the pancreas, the stomach, or other internal organs without an autopsy. At the other extreme, the worst reporting is evident where a large proportion of deaths is ascribed to "senility and ill-defined symptoms." As a general guide, the larger the proportion of deaths in this category the poorer the statistics in general.

Morbidity data are more difficult to obtain and less reliable. Many developed countries, including the United States, maintain national morbidity surveys in which samples of the population are given clinical exams and interviewed at length. It is from such surveys that the prevalence of diseases such as arthritis or lower back pain is estimated. Such surveys, however, are sampled on the basis of demographic and socioeconomic groupings across the nation. Data for specific geographic areas cannot be derived from them. National and usually state reports on morbidity are available for reported diseases. Such morbidity data, however, needs to be regarded with suspicion and interpreted cautiously. For some diseases, such as plague, the data are probably complete; but for many diseases, such as those that are sexually transmitted, they are notoriously inaccurate.

There are several levels of reportable diseases. Plague, yellow fever, and cholera are quarantinable diseases under the *International Health Regulations* and must be reported internationally. Smallpox used to be in this category but has been eradicated from the earth. Even mandatory reports are sometimes not sent because the economic cost can be so severe—from loss of tourism, for example. Case reporting is also required for the diseases under international surveillance: louse-borne typhus, relapsing fever, paralytic poliomyelitis, influenza, and malaria. Beyond these, each nation and usually its states or local communities determine what disease is important enough, or possibly controllable enough, to be reported. Thus, in the United States typhoid or rabies is nationally reportable, but individual cases of food poisoning or the common cold are not.

Geographers interested in specific diseases or in determining health care needs for a variety of diseases, sometimes have recourse to the records of private physicians or hospitals. For some diseases these records can be good sources, if the geographer can get past the confidentiality problems posed by the need for addresses. Some geographers conduct their own surveys and base their studies on the subjects' self-reporting, which is relatively accurate for some conditions and useless as a source for others.

Problems with Reported Rates

The casual reader as well as the researcher needs to be aware of the many sources of error within the neat, authoritative, published morbidity and mortality statistics. These are discussed below in terms of numerator problems, denominator problems, and scale problems. The major types of rates and ratios used for medical statistics are defined in Vignette 1-2.

The numerator of a rate is composed of events, for our purposes the number of deaths or cases of illness from specific diseases. Problems crop up because diagnosis varies over space and time. The types of laboratory tests or special equipment available to one physician and not to another can result in different diagnoses. The school at which a physician is trained inculcates its own procedures and definitions. The degree to which laws mandate or permit autopsies varies. All of these things have spatial bias. In some states the coroner must be a physician, but elsewhere coroners are popularly elected and may have no medical background at all. In some states reports of deaths due to ill-defined symptoms or senility are rejected and autopsies are ordered; in others even large numbers of such reports from certain counties will be accepted without notice.

There are also biases in diagnosis that result from changes in classification over time. The greatest classification change in the United States occurred in 1949. Until that time, cause of death was statistically classified according to the "lethal importance" of the conditions the physician reported to be present.

After that year, the "underlying cause" of death was specified by the physician. For example, after 1949 emphysema leaped into importance in the United States, in a seemingly terrible epidemic, when in fact its significance had been repressed by a nonlethal designation. Reporting of emphysema also exhibits spatial bias; there is great variation in its diagnosis internationally. Most British physicians diagnose "bronchitis" where United States physicans find "emphysema."

The denominator of a rate represents the population at risk of an event. Arriving at a fair count of people presents spatial and time biases. Consider a place in which 1,000 people live on the first of January, 80 die during the year, 50 are born, and 5,000 tourists come to spend the summer. If there were five deaths from motor vehicle accidents, what is the appropriate denominator for determining the death rate from that cause? Conventionally the midyear population is used to balance deaths and births, and this is reasonably accurate if deaths occurred evenly throughout the year and not in one terrible disaster, such as a tornado. But the tourists were part of the population at risk for several months, and a death rate attributed to the resident population alone would be inflated. Furthermore, even an excellent census has its undercounts, overcounts, and biases. A de jure census, such as most developed countries use, attempts to reassign people from where they were at the time of the census to where they habitually and legally reside. Inevitably some people get counted twice. Some people, especially minority ethnic groups in young, mobile categories, escape being counted altogether. Illegal aliens may be counted in one place and not another, but they are at risk of those automobile accidents.

The ages that people report to the census exaggerate the age of older people. Sometimes respondents forget to report the existence of people under five. Everyone, furthermore, seems to like to round to zero, and there is a general phenomenon of "age heaping" at 20, 30, 50, and so on. Death rates could be higher for 31-year-olds as a group because many of them declared themselves to be 30-year-olds, whose rates consequently go down (since the population of those who are 31 is reduced and that of those who are 30 is enlarged while the number of deaths remains unchanged). Demographers use a variety of multipliers and smoothing indexes to remove these distortions, but often for calculation of disease rates the reported census population is used without adjustment as a denominator.

The distortions of scale stem from small populations, geographical variations among areas of data collection, and modifiable units of observation. When a particular population is very small, such as the number of old black males in Appalachian counties, random events occurring to them result in enormous fluctuations and extremes. For example, consider a small town that has twenty 18-year-olds. Five of them are killed one year in a car accident. The accident rate is 250 per 1,000, enough to raise any mean or distort computerized map intervals. Yet, the next year the accident rate may be zero. For this reason it is important to aggregate enough data for small populations to stabilize the fluctuations. The data can be aggregated over time by using 5-, 10-, or 20-year periods or creating running averages. Alternatively, data can be aggregated over space by adding enough small units together to create a stable population. When there is a lot of spatial variation, aggregation over time is better; when there is rapid change, aggregation over space is better. When there is rapid change and diverse conditions spatially, the area must usually be excluded from the larger study.

Another problem with aggregation is related to the existence of spatial variation in conditions. Data are usually available for units that serve administrative convenience but that may mask material geographic differences (see Vignette 1-3). Imagine two adjacent counties through which a river flows. Encephalitis is epidemic to an equal extent in both counties among the people living in the river lowland. One county, however, has a large urban population and the other is mainly rural. The encephalitis cases and the populations are totaled by county unit, giving very different rates of occurrence with no etiological information and misleading information for public health intervention. The alternative to using such conveniently aggregated data, however, is often expensive fieldwork for which funding is scarce.

These terms, measures, and sources of data are the common property of all who investigate health-related topics. Geographers need to be thoroughly familiar with them and their pitfalls if sound research is to be promoted or appreciated.

THE CHALLENGE OF MEDICAL GEOGRAPHY

Place was important to medicine until the middle of the 19th century. For 2,000 years medicine was concerned with geographic variations in air, water, soil, vegetation, animals and insects, diet, habit and custom, clothing and house type, government, and economy. The 19th century saw a paradigm change, a change in the great overarching, all-orienting idea of how disease occurred and what questions were worth asking. In the last century *germ theory*, or what is known as the "doctrine of specific etiology," has resulted in revolutionary advances. The discovery that microbes invade human bodies and cause alterations that result in disease led to asepsis and sterilization, vaccination, antibiotics, chlorination of water and treatment of sewage, and at least 30 more years of life for the average person in a developed country. (See Vignette 1-4)

Specific etiology, or one cause (germ) that is both necessary and sufficient for each disease, is less relevant to a society where people die from heart disease, cancer, kidney failure, alcoholism, and murder. These are diseases of multiple, complex causes based as much in culture as in biology. Even for infectious diseases, germs are no longer considered to be the "sufficient" cause. The tubercle bacillus is necessary to cause tuberculosis, but the disease depends on nutrition, genetics, treatment, the presence of disease conditions, crowding and ventilation, spitting, and mental attitude. The etiology and control of strokes or infertility are even more complex. Yet the progress of specific etiology has been paralleled by the progress of specialization and the increasing divorce of body from mind and environment.

As the contradictions between the dominant biomedical orientation and the health needs of people increased, the social sciences became more involved. The last few decades have seen the development of flourishing concentrations in medical anthropology, medical sociology, medical economics, and psychol-.ogy. Historians are reconsidering the significance of disease for major social and economic changes. They are investigating the role that historical connections between empires and the development of trade has played in the spread of disease or concepts of disease causation. Changes in technology such as glass-making and the invention of the internal combustion engine have profoundly altered the disease maintenance and health care systems. Political factors influence government policy and determine the availability of sewerage systems and potable water as well as the mix of medical systems and the distribution of resources. Such organizations as universities, foundations, and insurance companies attract the attention of political scientists by affecting the quantity and goals of foreign aid and the standards and technology of domestic care. Medical economics has demonstrated the great monetary savings of prevention programs. The problems of technological development and ever-rising medical expenses on one hand and an aging population with uneven economic resources on the other demand that flexible and innovative strategies be developed. Sociologists have demonstrated the importance of class and ethnicity in everything from defining the sick role and choosing a doctor to diet and exercise habits. How far people are willing to travel for care and what care is accessible and acceptable to them are as sociologically relevant as occupational exposures to hazards and life-style changes. Anthropologists have illuminated the many cultural belief systems about disease causation and prevention and the pharmaceuticals and therapy strategies of traditional medical systems. Issues of diet and mental health have received special attention.

Medical geography draws on the concepts and uses the techniques of all these disciplines and adds spatial and ecological perspectives. It exemplifies the interdisciplinary nature of geography, bridging the gap between the social and the physical and biological sciences. Cognate disciplines for training a medical geographer include epidemiology, history, sociology, economics, anthropology, psychology, zoology, entymology, botany, parasitology, meteorology, geology, health administration, environmental engineering, and biostatistics.

The development of medical geography has been part of the response to society's inadequate theory, methodology, understanding, and response in relation to health and disease. The integrative perspective of geography and the questions and methodologies of its various traditions are needed. These words say it best:

The application of geographical concepts and techniques to health-related problems places medical geography, so defined, in the very heart or mainstream of the discipline of geography. I would suggest that there is no professional geographer, whatever his or her systematic bent or regional interest, who cannot effectively apply a measure of his or her particular skills or regional insights towards the understanding, or at least partial understanding, of a health problem. This is the essential challenge of medical geography. (Hunter, 1974, p 3–4)

REFERENCES

Amedeo, D., & Golledge, R. G. (1975). *An introduction to scientific reasoning in geography.* New York: Wiley.

Audy, J. R. (1971). Measurement and diagnosis of health. In P. Shepard & D. McKinley (Eds.), *Environ/mental: Essays on the planet as a home* (pp. 140–162). Boston: Houghton Mifflin.

Barrett, F. A. (1980). Medical geography as a foster child. In M. S. Meade. (Ed.), *Conceptual and methodological issues in medical geography* (pp. 1–15). Chapel Hill, NC: University of North Carolina, Department of Geography.

Berghaus, H. (1852). *Physikalischer Atlas.* Gotha: J. Perthes.

Blunden, J. R. (1983). Andrew Learmonth and the evolution of medical geography—A personal memoir of a career. In N. D. McGlashan & J. R. Blunden (Eds.), *Geographical aspects of health* (pp. 15–32). New York: Academic Press.

Chalmers, L. (1776). *An account of the weather and disease of South Carolina.* London: Charles Dilly.

Dickson, J. H. (1860). Report on the medical topography and epidemics of North Carolina. Philadelphia: Collins.

Dubos, R. (1965). *Man adapting.* New Haven, CT: Yale University Press.

Finke, L. L. (1792–1795). *Versuch einer allgeminen medicinisch—pratkischen geographie* (3 Vols.). Leipzig: Weidmannische Buchhandlung.

Fuchs, C. F. (1853). *Medizinische geographie.* Berlin: Duncker.

Geddes, A. (1978). Report to the Commission on Medical Geography. *Social Science and Medicine,* 12D, 227–237.

Gilbert, E. W. (1958). "Pioneer Maps of Health and Disease in England." *The Geographical Journal,* Vol. 124, part 2, (June, 1952), p. 178.

Hawley, A. (1979). Draft maps of syphilis in North Carolina. Data source: *Community Disease Mortality Statistics, North Carolina, 1979.* Raleigh, N.C.: Department of Human Resources, Division of Health Statistics, State of North Carolina.

Hippocrates. (1886). *The genuine works of Hippocrates* (F. Adams, Trans.). New York: William Wood.

Hirsch, A. (1883–1886). *Handbook of geographical and historical pathology.* (3 Vols., C. Creighton, trans.). London: The New Sydenham Society.

Howe, G. M. (1970). *National atlas of disease mortality in the United Kingdom* (2nd ed.). London: Nelson.

Howe, G. M. (1972). *Man, environment, and disease in Britain.* New York: Barnes and Noble.

Hunter, J. M. (1974). The challenge of medical geography. In J. M. Hunter (Ed.), *The geography of health and disease* (pp. 1–31). Chapel Hill, NC: University of North Carolina, Department of Geography.

Jusatz, H. J. (Ed.). (1968–1980). *Medizinische Landerkunde* (Vols. 1–6). Geomedical Monograph Series. Berlin: Springer-Verlag.

Learmonth, T. A. (1972). Medicine and medical geography. In N. D. McGlashan (Ed.), *Medical geography: Techniques and field studies* (pp. 17–42). London: Methuen.

Marble, D. F., Calkins, H. W., & Peuquet, D. J. (1984). *Basic readings in Geographic Information Systems*. Williamsville, NY: SPAD Systems.

May, J. M. (1950). Medical geography: Its methods and objectives. *Geographical Review, 40*, 9–41.

May, J. M. (1958). *The ecology of human disease*. New York: MD Publications.

May, J. M. (Ed.). (1961). *Studies in disease ecology*. New York: Hafner.

Ramsey, D. (1796). A sketch of the soil, climate, weather, and diseases of South Carolina. Charleston: W. P. Young.

Rodenwaldt, E., & Jusatz, H. J. (Eds.). (1952–1961). *Welt-Seuchen Atlas* (Vols. 1–3). Hamburg: Falk.

Shisematsu, I. (1981). *National atlas of major disease mortalities in Japan*. Tokyo: Japan Health Promotion Federation.

Shoshin, A. A. (1962). *Principles and methods of medical geography*. Moscow: Academy of Sciences.

Sorre, M. (1943). *Les fondements biologiques de la geographie humaine*. Paris: A. Colin.

Stamp, L. D. (1964). *The geography of life and death*. Ithaca, NY: Cornell University Press.

World Health Organization. (1978). *Basic tabulation list*. Geneva: World Health Organization.

Further Reading

Banks, A. L. (1959). The study of the geography of disease. *The Geographical Journal, 125*, 199–216.

Eyles, J., & Wood, K. J. (1983). *The social geography of medicine and health*. New York: St. Martin's Press.

Joseph, A. E., & Phillips, D. R. (1984). *Accessibility and utilization: Geographical perspectives on health care delivery*. New York: Harper & Row.

Learmonth, A. T. A. (1975). *Patterns of disease and hunger*. North Pomfret, VT: David & Charles.

Pyle, G. F. (1979). *Applied medical geography*. New York: Wiley.

Vignette 1-1

MEDICAL CARTOGRAPHY IN HISTORY

Geographers sometimes forget the power of a simple map. Yet much of the early geographic disease studies' impact on our understanding of health stems from the use of maps. To quote an 1852 cholera study,

> Geographical delineation is of the utmost value, and even indispensible; for while the symbols of the masses of statistical data in figures, however clearly they might be arranged in the Systematic Tables, present but a uniform appearance, the same data embodied in a Map, will convey at once, the relative bearing and proportion of the single data together with their position, extent, and distance, and thus, a Map will make visible to the eye the development and nature of any phenomenon in regard to its geographic distribution. (Petermann, 1852, cited in Gilbert, 1958, p. 178)

Modern medical geography (frequently called medical topography during the 1800s) began in Europe during the late 18th century. Investigators, who were usually medical practitioners and not geographers, described a place's topography and climate as they related to health and disease. While disease distributions were often described in detail and the reports sometimes did contain detailed topographic maps, they did not contain disease maps.

The yellow-fever epidemics of the late 18th and early 19th centuries and the cholera outbreaks of the 19th generated the first disease maps. Dot maps of the distribution of yellow-fever victims were used by both contagionists and anticontagionists in their argument over the nature of what caused that dread disease. Contagionists considered it a single disease brought by travelers from places already afflicted with yellow fever, while anticontagionists thought that it simply emerged from crowded, filthy urban areas. Apparently the first such map was produced by Dr. Valentine Seaman in his anticontagionist treatise on yellow fever in New York City in 1798. His work was continued over the next half century by many other people.

In 1852 Heinrich Berghaus published his *Physikalischer Atlas*. One of its eight sections included a number of medical maps and charts. These were the first medical maps included in an atlas, the first to show the distribution of a variety of epidemic and endemic diseases, and the first published by a major cartographer. The extraordinary quality of these maps represents a singularly important development in medical cartography.

The most famous 19th century disease map was John Snow's 1854 dot map of cholera around Broad Street water pump in London (Vignette Figure 1-1). The clustering of cholera in the vicinity of the well supported Snow's contention that cholera was a water-borne disease, with the pump the local source of infection. He urged that the pump be shut off. It was, and the local incidence of cholera declined quickly.

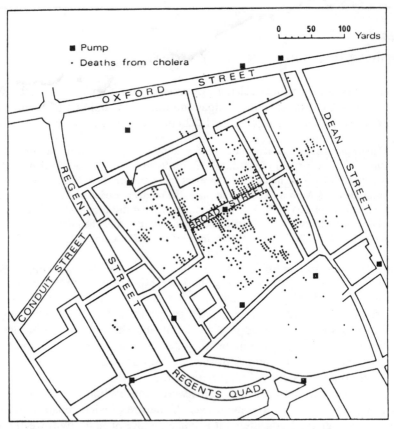

Vignette Figure 1-1. Snow's map of cholera. The affected well is clearly identified by the concentration of cases in its vicinity. Reprinted from Howe, G. M. (1972). *Man, environment, and disease in Britain.* New York: Barnes and Noble Books, p. 178. Copyright 1972. Reproduced by Permission. Original source: Snow, J. (1855). *On the mode of communication of cholera.* London, 1855.

Between 1950 and 1954 Jacques May produced a series of 17 maps of global disease and nutrition distributions. They were published, along with a limited commentary, as supplements to *The Geographical Review.* Many were later published in reduced size in May's *Ecology of Human Disease* (1958). Between 1952 and 1961 Rodenwaldt and Jusatz edited the more comprehensive, three-volume *Welt-Seuchen Atlas* (*World Atlas of Epidemic Disease*) in German and English. The atlas includes global and regional maps (with a special emphasis on Europe) and commentary on the diseases.

Vignette 1-2
RATES AND RATIOS

Rates give the frequency of one thing (numerator) relative to another (denominator) within a given period of time. In demographic (population) studies such as health studies, the numerator is the event happening to the population, and the denominator is the population at risk for the event. As an equation, rate is expressed in the following manner:

$$\text{rate} = \frac{\text{number of events in a given population for specified time and place}}{\text{total population at risk during the specified time in that place}}$$

The better one can specify the population actually at risk for the event, the more accurate and informative the results. When one must use the total population, the rate is referred to as *crude*. With reference to natality, for example, one may calculate the

$$\text{crude birth rate} = \frac{\text{number of births during year}}{\text{total midyear population}} \times K$$

or the better specified

$$\text{fertility rate} = \frac{\text{number of births during year}}{\text{midyear number of women aged 15–45}} \times K$$

Rates are often multiplied by a constant, K (usually 100, 1,000, or 100,000), to facilitate comparisons among places and time periods and, in some cases, to produce whole numbers. It is more comprehensible to speak of 11 babies dying per 1,000 born alive than of .011 deaths per baby.

Specific rates present a more accurate picture of what is occurring, but they require specific categories of data that are not always available. When considering birthrates, for example, it is obvious that men, children, and old people do not give birth—they are not at risk for the event. It was found in one country, for example, that the birthrate, calculated with total population as a denominator, was going down even though most women were improving in health and actually having more children than before: as the number of children increased, their proportion in the total population increased, and the birthrate decreased. In this case the fertility rate is better specified than the birthrate. Specific rates are always preferred.

Ratios describe the proportion of one absolute quantity as compared to another, at a given point in time. Usually the magnitude of the phenomenon is lost. If the standard mortality ratio of Somewhere is 80 and that of Somewhere Else is 160 for a disease, one knows that there is twice as much of it Somewhere Else, but one still does not know whether it is rare or common.

The following rates and ratios are among the most frequently used in medical geography.

RATES

Crude Death Rate (c.d.r.). The number of deaths in an area during some time period divided by the total midperiod population of that area (see Vignette Table 1-2).

Age-Specific Death Rate (a.s.d.r.). The number of deaths of people at a certain age (year or group of years) in an area during some time period divided by the population of that age category in that area (see Vignette Table 1-2).

Standard (age-adjusted) Death Rate. What the crude death rate would be if the age structure of the population being studied were the same as the age structure of the standard, or reference, population (see Vignette 2-1). The standard death rate can be adjusted not only for differences in the proportion of the popula-

Vignette Table 1-2. Rate Calculation of Age-Specific Rates Per 1,000

Age	Population	Diagnosed Cases	Deaths	Morbidity	Mortality
0–1	300	196	15	653.3	50.0
1–9	2,250	286	11	127.1	4.8
10–19	1,700	121	5	71.1	2.9
20–29	1,400	38	2	27.1	1.4
30–39	1,350	29	1	21.5	0.7
40–49	1,200	33	1	27.5	0.8
50–59	1,000	29	2	29.0	2.0
60+	800	59	6	73.8	7.5
Total	10,000	791	43		

	Per 1,000
Crude death rate $\dfrac{(43 \times 1000)}{10,000}$	4.3
Incidence rate $\dfrac{(791 \times 1000)}{10,000}$	79.1
Disease-specific infant mortality rate $\dfrac{(15 \times 1000)}{300}$	50.0
Fatality rate $\dfrac{(43 \times 100)}{791}$	5.4%

tion at various ages but also for different proportions of sex, ethnicity, income, or other classifications.

Infant Mortality Rate (i.m.r.). The number of deaths occurring to children under 1 year of age in an area during some time period, divided by the number of births during that period (see Vignette Table 1-2). The infant mortality rate is frequently used and is valid for comparison among places for three main reasons: (1) It is an age-specific rate and so is not affected by differences in the age structure of various populations; (2) The numbers of births and deaths to infants under 1 year are some of the most widely collected and available data; and (3) Infant mortality is highly sensitive to conditions of both the social and natural environment and rapidly reflects deterioration or improvement of health conditions.

Incidence Rate. The number of cases diagnosed in an area during some time period divided by the total midperiod population of that area (see Vignette Table 1-2).

Case Mortality (Fatality)Rate. The number of people who die from a disease divided by the number of people diagnosed as having that disease, all within a certain time period. It is usually expressed as a percentage and applied to a specific outbreak of a disease in which all patients have been followed for long enough to include all attributable deaths. A disease with a high fatality or case mortality rate (not general mortality rate) is often referred to as "virulent."

Attack Rate. The incidence rate for a disease in a particular group of people during a limited period of time, as during an epidemic. It tells what proportion of the population becomes sick with the disease. The *secondary attack rate* expresses the number of cases occurring within the incubation period of the first case among the sick person's contacts—a measure of how contagious it is.

RATIOS

Standard Mortality Ratio. The proportion of observed deaths in a unit compared to the number of deaths that would be expected there if the age distribution were the same as that of the standard population (see Vignette 2-1).

Relative Risk. The proportion of incidence of a study variable (disease) in a group when the group is classified according to certain exposures, such as those who smoke a pack of cigarettes a day or who do not, or those who live within one mile of a smelter and those who live further away.

Population per Bed. The number of people in an area, usually an administrative unit, for each bed in a hospital.

Population per Physician (or Specialist). The number of people in an area, usually an administrative unit, for each physician. This ratio is often calculated in reverse manner, that is, as physicians per (1,000 or 100,000) population.

LIFE EXPECTANCY

Neither a rate nor a ratio, life expectancy is a mathematical construct. It represents the mean number of years a person would live if he or she were in a hypothetical group (cohort) of people, born at the same time and proceeding through life subject to all the age-specific mortality rates existing at the time of birth. The cohort of population may be started at any age to determine the remaining years of life that can be expected, as in life expectancy after retirement. Because so much mortality befalls small children, people have a longer life expectancy when the cohort is started at 5 years of age rather than at birth. Life expectancy is also a valid measure for comparison among places, as it is based on age-specific rates. It is a hypothetical measure, however, because while real people live, the age-specific rates change, and no one person could actually experience all the present mortality conditions affecting people of every age.

Vignette 1-3

MODIFIABLE UNITS

Geographers often face the problem that the boundaries of the areas they use as their basic units of observation can be changed. Units may be split up or put together to make larger units, or unit boundaries may be shifted in some other way. For example, census tract boundaries may be altered from one decennial census to the next. A political party may attempt to realign voting districts in order to obtain more representation in government. Changes in boundaries often alter the level of analysis, make comparisons difficult, and affect the interpretation of results, as illustrated in the following examples.

The first example shows what might happen when one is looking for the causes of a disease. Suppose there is an industrial chemical being dumped into a river, which causes people who live along the river and drink its water to contract a certain disease. The disease can also be contracted by breathing car emissions or by breathing a heavy metal put into the air by a smelter. Because people who live near the smelter, along the river, or in the town tend to have common socioeconomic characteristics, ethnicity, income, and race may be associated with the disease. If one were to study this disease of unknown etiology at different scales or using different geographic units of analysis, different associations would be identified (see Vignette Figure 1-3a).

Study 1 focuses on a minor civil division where people breathe the air of a smelter (a) and car emissions (b), and so a and b are found to be associated with the disease. Study 2 examines another minor civil division where people drink water from a stream (c) and breathe car emissions (b). Study 3 is carried out in a rather rural, minor civil division wherein people work in a smelter and

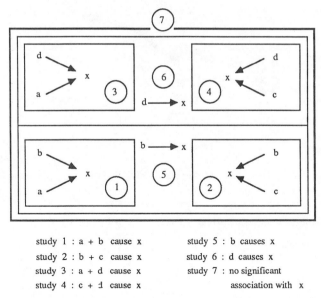

study 1 : a + b cause x   study 5 : b causes x

study 2 : b + c cause x   study 6 : d causes x

study 3 : a + d cause x   study 7 : no significant

study 4 : c + d cause x   association with x

Vignette Figure 1-3a. Disease causation factors at different scales of analysis.

breathe its air (a) and are German in ethnic origin (d). In study 4 the rural minor civil division has ethnic German people (d) who drink the stream water (c). These associations are found at the microscale, but they are puzzling because there is so little consistency. Suppose that study 5 looks at county data, combines areas 1 and 2 and finds that car emissions (b) are important. Study 6 examines county data, combines areas 3 and 4, and finds that being German (d) is important. Finally, study 7 uses state data and combines all these areas but finds no important associations.

The second example of modifiable units deals with health care delivery. Suppose a health systems agency (HSA) consists of 10 counties, as shown in a rough sketch in Vignette Figure 1-3b. Vignette Table 1-3 shows the number of physicians and total population for each county.

Planners in the HSA might wish to organize their counties in different ways and then compare groups of counties to determine if they were relatively well or poorly served by physicians. Suppose they compare a "northern tier" (counties A–E) with a "southern tier" (counties F–J). The northern tier would have an overall physician-to-population ratio of 128.8 and the southern tier one of 171.2. Now suppose that there is an interest in comparing the Standard Metropolitan Statistical Area (SMSA) (counties D, F, and G) with the other, more rural counties. The SMSA has a physician-to-population ratio of 187.5 and the other seven counties one of 84.9. Choosing the way one groups the counties clearly changes the assessment of the resource situation.

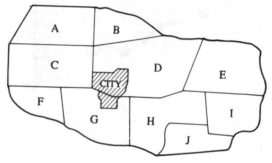

Vignette Figure 1-3b. Counties within a health systems agency jurisdiction.

Vignette Table 1-3. Health System Agency County Physician-to-Population Ratios

County	Number of Physicians	Population	Physician-to-Population Ratio (per 100,000)
A	12	19,122	62.8
B	15	25,639	58.5
C	2	4,553	43.9
D	126	62,798	200.6
E	10	16,014	62.4
F	40	29,528	135.5
G	74	35,670	207.5
H	22	9,608	229.0
I	5	5,926	84.4
J	8	6,321	126.6
Totals	314	215,179	145.9

Vignette 1-4

CARTOGRAPHIC CONSEQUENCES OF DATA CATEGORIZATION

Medical geographers often need to present an array, or statistical distribution, of data. It might be possible to present the entire set of data in a table or a map with the actual value written into each geographic unit. Indeed, if the primary goal is to convey precise statistics, that might be the correct approach. The reader may be overwhelmed by the volume of information, however. Additionally, interpretation of data presented in this fashion is difficult.

The far more acceptable alternative is to reduce the complexity of the data by collapsing it into a much smaller number of classes or categories that can

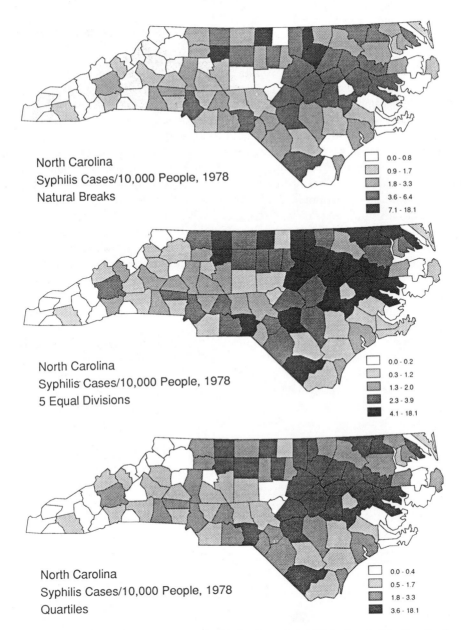

North Carolina
Syphilis Cases/10,000 People, 1978
Natural Breaks

☐	0.0 - 0.8
▨	0.9 - 1.7
▨	1.8 - 3.3
▨	3.6 - 6.4
■	7.1 - 18.1

North Carolina
Syphilis Cases/10,000 People, 1978
5 Equal Divisions

☐	0.0 - 0.2
▨	0.3 - 1.2
▨	1.3 - 2.0
▨	2.3 - 3.9
■	4.1 - 18.1

North Carolina
Syphilis Cases/10,000 People, 1978
Quartiles

☐	0.0 - 0.4
▨	0.5 - 1.7
▨	1.8 - 3.3
■	3.6 - 18.1

Vignette Figure 1-4. Three maps of the distribution of syphilis in North Carolina. Each map is based on the same set of morbidity data. Map A uses natural breaks in the data to define categories. Map B divides the data into five categories of 20 countries. Map C divides the data in quartiles of 25 counties. From maps drawn by A. Hawley, University of North Carolina, Department of Geography, Chapel Hill; and *Community Disease Morbidity Statistics, North Carolina 1978* (pp. 3-1–3-100). Raleigh, N.C.: Department of Human Resources, Division of Health Statistics, State of North Carolina, 1979.

then be easily mapped. Such simplified data sets result in maps that are far easier to digest and interpret.

Unfortunately, the problem of appropriate data classification is a vexing one. There are many different ways to categorize any set of data. If we are dealing with an ordered array of data, that is, data distributed sequentially from high to low, we might wish to categorize the data into four or five groups, each containing an equal number of data units. Alternatively, we could break the array into a number of categories, each containing an equal share of the total variation (for example, data ranging from 20 to 79 could be arranged into four groups of 20–34, 35–49, 50–64, and 65–79). Or, if the data meet statistical requirements of normality, we might identify the mean of the set and create categories by identifying standard deviations above and below the mean. Other statistical categorization schemes basically collapse the array one step at a time, first combining the two data units that are numerically closest and giving the joining a value that is the average of the two, then combining the next closest pair, and so on, until the data have been aggregated into an acceptable number of categories.

Any of these categorization approaches may be justifiably used, and yet each might lead to a mapped pattern quite different from the others. Consider, for example, the distribution of syphilis in North Carolina (Vignette Figures 1-4). Each of these three maps is based on exactly the same set of data. We can see certain underlying patterns, such as the concentration of generally higher values in the coastal plain of the state. Yet the result is three distinctly different maps.

Which map is correct? They all are. Which is the right one to use? That is a more complex question. Its answer is based on the investigator's clear understanding of data and the research problem. He or she must decide which map (or perhaps maps) best portrays the pattern in the data. Using the map that best matches a preconceived notion should be carefully considered against the alternatives suggested by the other maps.

2

The Human Ecology of Disease*

The *human ecology* of disease is concerned with the ways human behavior, in its cultural and socioeconomic context, interacts with environmental conditions to produce or prevent disease among susceptible people. This constitutes the etiology, or causal evolution, of health and disease. Population genetics, physiology, and immunological and nutritional status are important to disease processes and must be understood as a prerequisite to sound research into these processes. Geography is also very important, as its roots are firmly anchored in the study of cultural and environmental interactions.

Geographers have traditionally studied the creation of landscape, the mobility and composition of population, the determinants of economic activity and its location, and the diffusion of things, ideas, and technology. All of these are of consequence to medical geography. The landscape is composed of insects, medicinal herbs, and hospitals as well as topography, vegetation, animals, water sources, house types, and clothing. Mobility is important to exposure to and transmission of disease. Elements of population composition include not only age structure, ethnicity, and literacy, but also immunological and nutritional status and genetic susceptibility. Health service delivery relates to economic activity, as do occupational health hazards. Disease agents and medical technology are subject to diffusion.

The main purpose of this chapter is to establish a conceptual framework for understanding why human disease and health vary over the surface of the earth. Health is defined in terms of adaptability and is related to complex systems of interaction among habitat (environment), population and cultural behavior. These three dimensions form a triangular model of human ecology and underlie disease etiology, consequences, and prevention. Each dimension is considered in turn. A concrete example, ascariasis, and a complex field of study, the ecology of nutrition illustrate the functioning of the model.

---

*In order to avoid confusion, let us state succinctly the difference between human ecology and a term we used earlier, cultural ecology. Human ecology is a broad term used in anthropology and epidemiology, as well as geography, to denote the patterns of human interaction with the physical environment, such as behavior and the environmental interactions of physiological reaction to air pressure or trace elements in water. Cultural ecology is more specific and refers to behaviors and belief systems within a particular culture such as diet, house construction, or hygiene.

HEALTH

J. Ralph Audy (1971) defined health as a "continuing property" that could be measured by the "individual's ability to rally from a wide range and considerable amplitude of insults, the insults being chemical, physical, infectious, psychological, and social" (p. 140). One might prefer the term stimuli, or hazards, to "insults." Such stimuli may be either negative or positive: the crucial thing is that the individual must respond to them.

Stimuli that can be classified in Audy's way may be infectious, physical, chemical, or psychosocial. Infectious insults consist of the pathogens, agents that cause disease. Every person is infected at all times with many billions of viruses, bacteria, and protozoa that cause no harm, such as intestinal bacteria. Changes in health status can cause a normally benign relationship to alter and become pathogenic. We are also constantly stimulated by physical insults, such as electromagnetic radiation. The trauma of tissue damage and broken bones can result from falls and violence. We live in a chemical soup. Our bodies are chemical systems, quite literally composed of what we eat. Petroleum derivatives and nicotine are now part of our chemistry. The absence of an essential vitamin or excesses of a basic food component, such as cholesterol, can also be chemical insults. Mental and social insults further influence physiological functioning.

It is possible to map at a variety of scales every kind of insult. The areas of a town could be mapped based on noise, people's fear of walking down the street at night, air pollution, visual blight or beauty, mosquito density, or alcohol consumption. Such maps could be overlaid to show regions of health hazards. These regions of insults form the environments to which individuals are exposed at the microscale as they move around.

At the microscale are self-specific environments. Everyone is wrapped in an envelope of heat, humidity, bacteria, fungi, and mites and may host lice and fleas. The driver of a car encounters a set of insults that differ by section of the road and the other vehicles around. The infectious and other insults encountered on a bus are quite different from those in the car. Within buildings, one is insulted by microwaves from the walls, magnetism from the electricity, light, changes in humidity, infections from other people, and psychosocial challenges from books, television, and conversation. The exact nature and range of insults that an individual is exposed to during a day is unique. Behavioral roles associated with age, sex, and occupation serve to create some groupings of insults, however, and geographical location delimits other groupings. These differences in exposure to various health hazards can be modeled.

The idea that health is a "continuing property" and not a characteristic that is either present or absent, involves recognition that health exists at various levels. The only absence of health is death. Health can exist at a threshold, marginally, or it can exist amply with great reserves. An insult has a "training" impact. That is, after the body has successfully rallied from the insult, the body is better able to cope with future insults of that kind. One's first public talk, first date, or first exam in graduate school is more difficult to cope with than the 20th. While a person is reacting to a stimulus, however, the level of health is decreased, and that person becomes less able to cope with another insult.

The way insults affect the level of health can most easily be diagramed for immunologic health. In Figure 2-1, two individuals are conceived and born and experience infectious insults. The first becomes infected with a cold virus, and while she is coughing, sneezing, and slightly feverish her health level declines a little. Soon, however, she is immune to that cold virus, and her health rebounds to a higher level. In this way she proceeds through a succession of infectious episodes. Through time, her level of health continues to increase until her early 20s and then gradually declines over the next several decades. The second individual also survives the massive insults that attend birth, but poor maternal nutrition has given her a lower birth weight and level of health. Soon after she rebounds from the cold virus with increased health, she is infected with bacteria that give her diarrhea. She is removed from the food supplements believed to be the cause and even from water, in an effort to stop the diarrhea. Her health level increases slightly as she masters the bacteria, but the episode has precipitated malnutrition because her diet has been marginal. With health lowered by malnutrition, before she can rally and restore herself, she is again assaulted by enteric (intestinal) bacteria. Malnourished, dehy-

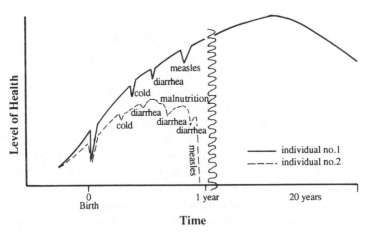

Figure 2-1. Health level and insult stress. The discontinuous solid line indicates health level over a lifetime, with a peak of health around age 20. The part before the discontinuity indicates levels of health for two individuals within their first year.

drated, her health level is greatly decreased. As she struggles to rally and grow, she becomes infected with the measles virus. Depleted of health reserves, she cannot rally and she dies. This latter scenario of multiple insults piling up and cumulatively lowering the health level below the vital threshold occurs frequently in Thirld World countries.

The stress of life events such as marriage and divorce, promotion and being fired, moving, losing a spouse, or having a baby can be scored to predict the likelihood of illness events. It has also been noticed that employee absences due to sickness tend to cluster. Both of these examples of the timing of illness events can be addressed in the terms of the framework that we have developed. Insults require adaptation, and health is lowered during that process, making a person temporarily less able to adapt to the next insult. Around final examination time, for example, many students get sore throats, as the cumulated stresses of little sleep, poor diet, anxiety, and other factors lower levels of health until throat bacteria that have been well controlled for weeks suddenly cause clinical illness.

THE TRIANGLE OF HUMAN ECOLOGY

Habitat, population, and behavior form the vertices of a triangle that encloses the state of human health. *Habitat* is that part of the environment within which people live, that which directly affects them. Houses and workplaces, settlement patterns, naturally occurring biotic and physical phenomena, health care services, transportation systems, and government are parts of the habitat. The following chapters develop aspects of the biotic and physical environment; in this chapter the discussion is limited to the constructed part of the habitat, where humans live and work.

Population is concerned with humans as organisms, as the potential hosts of disease. The ability of a population to cope with insults of all kinds depends on its genetic susceptibility or resistance, its nutritional status, its immunological status, and its immediate physiological status with regard to time of day or year. The effects of age, genetics, and other population components are largely implicit in the remainder of this book, but are essential to all its presentations.

Behavior is the observable aspect of culture. It springs from cultural precepts, economic constraints, social norms, and individual psychology. It includes mobility, roles, cultural practices, and technological interventions. The triangular ecological model differs from sociological models in its separate consideration of behavior and population. Education, for example, is an element of behavior rather than of population status. Education involves behavioral exposures to an opportunity in the habitat, an experience that can influence behavior in a way that improves health status by reducing harmful exposures, increasing protective buffering, and inducing alteration of the habitat itself.

Through their behavior people create habitat conditions, expose themselves to or protect themselves from habitat conditions, and move elements of the habitat from place to place. The habitat presents opportunities and hazards to the population, which can modify its behavior. In doing so, habitat affects population genetics, nutrition, and immunology. The status of the population affects the health outcome from the habitat stimuli and the energy and collective vigor needed to alter behavior and habitat.

It is possible to elaborate upon this basic triangle in many ways. Subsystems can be created for economics, politics, or religion. Motivation systems for behavioral alteration can be analyzed. The effects of changes in major cultural paradigms, such as the purpose and role of humans in the environment, can be isolated and spotlighted as insets.

Habitat

One leg of the ecologic triangle is habitat. It contains sunshine, insects, distance, and many other elements of the physical and biotic environment. Here we consider the environment built by humans.

The Built Environment

Asleep or awake, humans spend most of their lives inside their houses or other buildings. Dubos (1965) points out that evolutionary stimuli now come more from the environment we have constructed than from nature.

What stimuli do you receive in your house? Is the house heated and cooled, so that the humidity is also affected? Is it well ventilated? Do any insects live in the woodwork or basement? Does your dog or cat sleep in the house? Are there windows to let in light? Are there dark corners and rooms? Is there perhaps formaldehyde in the insulation, or lead solder in the pipe joints, or asbestos in the shingles? Do you have radiation sources such as a television or a microwave oven? Or do you have radiation only from electric wiring in the walls? Is the concentration of dandruff, hair, dust, and allergenic materials higher inside or outside your house? What is the noise level? Do you feel crowded or isolated there?

The type of house, the presence of domestic animals, and the kinds of pens and buildings within which they are confined are all of consequence to health. It matters to insect ecology (and hence to disease transmission) whether roofs are made out of thatch or ceramic tile and whether windows are screened. It matters for the survival time and contagion of bacteria whether architecture is oriented toward private, shaded, inner courtyards and interior darkness, or is open with the interior almost continuous with the outdoors. It matters whether kitchens are inside or outside the dwellings and whether there are piped water and flush toilets.

Details matter in the ecology of disease. A certain kind of chimney can cause a room to be smoky, so that mosquitoes are repelled, but eyes are chronically irritated and eventually blinded. Floorboards can be spaced so that food and dirt fall through to be scavenged by pigs and chickens below, or they can be placed tightly together so that one needs to learn to use a dustpan and brush, or they can be carpeted so that fleas can spend their entire life cycle in the living room. Houses can be built out of cold, damp stone and be full of drafts. Alternatively, they can be constructed from mud and straw and provide good nesting sites for insects. Humans create much of their disease environment.

Changes in the built environment can result in profound alterations of disease conditions. We do not know why some diseases disappeared from Europe and others increased during the last few centuries. Leprosy, for example, used to be common in Europe but no longer is endemic. Dubos (1965) and others argue that changes in the built environment, such as cheaply produced window glass and architectural principles that allowed construction of multiple chimneys, flooded even the houses of the poor with light and warmth, drastically changing the habitat for disease agents. In the cities of industrial Europe the construction of dark, unventilated, and crowded tenements provided an ideal habitat for tuberculosis bacteria.

Settlement patterns, the way people are clustered and distributed on the land, also influence health conditions. On a microscale, geographers look within settlements at the spatial arrangement of residences and land uses. Usually three general settlement *forms* are distinguished: nuclear, dispersed, and linear. At the macroscale, geographers look at how the settlements are distributed with regard to each other. A settlement *system* is comprised of various sizes of settlements, including large cities and the distances, directions, and connections among them which form the structure of a functional region of trade, ideas, and other interactions.

The most common setlement form is *nuclear*. Most rural people in the world live in houses clustered in a village from which they walk out to the surrounding agricultural fields, with forest and grassland lying beyond. The settlement land use buffers most households from any insect-transmitted diseases from woods and fields, but the nuclear form facilitates the fecal contamination of water sources and the spread of contagious diseases. Houses in a *dispersed* settlement form are located on the farm land of their owners, and neither air nor water provides much focus of contagion for the scattered population. Each household, however, is exposed to diseases originating in the natural surroundings. A *linear* settlement, in which houses are lined up along both sides of a river, canal, or road, has an intermediary position and often is characterized by the worst conditions of the other settlement forms. People are only partially buffered from insect-transmitted diseases because the rear of the dwelling is exposed; and the clustering of houses provides a focus for contagion, especially for those households downstream from other dwellings.

When a disease agent either produces immunity or kills, there must be a population large enough to support its continuous circulation among susceptible people. Otherwise the disease cannot be maintained, as can be seen when infectious diseases such as colds or influenza are introduced to small island populations today. A few diseases, such as chicken pox (whose viral agent can lie dormant in an immune person for decades before being shed by the disease known as shingles to infect a new generation) have evolved special strategies for surviving in small human groups. Most human infectious diseases could not have existed before the creation of cities transformed the settlement system. Disease agents that mutated from animal diseases, as must have happened many times, found with urbanization a large enough interacting population among the cities in a settlement system to support their continuing circulation. The apparent transfer and adaptation to humans of the AIDS virus from monkeys in Africa is but the latest manifestation of a new, virulent disease being created. A settlement system of several hundred thousand interacting people is necessary to maintain the circulation of measles, for example. That disease is thought to have evolved from canine distemper and to have been a very virulent disease in the ancient Mediterranean world.

Various infectious diseases evolved in separate settlement systems, within which the population over centuries grew resistant. Diseases remained regionalized for long periods of time because people infected with a disease agent either died on the journey or became immune before they could reach another region. Throughout history, the progressive development of transportation has resulted in exchange of disease agents between formerly separated populations. The contact of the Roman and Chinese empires brought a deadly exchange of disease agents, as did the later contact of Europeans and Native Americans.

Today disease agents can easily cross the ocean on a jet airplane. The accelerating mobility of the human population also seems to have created different disease entities by the sheer intensity of transmission that has been made possible, as is illustrated in the discussion of the development of dengue hemorrhagic fever in Chapter 3.

Health services are an integral part of the human habitat. Whether a disease is diagnosed and how it is treated depend partly on whether facilities for urine analysis, X rays, brain scans, and various blood tests are available to the physician. Some health facilities lack electricity and in many ways constitute a very different health habitat than is presented by the presence of a university research hospital. Factors as diverse as international economics and roads washed out by monsoon rains can affect the availability of antibiotics or blood for transfusion. The availability and accessibility of health facilities and health personnel comprise a critical part of the health habitat.

Population

The "nature" of the population, that is, the characteristics, status, and conditions of individuals as organisms, does much to determine the health consequences of any stimulation. Whether the stimulus is a bacterium, light, drug, sound, or thought, the reaction will differ according to the body's biochemical state. This physiology is in part inborn through the genetic code, but it is also influenced by such factors as weather, nutrition, previous experience, and age.

Genetics

Once it became possible to "read" the DNA (deoxyribonucleic acid) sequence of acids and bases that encodes the structure and processes of life, the sciences of genetics and biochemistry expanded explosively. Our knowledge of how genetic information is stored, transmitted, and activated at the appropriate time is undergoing almost daily revision. The DNA chain can be broken in some places but apparently not in others. Pieces can be switched, overridden, deleted. Some information has persisted from ancient times, with no known function today. Some is activated to govern the production of enzymes, hormones, or other proteins only at certain times in life and then is "turned off." The functioning of the genetic code is much more complex, and the encoded instructions much more varied, than had been imagined.

Research in biochemistry and genetics has been overwhelmed by the recent findings on the variability and plasticity of human inheritance. The paired, rod-shaped chromosomes in the nucleus of each body cell contain the paired genes that control heredity. There are specific points, known as structural loci, on the chromosomes for genes governing each characteristic (trait). Genes that occupy the same locus on a specific pair of chromosomes and control the heredity of a particular characteristic, such as blood type, are known as *alleles*. When more than one version of the same trait is common, such as blue and brown eyes or Type A and Type B blood, the population is said to be *polymorphic* for that trait. Humans have long been recognized to be polymorphic for blood type, skin color, hair texture, stature, and other traits that used to be categorized by the concept of "race." There are, at a conservative estimate, more than 50,000 structural loci, and about one third of these are polymorphic. Each individual has two different alleles for about one third of these polymorphic structural loci, or about one tenth of his or her entire genetic inheritance.

One way to illustrate the importance of polymorphism for human health is to consider the histocompatability system (HLA; "histo" refers to tissue) of human leukocyte antigens (white blood cell substances that induce production of antibodies): in short, the genetic control of the body's immune system. The HLA region of the chromosomes has at least four loci. More than 20 alleles may occur at one of these, 40 at another. In total, more than 80 alleles are involved at these four loci, and the possible number of reproductive combinations exceeds 20 million. It has long been known that people have different types of blood, A, B, AB, or O, and that the Rh factor is positive or negative. Currently, more than 160 red blood cell antigens are identifiable. Most have been implicated in blood transfusion reactions and presumably are involved in mother–fetus exchanges. More than a hundred variants of human hemoglobin are also known. This is the type of genetic variability involved in acceptance or rejection of organ transplants, defense against cancer, and resistance to diseases such as malaria or measles.

One can visualize thousands, even millions, of maps of the relative frequencies of genetic traits for antigens and enzymes, hormones, bone structure, skin pigmentation, and other coded instructions for forming human beings. Each mapped surface is continuous over the earth, for almost all of these variants occur everywhere. The frequency of most genetic variations in a population has a spatial gradient (slope), however. The gradients for types of blood antigens, for example, run east–west in Europe, with Type A and Rh factor decreasing to the east. Skin pigmentation and body size, however, have a north–south gradient in Europe.

The manner in which each genetic distribution compares with environmental, cultural, or disease patterns is of great interest. Physical environmental factors such as solar radiation, cultural factors such as livelihoods and marriage ideals, and migration histories have all affected genetic patterns. Geographers have joined the search for associations between blood type and disease susceptibility and between metabolic differences and cultural evolution. Any effort to categorize all distributions, however, has to involve rather arbitrary criteria and great simplification.

The concept of race has become obsolete insofar as it denotes a classification with concrete existence, rather than a convenient categorization for some particular purpose. In social terms it has been replaced by the categorization of "ethnicity," for it has been many decades since behavioral, linguistic, or mental characteristics were scientifically associated with the genetic inheritance of physical traits. Biologically the concept of race is still used occasionally as a convenient categorization for the relative frequencies of many alleles, in full recognition that group boundaries and the traits included are arbitrary.

When race is used as a grouping of alleles to indicate the overall genetic distance or closeness of populations, it sometimes establishes a useful research framework. The Japanese, for example, have different health problems than Europeans and Americans also living in industrialized countries at the same latitudes. The disease experience of successive generations of Japanese in the United States has been studied to help untangle genetic, cultural, and environmental factors in cancer etiology. Anyone doing or reading about research that involves a racial classification, however, needs always to question closely the nature, purpose, and appropriateness of the characteristic used.

Most diseases that tend to be transmitted in families or that occur more frequently in particular population groups result from *genetic susceptibility* rather than genetic causation. That is, the disease requires a stimulus, or "cause," to occur, but another person who is not genetically susceptible will not respond to the same stimulus with disease. Some types of cancer and the virulence of diseases such as tuberculosis, measles, and malaria, are thought to be related to genetic susceptibility. Genetically *caused* disease tends to be associated with rare recessive traits which, while terribly important to the individuals and families involved, are not a concern at the population level except under conditions of inbreeding. Hemophilia among the royal families of Europe is the most famous example of such a disease. Harmful traits in a population that is not inbred tend to be eliminated rather than reproduced, and when they are found to occur frequently in a population, one looks for some advantage that they bestow.

Genetically Based Differences in Metabolism

There is considerable interest today in investigating the genetic base of differences in human metabolism (energy and material transformation within cells) as they interact with culture and health. Lactose intolerance and alcoholism are briefly discussed here.

Lactose intolerance is a classic geographic puzzle in human ecology. Lactose is the only carbohydrate in milk. The enzyme lactase splits lactose into glucose and galactose, which are absorbed into the blood stream as nutrients. Lactase appears in the human fetus in the third trimester of pregnancy, reaches a peak at birth, and falls gradually in childhood. The condition of lactose intolerance, or being unable to digest milk because of an inability to produce lactase, is the usual condition of most adult mammals, including human beings. Among some populations, notably Europeans, the gene for producing lactase does not shut off, and most adults are able to digest milk. People of European descent used to consider this the normal human condition.

Simoons (1970) has proposed a geographical hypothesis of biological and cultural interaction to explain the distribution of lactose intolerance around the world. He estimates that among Asians, about 90% of the population is lactose intolerant. Among Africans the prevalence of lactose intolerance shifts from tribe to tribe, even from village to village, but in general, intolerance prevails. The question raised by his research is whether, for example, the Chinese lost their ability to digest lactose because they defined cattle-keeping as barbaric and consequently excluded from their diet milk and the meat of herd animals (eating the meat of only dooryard scavengers like chickens and pigs), or whether they developed their cuisine because they were unable to digest milk. Similarly, did Caucasians develop a high frequency of lactose tolerance because they became herders of cattle and goats, with mixed farming systems to support their draft animals, or did they take to dairy foods and raise the animals that produced milk because they could digest lactose? In Africa the ethnic groups that herd cattle generally are much more lactose tolerant than the farming groups that do not, but the genetic pattern is very complex as a result of invasions, migrations, and intermarriage. Lactose intolerance has important ramifications for worldwide emergency relief (usually involving powdered milk), agricultural extension and development aid of all kinds, and vitamin D and calcium supplementation.

Studies involving the metabolism of alcohol have shown that Chinese, even when acculturated to United States drinking habits, metabolize alcohol differently than European ethnic groups. A single glass of wine may produce the dizziness, flushing, and nausea that usually characterize much higher levels of alcohol consumption among Europeans. Some ethnic groups, such as Jews, have very low levels of alcholism while others, such as Russians, have rather high levels. Most such differences have been explained in sociocultural terms, as alcohol plays different roles in different societies. But the question arises: Is the different role of alcohol in European and Chinese cultures a result of, or a cause of, differences in metabolism of alcohol? As the study of alcoholism progresses, it is being found that upbringing, life stress, familiarity, and social custom are important, but alcoholism also clusters in families. Even when adopted in infancy the children of alcoholic fathers are more likely to become alcoholic, and identical twins reared apart and brought up differently seem to show similar susceptibility (studies have involved small numbers). Alcohol has been produced for millenia in some cultures to store excess grain and to provide an alternative to polluted water; in other societies it has been little used in these ways. As in the case of lactose intolerance, geographical associations of culture, alcohol, and population may help establish patterns and promote understanding of a serious human health problem.

Other Influences on Population Status

Beyond genetics there are many other inputs into the functioning of the human biochemical system. A most important geographical influence is weather. Humans adjust to differences of temperature and humidity, air pressure, altitude and oxygen level, and hours of daylight by altering their body chemistry, physiology, and metabolism. These changes create different nutritional needs and different internal conditions for infectious agents or for drugs (see Chapter 5).

Age is also a critical factor for health status. Those who study geriatrics (diseases of old age) are just beginning to understand the aging process, but many of the biochemical changes are obvious. Metabolism changes in response to different energy requirements when behavioral roles change with age and growth, maturation, and reproduction are completed. Hormone and enzyme production, organ function, and deposition and storage of fats and chemicals in the bones, liver, and blood vessels all change with age. Experience accumulates, and the immune system recognizes and copes with a greater variety of infectious agents—and is sometimes more sensitized from long exposure to a variety of allergenic substances. The regulation of some homeostatic systems, such as temperature control, tends to become less efficient with age; other systems gain in efficiency, from practice.

The age structure of a population in large part will determine consequences as diverse as the spread of an infectious agent and the severity of the illnesses it causes, the effects of changes in the weather or of an air pollution episode, the expression of carcinogenic agents in clinical disease, and the need for health services. Because age affects so many dimensions of health status, it needs to be accounted for in virtually every study of disease etiology (see Vignette 2-1).

Behavior

Cultural behavior supports the interactive triangle of disease ecology in four ways: (1) Humans create many habitat conditions; (2) Behavior exposes individuals and populations to some hazards and protects them from others; (3) People move not only themselves from place to place but also other elements of disease systems; and (4) Behavior affects the quality of the population by controlling genetics through marriage customs, nutritional status through food customs, and immunologic status through the technology of vaccination and customs of deliberate childhood exposure. Each of these aspects of behavior will be discussed.

Very little of the earth's surface is unaltered by human activity. Vegetation has been burned, removed, and planted. Water has been withdrawn from below the earth's surface and stored and distributed over hundreds of miles on the surface. The chemical composition of the air and water has been altered. The kind of buildings people construct, the kind of industries they work in, the kind of vehicles and roads they build, and the kind and scale of the settlement systems they live in, all form the hazards to which they are exposed. This cultural creation of disease is as true for malaria as for lung cancer.

Besides creating hazards, cultural practices also function to protect people. Many protective customs have been developed. In the Chinese culture realm people habitually drink tea, rather than unboiled water. Many of the Jewish and Moslem dietary practices are protective, and the burial of feces is prescribed in the Old Testament. The European custom of using a handkerchief for blowing the nose is as protective as the Chinese custom of blowing the nose onto and spitting upon the street is endangering. Wearing shoes provides almost total protection against hookworm. Frequent washing of hands, especially after defecation or before handling food, is effective protection from diseases as diverse as the common cold and hydatidosis (invasion of tissue by tapeworms; see Table 3-5).

We have learned to construct buffers against disease deliberately: consider the chlorination of water supplies, the use of seat belts, and the prevalence of the water-sealed toilet. There are cultural origins and diffusion paths for protective buffers, and sometimes diffusion and adoption take time. There are also economic and social class barriers to their occurrence. For example, diarrhea is one of the greatest causes of death in the world. Millions of children die of diarrhea because, although it is known how to protect them, the appropriate knowledge and means never reach them.

Behavioral roles, varying by age and sex and often by ethnicity, largely determine who is exposed to what. There is great cultural variation in social norms. In one culture men are exposed to the infections of the crowded marketplace because women are thought not to understand money, while in another culture women are exposed because men are considered far too unreliable to handle the family money and women are considered superior at bargaining. Who herds animals in pastures and gets hookworm? Who handles dangerous industrial chemicals and gets cancer? Who washes laundry in the morning when *Mansonia* mosquitoes are biting and gets filariasis (a worm infestation; see Table 3-2)? Who tends the orchard in the evening when *Aedes* mosquitoes are biting and gets dengue fever? Who goes to school and gets mumps? Who drives a vehicle and suffers a concussion? Who fights a war and is maimed? Who goes to the city for wage labor and who leaves the home village for marriage?

Mobility patterns are of critical importance for the diffusion and incidence of disease, and they are discussed at some length in Chapter 4. Medicinal plants as well as insects that transmit disease have been spread around the world by humans. When people migrate, they carry customs, adaptive in the place of origin, that are often inappropriate in the new place, such as building in Hawaii a house with a steeply pitched roof to shed snow. Such "cultural baggage" may include ideas about poisonous or nutritious plants. Formerly protective behavior may even be harmful in the new place. A good example is geophagy, the practice of earth eating. In parts of Africa pregnant women eat earth formed into molds and sold in markets. The earth comes from special sites and is often high in calcium and other minerals. It is believed to provide critical nutrients missing in the diet. The practice persists in the United States among a few people of African descent. An old woman in rural Georgia may keep a jar of her favorite clay in her pantry, although she does not need its minerals (inappropriate behavior); and her daughter who has migrated to a northern city may consume laundry starch as a substitute for earth (harmful behavior). With mobility, custom often becomes disassociated from purpose.

It is easy to understand how behavior can affect the genetics of a population. Marriage customs have evolved partly to control genetics. All societies appear to forbid incest, although it is defined in different ways. Some people prohibit the marriage of cousins, but others prescribe it. Some peoples prohibit racial intermarriage, whereas others are far more concerned with class status. Technology, a cultural creation, also affects the quality of the population. In one society people with poor vision make poor hunters and providers; in another, eyeglasses have rendered eyesight an irrelevant criterion for marriage. Health services and technology enable people with genetic traits for diabetes, hemophilia, or deficient immune systems to live and reproduce. The immunological status of a population used to depend almost entirely on its age structure, because experience with infectious agents accumulates over time. Now technology allows even newborns to be vaccinated for diphtheria, whooping cough, and other frequently fatal diseases. The degree of disease resistance of the population thus has become a cultural construct.

The Cultural Ecology of Ascariasis

Hundreds of millions of people are unaware that they have ascariasis. The disease agent is the intestinal roundworm that is adapted to humans, *Ascaris lumbricoides.* The adult worm in the intestine passes eggs into the feces. The eggs embryonate in the soil for about 3 weeks before they are infective. They are eaten by people, perhaps through soil contamination of vegetables, dust contamination of eating utensils, or a child's eating of dirt. The eggs hatch, and the larvae penetrate the gut wall and make their way to the lungs, where they develop in the rich oxygenated tissue for more than a week. They break through into the lung and are coughed up into the mouth and swallowed once again. This time, as adults, they dwell in the small intestine and pass eggs into the feces. A light infection is usually asymptomatic (lacking symptoms). The first awareness of infection may come when live adult worms are seen in the stool or perhaps are passed through the mouth. Heavier infection causes digestive disturbances, abdominal pain, and loss of sleep. Death can (rarely) result from intestinal obstruction.

This life cycle and transmission chain is complex, but well known and medically straightforward. Epidemiologically, small children have the highest rates of infection, followed by mothers who tend small children. It is a disease acquired in the home and its environs. Those who work in the fields or herd livestock may get hookworm, but seldom roundworm.

The cultural and environmental interactions involved in ascariasis are actually quite complex, as Figure 2-2 illustrates. A person infected with the roundworm can be treated with antihelminthic drugs. There may be side effects, but unlike the case with other helminthic (worm) infections of the lungs, liver, or blood vessels, the killed intestinal worms are easily eliminated from the body. Sanitary disposal of the eggs prevents humans from ever contacting them. In some cultures, however, even if the adults construct and use latrines or are wealthy enough to have a flush toilet, the home compound may be contaminated by toddlers because they are not diapered. Although embryonated eggs may remain viable for years, the eggs often encounter adverse environmental conditions in the soil. Temperature affects the time they

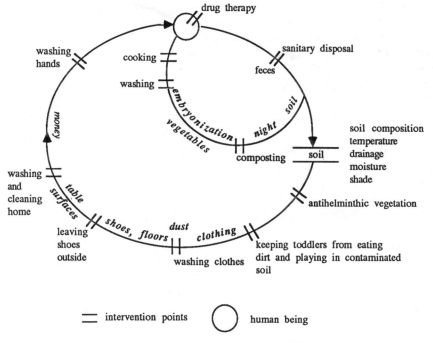

Figure 2-2. The ecology of ascariasis.

take to embryonate. Physical conditions like sun or shade, moisture, acidity or alkalinity affect whether the embryos can develop and how long they survive. There are predators that eat the eggs, and plants that produce chemicals to protect themselves from helminths sometimes affect the human parasites as well. Customs regarding defecation habits are very important. A certain stream bank may be used by everyone, or people may seek privacy in their own orchard or woodlot. Customs differ about how far from the house one must go, and about where small children can play. In some cultures human waste is collected and used for fertilizer. The Chinese were once said to be the most heavily parasitized people in the world because of this practice. Apparently composting at high temperatures will sterilize the eggs; this is done in some places, but not in others. The embryonated eggs get into the human mouth through a variety of paths that involve dust as well as direct contamination. Money or dusty fruit may carry the eggs. The eggs often enter the house on dirty feet and contaminate counters, dishes, and hands. All the customs about not wearing shoes into the house, frequency of washing clothes, floors, and furniture, and habitual washing of hands become important. Sometimes these habits depend not only on custom, but on the physical availability of abundant water for hygiene. When the nearest well in the dry season is more than a mile distant, when the stream used for water is 5,000 feet down the mountainside, or when water must be melted from snow or obtained by hacking through thick ice, floors tend not to be washed very often.

NUTRITION AND HEALTH

The interaction between nutrition and disease is two-way. In themselves deficiencies of minerals and vitamins constitute diseases, but malnutrition also makes the body more susceptible to disease. Antibodies are not produced quickly or abundantly, tissues heal slowly, bones are more easily broken, membranes become permeable. A malnourished person is more likely to get sick, and the course of the sickness is more likely to be severe and drawn out; the level of health is lowered by the malnutrition, and the person is less able to rally from new insults. Sickness may cause malnutrition. The body commonly responds to fever and nausea with a loss of appetite. Feverish people tend to sleep for long periods and not eat regular meals. If diarrhea or vomiting occurs, the body's absorption of nutrients is hindered. Prolonged diarrhea drains the body of salts and disturbs the electrolyte balance. When the body copes with infection, high levels of nitrogen are eliminated in the urine. Even the mild reaction resulting from vaccination will cause nitrogen to be eliminated, so that the body requires protein supplementation for recovery. It is easy to understand how a newly weaned child with borderline nutrition can develop kwashiorkor (a form of severe protein deficiency) after measles, or pneumonia after malnutrition.

Dietary excess can also result in disease conditions. Sugar consumption, for example, is related to the occurrence of diabetes and dental caries. Cholesterol, an animal fat, is related to atherosclerosis. Several salts can cause hypertension.

Regionalization of Diet and Deficiencies

Nutritional hazards can often be associated with cropping pattern, season, or culture realm. Jacques May devoted the last 20 years of his life to studying the ecology of malnutrition and left us a voluminous literature on the subject. He showed appreciation of both the enviromental and cultural side of nutritional patterns.

Environmental parameters of rainfall and evapotranspiration potentials, freezing, heat, soil, evolution, and topography put limits on what can be cultivated. Where droughts are frequent and soils tend to be saline, the various kinds of millets are reliable and yield good protein. Where water is abundant and temperatures warm, rice returns the highest number of calories per acre and provides good nutrition. Winter wheat yields more than spring wheat, but winter wheat usually cannot be grown in Siberia, Manchuria, and much of Canada. Many of our grains evolved in places with long hours of sunlight during the growing season. They either fail to ripen or yield poorly under tropical conditions of an unvarying 12 hours of sunlight. The agricultural patterns of the world formed within these parameters. These are generalizations; rice is grown even in northern Japan, using specially adapted strains. For most of human history 20 inches of rainfall marked the limit of cultivation and the beginning of pastoralism, but in recent years irrigation has greatly extended the amount of arable land.

To these physical parameters must be added the cultural patterns that have formed broad dietary regions (mapped in Figure 2-3). Food crops were domesticated in specific cultural hearths and then spread over paths of cultural contact. The natives of Middle America domesticated corn, squash, beans, tomatoes, chili peppers, and numerous vegetables. They did not domesticate many animals and had no equivalent for the cow, but the collective amino acids in corn, squash, and beans are well balanced and provide complete protein. In the Middle East wheat was domesticated and became the staff of life. It was balanced by the domestication of cattle, sheep, goats, and other dairy and meat sources. In Southeast Asia rice was domesticated, and the fish cultivated in the rice fields provided the balancing protein. In West Africa the cocoyam, oil palm, and a variety of vegetables, such as okra and watermelon, were domesticated.

The original domesticates continue to constitute the core regional diets. Each diet is balanced and potentially sufficient for people who can afford to get enough of it. Only the domesticated starch roots, especially manioc (otherwise known as tapioca or cassava), are so deficient in anything but calories that they constitute a poor diet. For most people the cereal grains and vegetables such as soybeans or chickpeas are the major sources of protein, although medieval Europeans and their Mediterranean predecessors developed a mixed farming system that used grain cultivation and animal husbandry to mutually support each other and yield high-quality protein. Diversity and variety in diet, however, are everywhere necessary to provide the full range of nutrients. A diet limited to staple foods is often the lot of the poor. Regional diets are thus often associated with specific nutritional deficiencies. Table 2-1 lists human nutritional needs, sources of nutrients, and deficiency diseases.

The best known of the staple-associated deficiency diseases is *pellagra*. The disease is first characterized by a distinctive butterfly-shaped rash on the back, which is easily reversed by consuming the vitamin niacin. At advanced stages it results in nerve damage and insanity. Maize, or corn, has little niacin. The way the Native Americans prepared maize made the most of its niacin, but it needs to be supplemented with niacin-rich foods. In the United States South, settlers' contact with the Cherokees was prolonged. Maize came to have a very

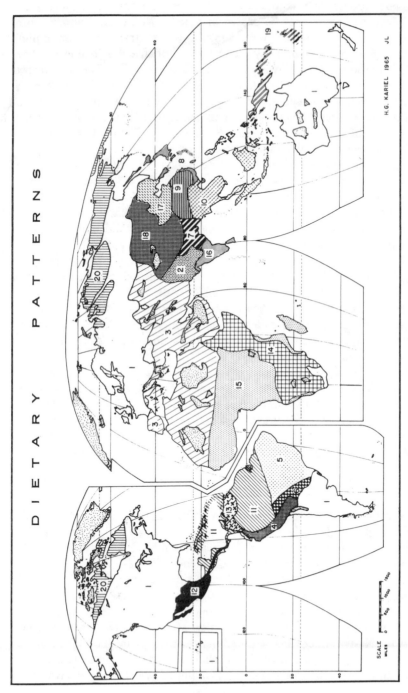

Figure 2-3. Worldwide dietary patterns. Classification of dietary regions is by the major sources of both calories and protein. From "A Proposed Classification of Diet" by H. G. Kariel, 1966. *Annals of the Association of American Geographers, 56,* pp. 74–75. Copyright 1966 by Association of American Geography. Reprinted by permission. (Key is on facing page.)

important role in the immigrants' diet. By the end of winter there was often little left for the poor to eat except cornmeal, a situation that became more common in the late 19th century when King Cotton became even more important than vegetables in the home garden. A niacin-deficient diet resulted. The majority of patients in southern mental institutions in the 1930s were pellagra victims. As poor people in parts of Africa, South America, and Asia turn to maize for most of their calories, pellagra is predictable.

A more recent deficiency association is that of *beriberi* with rice. Beriberi, a disease involving weakness, paralysis, and eventual heart failure, results from the lack of thiamine, needed for carbohydrate metabolism. Rice has some

DIETARY PATTERNS

	Major source(s) of calories[1]	Important source(s) of protein[2]
1	Wheat, potato, sugar, meat, fats and oils	Beef, pork, mutton, dairy products
2	Wheat, millet and sorghum, barley, rice	Dry beans, dry peas, chickpeas, lentils
3	Wheat, maize, barley, rice, fats and oils	Beef, pork, mutton, dry beans, dry broad beans, chickpeas
4	Wheat, maize, barley, potato	Dry beans
5	Wheat, maize, rice, sugar	Beef, dry beans
6	Wheat, maize, cassava	Beef, dry beans
7	Rice	Dry beans, dry peas
8	Rice, wheat	Fish, soybeans
9	Rice, maize, sweet potato	Pork, fish, soybeans, peanuts
10	Rice, maize, sweet potato, coconuts, cassava	Fish, soybeans, peanuts, dry beans
11	Rice, maize, bananas, yams, cassava, sugar	Dry beans, dry peas
12	Maize	Dry beans
13	Maize, wheat, potato	Beef, dry beans
14	Maize, millet and sorghum	Dry beans, dry peas, dry broad beans, chickpeas, lentils
15	Millet and sorghum, maize, rice, yams, cocoyams, sweet potato, cassava, bananas	Dry beans, dry peas, peanuts
16	Millet and sorghum, rice, cassava, coconuts	Fish, dry beans, lentils, peanuts
17	Millet and sorghum, wheat, maize, potato	Pork, mutton, soybeans, peanuts
18	Barley	Dairy products, mutton, goat
19	Cassava, yams, taro, bananas, coconuts	Fish, pork
20	Animal fats, wheat	Fish, local game animals

[1] Major sources derived from starchy staples and other sources where applicable.
[2] In addition to that protein which is derived from grains.

Table 2-1. Human Nutritional Needs, Sources, and Deficiencies

Nutrient	Major Body Function	Major Sources	Deficiency Disease
Balanced amino acids	Precursors of structural protein, enzymes, antibodies, hormones	Good sources: meat, fish, dairy products, legumes Adequate sources: rice, corn, wheat Poor sources: cassava, yams	Kwashiorkor
Fatty acids	Cell membrame function, regulation of digestion and hormones	Vegetable oils	Skin lesions
WATER-SOLUBLE VITAMINS			
B-1 (thiamine)	Removal of carbon dioxide	Organ meats, whole grains, legumes	Beriberi (nerve changes, weakness, heart failure)
B-2 (riboflavin)	Energy metabolism	Widely available	Lip cracks, eye lesions
Niacin	Oxidation-reduction reactions	Liver, lean meats legumes (can be formed from amino tryptophan)	Pellagra (skin rash, nerve and mental disorders)
B-6 (pyridoxine)	Amino acid metabolism	Meats, vegetables whole grains	Irritability, convulsions, kidney stones
Pantothenic acid	Energy metabolism	Widely available	Fatigue, sleep disturbances
Folacin	Nucleic acid and amino acid metabolism	Legumes, green vegetables, whole grains	anemia, diarrhea
B-12	Nucleic acid metabolism	Muscle meats, eggs, dairy products (not in plant foods)	Pernicious anemia, nerve disorder
Biotin	Fat synthesis, amino acid metabolism	Legumes, vegetables, meats	Fatigue, depression
Choline	Precursor of neurotransmitter	Yolk, liver, grains, legumes	Not known
C (ascorbic acid)	Maintains cartilage, bone, dentine; collagen synthesis (= about 30% of body protein)	Citrus fruits, tomatoes, green vegetables	Scurvy (degeneration of skin, teeth, blood vessels)
FAT-SOLUBLE VITAMINS			
A (retinol)	Constituent of visual pigment; maintenance of epithelial tissue	A or pro-A (beta-carotense) in green vegetables, dairy products	Xerophthalmia, night blindness, blindness
D	Growth and mineralization of bone, absorption of calcium	Cod-liver oil, eggs, fortified dairy products	Rickets (bone deformity), osteomalacia in adults

Table 2-1. (*continued*)

Nutrient	Major Body Function	Major Sources	Deficiency Disease
E	Antioxidant to prevent cell-membrane change	Green leafy vegetables, seeds, shortenings	Anemia
K	Blood clotting	Green leafy vegetables	Internal hemorrhages, severe bleeding
MAJOR MINERALS			
Calcium	Bones and teeth, blood clotting, nerve transmission	Dairy products, dark green vegetables, legumes	Stunted growth, rickets, osteoporosis, convulsions
Phosphorus	Bones and teeth, body pH balance	Dairy products, meat, poultry, grains	Weakness, demineralization of bones
Potassium	Body pH balance, water balance, nerve function	Meats, milk, fruits	Muscular weakness, paralysis
Choline and sodium	pH balance, water balance, formation of gastric juices	Common salt	Muscle cramps, apathy
Magnesium	Activates enzymes, protein synthesis and nerve transmission	Whole grains, green leafy vegetables	Growth failure, weakness, spasms behavioral disturbances
Iron	Constituent of hemoglobin and enzymes in energy metabolism	Eggs, organ meat, meats, legumes, whole grains, green leafy vegetables	Iron-deficiency anemia
Fluorine	Bony structure and teeth	Water, seafood	Higher frequency of tooth decay
Zinc	Constituent of many enzymes, tissue healing, digestion	Widely distributed	Growth failure, small sex glands, slow healing
Copper	Constituent of enzymes of iron metabolism	Meats, water	Anemia
TRACE MINERALS			
Iodine	Constituent of thyroid hormones	Marine fish, dairy products, vegetables (depending on soil)	Goiter
Selenium	Association with vitamin E	Seafood, meat, grains	Anemia, cardiovascular disease
Chromium	Glucose and energy metabolism	Fats, oils	Impaired ability to metabolize glucose
Manganese, molybdenum, cobalt, silicon, vanadium, tin, nickel	Constituents of enzymes; functions poorly known in humans; essential to animal nutrition	Organ meats, fats; some widely available	Not reported in humans

Note. Adapted from "The Requirements of Human Nutrition" by N. S. Scrimshaw and V. R. Young, 1976. *Scientific American, 235,* p. 168–170. Copyright 1976 by *Scientific American.* Adapted by permission.

thiamine in the husk. When rice is milled at home with mortar and pestle, some husk remains in the finished product. As part of economic development, power mills have spread along new railroad lines and roads into formerly rural and isolated areas. Milling rice by hand is arduous and time consuming, and the pretty, white, polished mill-rice is greatly preferred. The white polished rice also stores better. The polishing almost completely removes the thiamine, however, so that with no other changes in the population's diet, beriberi has recently spread in almost epidemic form. Just at plowing time, when the need for carbohydrate energy is high, beriberi appears. In Thailand, enrichment of rice at the mills has proved to be an answer to the deterioration of nutritional ecology.

Pastoralists tend to have ample protein but inadequate B vitamins. The pastoralist who eats few green leafy or yellow vegetables and no butterfat is often short on vitamin A. For children one common result of vitamin A deficiency is blindness, because the cornea becomes softened (keratomalacia). Historically, African pastoralists traded meat and milk for vitamin B–rich grains from sedentary farmers and for vitamin A–rich palm oil from the inhabitants of chronically protein-deficient rain forest areas.

Kwashiorkor, an African word for a protein deficiency disease, is most common in the wet tropics. Kwashiorkor occurs even when calories are adequate. There are many reasons for the protein deficiency of the wet tropics, including leaching and low nitrogen content of the soils and the dietary importance of starchy root crops like cassava and taro. Although some dwarf, resistant breeds of cattle have been developed and there are vigorous internationally supported breeding programs, over the centuries trypanosomiasis (sleeping sickness) made it impossible to raise livestock in large areas of Tropical Africa. This has meant loss of animal protein from milk and meat. Fish are an important source of protein, especially for places near the sea. Dried fish made into a sauce is a high-protein condiment used across tropical Asia and traded far inland.

Rickets, a bone-deformity disease, used to be common in high-latitude countries. The sun's ultraviolet radiation stimulates the body to produce vitamin D, which is necessary for calcium metabolism. Wrapped in heavy clothing when outdoors and spending most of the low-light winter period indoors, growing children did not get enough sunlight. The situation became worse in the high-rise, coal-polluted, dark industrial cities. Fish liver is one of the few dietary sources of vitamin D, and so cod liver oil became a common tonic. The vitamin is fat soluble, and eventually the happy solution (for lactose-tolerant populations) was hit upon of adding the vitamin to milk, where it could be ingested with the calcium that needed metabolizing.

Famine also has its regional patterns. Starvation can claim the poor even in well-fed lands, and crop failure, flood, earthquake, and hurricane can strike anywhere. Hundreds of local famines have been recorded in Europe in the last few centuries. Repeatedly during the 20th century China's sorrow, the Huang

(Yellow) River, escaped its levees to spread across the North China plain and bring famine to tens of millions. Famines tend to occur in areas of unreliable rainfall. Cultures can adapt to almost any level of rainfall as long as it is reliable. The great percentage variations in annual rainfall on the arid margins of cultivable land have long been associated with periodic famine. Ever worse famines are predictable as agriculture in these areas intensifies and their population grows. The grasslands and marginal agricultural lands of the African savanna (Vignette 3-1), land under the rain shadow of the Western Ghats and margins of monsoon penetration in India, and the degraded scrublands of northeastern Brazil are all famine-prone.

Beyond any environmental influences on food supply, however, looms the power of cultural preference, prescription, and prohibition. Swine would fit in well with Middle Eastern agriculture, but there are religious injunctions against eating pork. Valuable though cattle are for draft power, milk, dung, and hides in India, the herds would be better fed and stronger if they were culled for food, but religion proscribes it. Even the poorest slum dwellers in the United States could get quality protein from the excess dogs killed daily by the animal pounds. The majority of the human race eats tasty dog meat, but in Western civilization it is taboo. Many peoples have vitamin A and iron deficiencies simply because they do not want to grow vegetables, not because of inadequate land. Some people refuse their children eggs because they are believed to cause worms. Most cultures have strict rules about what pregnant women, sick people, or infants can and cannot eat. Sometimes these prescriptions promote good nutrition, and sometimes they hinder it. There is the widest possible range of opinions about what constitutes good food supplementation for a nursing infant, and what children should eat after being weaned. Some infant mortality and kwashiorkor are caused by the lack of income to purchase anything but the local starchy staple; but much of it is caused by ignorance of what an infant needs for nourishment.

World dietary patterns are undergoing many changes. In the last few centuries, crops and animals have spread from their hearths to be incorporated into other dietary regimes. Potatoes have gone from South America to Europe, wheat and soybeans to the Americas, maize to China. Food storage and processing, transportation and finance, and types of ovens and cooking fuels have changed. Local famines resulting from isolated valley floods, hurricanes, or insect outbreaks have largely been eliminated by regional and national economic and political cooperation as new scales of organization have developed. International trade, refrigeration, veterinary science, and the development of new protein drinks, supplements, and enrichment targets have changed world nutrition. All of these nutritional changes affect the health status of populations: because of radical changes in diet, the Japanese have grown taller, and as people in the United States change their diet, they die less frequently from heart disease.

Distribution

The issues of food distribution can be studied and dealt with at every scale. The purchasing power of the Netherlands and Japan allows them to buy the food they need in the international market, while countries like Bangladesh, which are far more self-sufficient in food, are considered to have a population problem. In the 1980s issues of social equity have become important at the national scale as countries struggle with the worldwide recession and occupational restructuring. For example, the United States exports food to earn a large part of its foreign exchange yet has malnourished poor and elderly. At the family scale, a household may have adequate food, but if it is customary for the male head of the family to eat first and get his choice of any eggs, fish, or special foods, then the dietary intake of pregnant women and small children may be inadequate.

Measurement of the availability of nutrients is difficult. Methods used include 24-hour recall, maintenance of food diaries, the weighing and sampling of foods as they are cooked, and the less intrusive feeding of a "ghost" visitor— an additional plate at table, which can be emptied daily by an investigator. All methods have their drawbacks and deficiencies. Individual and household studies are expensive. One result is that small-area data for nutrition are almost nonexistent. For some purposes state-level marketing data may be adequate, but few data of value are available for medical geographical research in nutrition.

CONCLUSION

Morbidity and mortality vary greatly around the world, not only in quantity but also in the particular diseases involved. Women in Ghana die of malaria, women in the United States die of breast cancer; children in Bolivia die of diarrhea and pneumonia, children in Italy die of automobile accidents. It is more valuable to understand disease processes however, than to construct great mental catalogues and memorize disease patterns as though they were so many imports and exports.

The customs, beliefs, and behavior that characterize each global culture realm and local ethnic group create the environmental conditions and exposure patterns that result in geographic distribution of health and disease. Genetics, to be sure, often underlies susceptibility and resistance, but the distribution of genes is also a result of adaptation to environment, population mobility, and cultural selection. Influences of the physical environment, such as radiation and trace elements, and insect and animal communities and their habitat are critical; but they are buffered and even formed by human agency. Entirely natural landscapes scarcely exist.

Medical geography *explains* the distribution of health and disease and identifies efficient ways to intervene and distribute trained personnel and technology. Every disease has its cultural ecology, its geographic regionalization, and its patterns of diffusion and change.

REFERENCES

Alland, A., Jr. (1970). *Adaptation in cultural evolution: An approach to medical anthropology.* New York: Columbia University Press.

Allison, A. C. (1954). Protection afforded by sickle-cell trait against subtertian malarial infection. *British Medical Journal, 1,* 290–294.

Audy, J. R. (1971). Measurement and diagnosis of health. In P. Shepard & D. McKinley (Eds.), *Environ/mental: Essays on the planet as a home* (pp. 140–162). Boston: Houghton Mifflin.

Bias, W. B. (1981). Genetic polymorphisms and human disease. In H. Rothschild (Ed.), *Biocultural aspects of disease,* (pp. 95–131). New York: Academic Press.

Dubos, R. (1965). *Man adapting.* New Haven, CT: Yale University Press.

Florin, J. W. (1971). *Death in New England: Regional variations in mortality.* Studies in Geography No. 3. Chapel Hill, NC: University of North Carolina, Department of Geography.

Howe, G. M. (1972). *Man, environment, and disease in Britain.* New York: Barnes & Noble.

Hunter, J. M. (1973). Geophagy in Africa and in the United States. *Geographical Review, 63,* 170–195.

Kariel, H. G. (1966). A proposed classification of diet. *Annals of the Association of American Geographers, 56,* 68–79.

May, J. M. (1958). *The ecology of human disease.* New York: MD Publications.

May, J. M. (1977). Deficiency diseases. In G. M. Howe (Ed.), *A world geography of human diseases* (pp. 535–575). London: Academic Press.

Miller, L. H. & Carter, R. (1978). Innate resistance in malaria: a review. *Experimental Parasitology, 40,* 132–146.

Newman, J. L. (1977). Some considerations on the field measurement of diet. *The Professional Geographer, 29,* 171–176.

Race, R. R., & Sanger, R. (1975) *Blood groups in man* (6th ed.). Oxford: Blackwell.

Scrimshaw, N. S., & Young, V. R. (1976). The requirements of human nutrition. *Scientific American, 235,* 50–64.

Simoons, F. J. (1969). Primary adult lactose intake and the milking habit: A problem in biological and cultural interrelations. *American Journal of Digestive Diseases, 14,* 819–836.

Simoons, F. J. (1970). Primary adult lactose intolerance and the milking habit: Part 2. A cultural historical hypothesis. *American Journal of Digestive Diseases, 15,* 695–710.

Simoons, F. J. (1974). Rejection of fish as human food in Africa: A problem in history and ecology. *Ecology of Food and Nutrition, 3,* 89–105.

Simoons, F. J. (1976). Geographic perspective on man's food quest. In D. Walcher, N. Kretchmer, & H. Barnett (Eds.), *Food, man, and society* (pp. 31–53). New York: Plenum Press.

Smith, C. J., & Hanham, R. Q. (1982). *Alcohol abuse: Geographical perspectives.* Resource Publications in Geography. Washington, DC: Association of American Geographers.

Stock, R. (1976). *Cholera in Africa.* London: International African Institute.

Vermeer, D. E. (1966). Geophagy among the Tiv of Nigeria. *Annals of the Association of American Geographers, 56,* 197–204.

Waterhouse, J., Correa P., Muir, C., & Powell, J. (Eds.). (1976). *Cancer incidence in five continents.* Lyon: International Agency for Research on Cancer.

Watterlond, M. (1983). The telltale metabolism of alcholics. *Science 83*, 4, 72–76.
Watts, E. (1981). The biological race concept and diseases of modern man. In H. Rothschild (Ed.), *Biocultural aspects of disease* (pp. 1–25). New York: Academic Press.

Further Reading

Dando, W. A. (1976a). Man-made famines: Some geographical insights from an exploratory study of a millennium of Russian famines. *Ecology of Food and Nutrition. 4*, 219–234.

Dando, W. A. (1976b). Six millennia of famine: Map and model. *Proceedings of the Association of American Geographers, 8*, 29–32.

Fabrega, H., Jr. (1974). *Disease and social behavior: An interdisciplinary perspective.* Cambridge: MIT Press.

Gary, L. E. (1977). Sickle-cell controversy. In A. S. Baer (Ed.), *Heredity and society: Readings in social genetics* (2nd ed., pp. 361–373). New York: Macmillan.

Knight, C. G., & Wilcox, R. P. (1976). *Triumph or triage: The world food problem in geographical perspective.* (Resource Paper No. 75-3). Washington, DC: Association of American Geographers.

Logan, M. H., & Hunt, E. E., Jr. (Eds.). (1978). *Health and the human condition: Perspectives on medical anthropology.* North Scituate, MA: Duxbury Press.

May, J. M. (1961). *The ecology of malnutrition in the Far and Near East.* New York: Hafner.

May, J. M. (1963). *The ecology of malnutrition in five countries of eastern and central Europe.* New York: Hafner.

May, J. M. (1965). *The ecology of malnutrition in middle Africa.* New York: Hafner.

May, J. M. (1966). *The ecology of malnutrition in central and south-eastern Europe.* New York: Hafner.

May, J. M. (1967). *The ecology of malnutrition in northern Africa.* New York: Hafner.

May, J. M. (1968). *The ecology of malnutrition in French-speaking countries of West Africa and Madagascar.* New York: Hafner.

May, J. M. & McLellan, D. L. (1970). *The ecology of malnutrition in eastern Africa and four countries of western Africa.* New York: Hafner.

May, J. M., & McLellan, D. L. (1971). *The ecology of malnutrition in seven countries of southern Africa and in Portuguese Guinea.* New York: Hafner.

May, J. M., & McLellan, D. L. (1972). *The ecology of malnutrition in Mexico and Central America.* New York: Hafner.

May, J. M., & McLellan, D. L. (1973). *The ecology of malnutrition in the Caribbean.* New York: Hafner.

May, J. M., & McLellan, D. L. (1974). *The ecology of malnutrition in eastern South America.* New York: Hafner.

May, J. M., & McLellan, D. L. (1974). *The ecology of malnutrition in western South America.* New York: Hafner.

Vignette 2-1

AGE STANDARDIZATION

Few things can be so misleading as comparisons between populations with different age structures. Because the very old and the very young have particularly high mortality rates, differences in the proportion of the population in these categories alter the number of deaths occurring in the population and so alter the crude death rates. Differences in the possibility of disease occurring are not accurately reflected.

Consider the crude death rates in three countries: Country A has 4.3 deaths per 1,000, B has 6.0 per 1,000, and C has 4.3 per 1,000. It appears that the population of country B is far worse off. Vignette Table 2-1a shows the age

Vignette Table 2-1a. Age-Specific Death Rates for Three Counties

County A				County B			
Age	Population	Deaths	Age-Specific Death Rate (per 1,000)	Age	Population	Deaths	Age-Specific Death Rate (per 1,000)
0–1	300	15	50.0	0–1	600	30	50.0
1–9	2,250	11	4.8	1–9	3,500	17	4.8
10–19	1,700	5	2.9	10–19	1,950	6	3.0
20–29	1,400	2	1.4	20–29	1,400	2	1.4
30–39	1,350	1	0.7	30–39	1,050	1	0.9
40–49	1,200	1	0.8	40–49	700	1	1.4
50–59	1,000	2	2.0	50–59	500	1	2.0
60+	800	6	7.5	60+	300	2	6.7
Total	10,000	43	c.d.r. = 4.3/1000	Total	10,000	60	c.d.r. = 6.0/1000

County C			
Age	Population	Deaths	Age-Specific Death Rate (per 1,000)
0–1	300	9	30.0
1–9	2,250	7	3.1
10–19	1,700	4	2.4
20–29	1,400	3	2.1
30–39	1,350	3	2.2
40–49	1,200	2	1.7
50–59	1,000	4	4.0
60+	800	11	13.8
Total	10,000	43	c.d.r. = 4.3/1000

structure and the age-specific death rates, however, and it can be seen that country B has the same mortality experience as country A. The difference is the proportion of its population that is at risk of the higher rates. In contrast, country C has the same crude death rate as country A, and an identical population structure, but its health conditions are characterized by higher mortality rates for the elderly and lower ones for infants.

One can compare mortality among these three countries by using the age-specific death rates. If individual years of age were used instead of 10-year groupings, however, or if more countries were involved, the tables would soon become too complex for comprehension. A single, summary figure is needed, one similar to the crude death rate but which takes into account differences in age structure.

This summary, comparative figure is the *standard death rate*. To calculate it, a standard population is needed. The age-specific death rates for all the populations are then applied to the age structure of the standard population. This answers the question, "How many deaths would there be if the study populations, given their mortality experiences, had the same age structure?" The total number of expected deaths is then divided by the total standard population to create a summary death rate that is valid for comparison.

Sometimes it is difficult to identify an appropriate standard population. Usually if one is comparing subdivisions of a whole, such as states of the union or counties of a state, the total population by race or sex, as appropriate, is used. A world standard and separate standards for Europe and Africa are available (Waterhouse *et al.*, 1976, p. 456). Sometimes in fieldwork there is no appropriate standard population available, and recourse is made to adding the study populations together and using the total as a standard. The choice of a standard population needs to be carefully considered with respect to the research question.

Vignette Table 2-1b illustrates the calculation of standard death rates. In this example the three populations were added together to make the standard. If B had been standardized on C's population or by some other combination, the result would have been different. The general formula for direct age adjustment is

$$\frac{\Sigma \; M_x \times P_x^s}{\Sigma \; P_x^s},$$

Where M = mortality rate, P = population at risk, x = age category, and s = standard populations, so that M_x is the mortality rate of population age x of the study population, and P_x^s is the population of age x in the standard population.

To calculate the directly adjusted death rate, one needs to know the age-specific death rates of the study population. It sometimes happens that one knows the total number of deaths, but not the ages of the deceased. If age-specific deaths are available for a standard population, the death rates may be

Vignette Table 2-1b. Direct Age Standardization

Age	Standard Population (A + B + C)	Population A		Population C	
		Age-Specific Death Rate	Expected Deaths	Age-Specific Death Rate	Expected Deaths
0–1	1,200	.0500	60.0	.0300	36.0
1–9	8,000	.0048	38.4	.0031	24.8
10–19	5,350	.0029	15.5	.0024	12.8
20–29	4,200	.0014	5.9	.0021	8.8
30–39	3,750	.0007	2.6	.0022	8.3
40–49	3,100	.0008	2.5	.0017	5.3
50–59	2,500	.0020	5.0	.0040	10.0
60+	1,900	.0075	14.3	.0138	26.2
Total	30,000		144		132

Crude death rate per 1,000 = 4.3
Standard death rate per 1,000: population A = 4.8, population C = 4.4

indirectly standardized. The direct process is reversed. The age-specific death rates of the standard population are applied to the study population's age structure to answer the question, "How many deaths would there be if the study population, given its age distribution, had the mortality experience of the standard?" The observed deaths are then compared with the expected deaths to create the standard mortality ratio. If this ratio of mortality experience is multiplied by the crude death rate of the standard population, a standardized death rate results. The indirectly standardized death rate is frequently used by geographers because age information for death is often not available at the microscale. The standard mortality ratio that is calculated by the indirect method is convenient for comparing the spatial patterns of common and rare diseases in the same place. The direct method of age standardization is statistically more valid, however, and should be used whenever practicable.

In the example in Vignette Table 2-1c, P is a part of country C. There was a total of 11 deaths observed in P, but the ages of the deceased are unknown. The age distribution of P's population is known, however, and there are vital rates available at the national level. It should be noted that 11 deaths is too small a number to justify use in further analysis, as it is subject to random fluctuation. Data for P would have to be aggregated over a longer period of time.

The general formula for indirect age adjustment is

$$\frac{D}{\Sigma\ M_x^s \times P_x} \times \frac{\Sigma\ M_x^s \times P_x^s}{\Sigma\ P_x^s},$$

where unknowns are as above and D = observed deaths in study population.

Vignette Table 2-1c. Indirect Age Standardization

Age	Population P	Population C Age-Specific Death Rate	Population P Expected Deaths
0–9	70	.0300	2.10
1–9	250	.0031	.78
10–19	225	.0024	.54
20–29	360	.0021	.76
30–39	220	.0022	.48
40–49	150	.0017	.26
50–59	120	.0040	.48
60+	105	.0138	1.45
Total	1,500		6.85

Deaths occurring in population P = 11

$$\text{Standard mortality ratio (s.m.r.)} = \frac{\text{observed deaths}}{\text{expected deaths}} = \frac{11}{6.85} = 1.6$$

Standard death rate per 1,000 = s.m.r. × c.d.r. of standard = 1.6 × 4.3 = 6.9

Vignette 2-2

DEMOGRAPHIC BASE MAPS

Populations are seldom evenly distributed across geographic space. Variations in quality and quantity of agricultural land, the intensity of its utilization, and the irregular pattern of urbanization are a few of the influences that create often extremely uneven population distributions. Geographic base maps (maps whose areas are proportionate to land area) may thus give undue weight to sparsely settled areas and insufficient emphasis to areas of dense population, creating an incorrect expression of a geographic distribution.

This problem can be alleviated by using demographic base maps, or population-by-area cartograms. Cartograms are maps in which some of the usual geographic qualities, such as size, shape, or contiguity, have been ignored so that the areal units can be transformed to be proportional to some other quality. Thus, if size were proportional to total population, maps of mortality, morbidity, or health-resource availability would more accurately reflect the underlying population base of those distributions. If the research focus is on a condition affecting some subset of the population, for example, the availability of obstetric-gynecologic specialists for females of childbearing age, then that subject population might be the more appropriate cartogram data base.

A major problem associated with nearly all demographic base maps, especially for the cartographic neophyte, is difficulty in identifying specific locations. The map simply does not look right. We are so used to the geographic base map that any different base creates confusion. Howe (1972) and others have met this problem by maintaining the geographically "correct" boundary for the entire study area and mapping the data through a series of geometric shapes located over the appropriate data sites and varying in size depending on the base population (Vignette Figure 2-2a). Different geometric

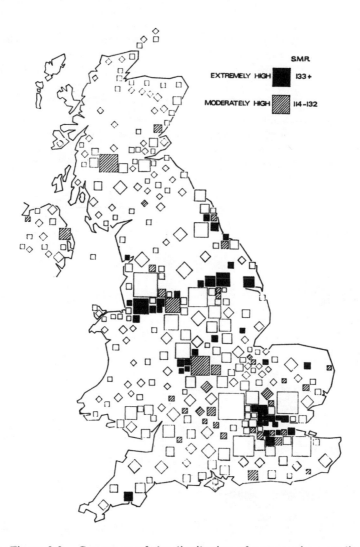

Vignette Figure 2-2a. Cartogram of the distribution of pneumonia mortality in the United Kingdom, 1959–1963. From *Man, Environment, and Disease in Britain*, p. 237. New York: Barnes & Noble Books, Nelson. Copyright 1972 by G. M. Howe. Reprinted by permission.

shapes allow inclusion of another piece of data, such as average income. An alternate approach is to maintain the proper geographic shape of each data unit but vary its size. This might be appropriate for a map of an area familiar to the audience, such as a map of the distribution of a state's infant mortality (where the demographic base could be the total number of children under age 1), to be presented to a state agency. Both of these map types leave large blank spaces (which many cartographers dislike) and may make identification of contiguities difficult.

A more widely used alternative adjusts the area of each data unit while maintaining proper contiguities and attempting to maintain relative shape and location (Vignette Figure 2-2b). Such maps are usually constructed by hand using either graph paper or a large set of small blocks. By manipulating these, visually satisfying maps can be created. The value of this form of cartogram may simply be that it eliminates the blank spaces. These cartograms are most appropriate for studies of movement, such as travel to distant medical specialists, or the diffusion of disease, where the identification of relative location and contiguities is important.

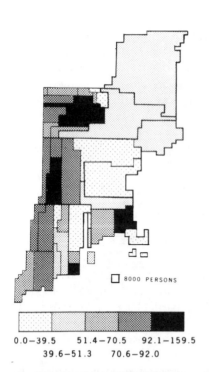

Vignette Figure 2-2b. Cartogram of New England diphtheria mortality rates, 1859–1861. From *Death in New England* (p. 97) by J. W. Florin, 1971, Chapel Hill, NC: University of North Carolina. Copyright 1971 by J. W. Florin. Reprinted by permission.

☐ 8000 PERSONS

| 0.0−39.5 | 51.4−70.5 | 92.1−159.5 |
| 39.6−51.3 | 70.6−92.0 | |

RATE PER 100,000

3

Landscape Epidemiology

The landscape that distinguishes a place is a complex expression of physical, biotic, and cultural processes. When one knows how to analyze its elements and patterns, one can usually determine what diseases can occur. This is true at every scale, from a house and its backyard to the transcontinental migratory paths of birds. As world population grows and the world economy changes, landscapes are being altered in ways that increase risk of disease or enhance protection from it.

This chapter is about the human ecology of transmissible disease, or the ways in which regions impart pattern to disease distribution. Before taking up the elaboration of disease systems and their regionalization, however, basic terminology and system parameters must be established. Then the human ecology of several diseases—yellow fever, bubonic plague, malaria, dengue hemorrhagic fever, and schistosomiasis—will be discussed at length. Each of these diseases is of major importance not only to human health but also to the advance of scientific understanding about the nature and occurrence of different types of transmission systems. Finally, the use of landscape epidemiology to regionalize the occurrence of, and locate points of intervention in, vector-borne disease systems will be summarized.

REGIONS

There are four types of regions involved in the study of landscape epidemiology. The first is biotic. Climate, altitude, and latitude interact to create broad biotic regions, *biomes*, with predictable locations (see Vignette 3-1). These include not only plants and animals, but also particular types of insects and microbes. If the biome's natural fauna and flora are displaced through land development, their replacements tend to be crops and domestic animals equally as distinctive to that biome.

The second type of region is also biotic: *realms of evolution*. Because oceans, deserts, and great mountain ranges formed barriers to the exchange of genetic information as the continents drifted apart, complexes of plants and animals followed separate paths of evolution in South America, Africa south of the Sahara, Asia south of the Himalayas and Yangzi (Yangtze) River, Australia, and in the remainder of Eurasia (Palearctic) and North America (Nearctic). These realms of evolution include microbes and insects as well as the more visible mammals and birds that popularly demarcate the regions. As trade and political empires established contacts beyond their culture realms, the biotas of the evolutionary realms were exchanged. Deliberately, in the case of food crops, and unintentionally, in the case of disease agents, sweeping processes of global readaptation were set in motion. Historians have only relatively recently paid attention to the health consequences that followed upon connection of these realms (McNeill, 1977). When the maritime and land-based transportation and trade networks of the Roman empire connected with those of the Chinese empire, for example, terrible epidemics swept the Mediterranean world. The best studied connection of evolutionary realms is that between Eurasia and Africa and the Americas. Sometimes called the Columbian Exchange, the connection involved the exchange of scores of crops and animals and the introduction to the New World of malaria, yellow fever, typhus, bubonic plague, cholera, typhoid, smallpox, scarlet fever, schistosomiasis, river blindness, and other pestilence. The consequences for the Native Americans were catastrophic. The AmerIndians who greeted Columbus at Santo Domingo were extinct within a century. Although it has proved difficult to estimate the population of the Americas before European contact, it is likely that before the first English settlers arrived more than half the Native American population had been exterminated by European diseases, spreading rapidly across the continents. It is probable that one disease traveled the other way: syphilis was introduced to Europe after the first voyage of Columbus.

The third type of region, *culture realm*, is a broad cultural area delimited by the extent of particular cultural practices and beliefs that largely originated in the primary culture hearths (see Vignette 3-2). China and the United States, for example, occupy vast territories with the same latitude and often similar climatic patterns, but because their inhabitants use the land differently and have different settlement forms, house types, and technology, their landscapes are distinctive. The use of human waste for fertilizer in the Chinese culture realm (which includes Korea, Japan, and Vietnam) and the use of dogs for herding sheep in the Euro-American cultural spheres are examples of regional patterns associated with health consequences: intestinal parasites and dysentery in the former, and hydatid disease (see Table 3-5), tapeworm, and numerous flea- and tick-transmitted diseases in the latter.

The fourth type of region, *natural nidus* (focus or cluster), has been defined by Soviet and German geographers. A natural nidus is a microscale region constituted of a living community, among the members of which a disease agent continually circulates, and the habitat conditions necessary to maintain the disease system. The life cycles and transmission chains of nidal diseases are complex. Several examples of different types of systems are described in this chapter.

TRANSMISSIBLE DISEASE SYSTEMS

In order of complexity, a disease's causative organism, the *agent* (pathogen, germ, or parasite), may be a virus, a rickettsia, a bacterium, a protozoan, or a worm. The type of disease agent is important for understanding the possibilities of intervention in the disease cycle. Outside of living tissue viruses cannot reproduce and have limited viability (life expectancy). Pharmaceutical drugs have little effect on them. The best weapon against viruses is a vaccine. Rickettsias cannot live free in the environment, but in other respects they are similar to bacteria. Rickettsias and bacteria both are sensitive to antibiotics, which means that infected people can be treated. Mass treatment programs based on a single injection are sometimes possible. The more complicated organisms, protozoa and especially helminths (flukes, filaria, and intestinal worms), are difficult to treat on a mass basis because powerful drugs that have many adverse side effects must be used. Treatment of individuals, such as returned United States tourists, is usually possible.

The organism infected by a disease agent is called the *host*. When animals are the ordinary hosts, the disease is known as a *zoonosis*. A disease that often infects both people and animals is an *anthropo-zoonosis*. When animal hosts serve as a continuing source of possible infection for human beings, our anthropocentric name for them is a *reservoir*. A disease endemic to an animal population is said to be *enzootic*, and one epidemic among animals is said to be *epizootic*. Some diseases are transmitted directly by contact or inhalation. Others are transmitted through contaminated food or water. This chapter is mainly concerned with disease agents that are transmitted by an arthropod.

The arthropod that transmits the disease agent between hosts, and in which the agent multiplies and often goes through life-cycle changes in form, is known as a biological *vector*. "Biological" distinguishes the vectors from flies or inanimate objects that may merely transport the agents mechanically (as vehicles or fomites). Many people speak of vectors as "insects," but ticks, mites, and other arachnids are not, strictly speaking, insects. The more inclusive category is the phylum, arthropoda. *Arbovirus* includes all arthropod-borne viruses. The major vectors are mosquitoes, biting blackflies, ticks, mites, sandflies, fleas, and lice; but gnats, midges, and other arthropods are sometimes involved. Most disease agents are strictly limited to transmission by a single species, or at most a genus, of vector. This establishes limits to their distribution because vectors have specific habitat requirements.

Intermediate hosts are organisms that are necessary to some stage of the agent's life cycle. Hosts in which the agent attains maturity or its sexual life stage are the *primary* or *definitive* hosts; hosts for the agent's asexual or larval stage are intermediate hosts. The fluke that causes schistosomiasis, for example, must alternately infect people and snails. The snails do not transmit the agent to people; they are not vectors. Nevertheless, it is often convenient to treat intermediate hosts as vectors, because the same kinds of population dynamics and intervention strategies are involved. Sometimes this concept is even further enlarged, as when dogs are treated as "vectors" of rabies to people. Such careless usage may be helpful for mathematical modeling of a disease system from our anthropocentric viewpoint.

Transmission Chains

The first chain of transmission diagrammed in Figure 3-1 involves direct communication of the agent by contact between people. Animals also communicate contagion (chain 2). It is possible for people to get a zoonotic infection, for example, through contact with skin fungi or consumption of raw milk.

Chain 3 illustrates a vectored human disease having no animal reservoir. The absence of a reservoir raises a possibility that does not exist for most anthropo-zoonoses, disease eradication. Although birds, reptiles, and monkeys have their own forms of malaria, for example, human malaria is solely a human disease transmitted by *Anopheles* mosquitoes. There was therefore

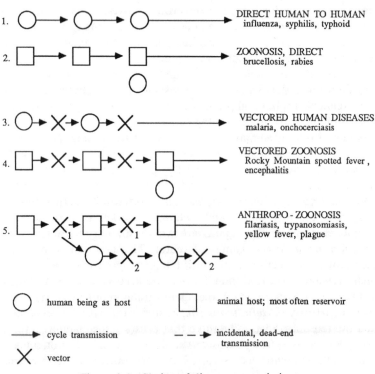

Figure 3-1. Chains of disease transmission.

some hope in the 1950s that it could be eradicated. We shall consider later some of the reasons that it could not.

Most vectored diseases involve transmission of agents among animals (chain 4). Many of these cycles are so ancient that agent and host have mutually adapted and no disease symptoms appear in the infected hosts. Sometimes people are accidentally infected when they intrude into the animal habitat. Usually they will develop no disease at all, as they will not be a compatible host. Sometimes they will be so susceptible to the rare infection that case mortality rates will be high. People are usually "dead-end" hosts, meaning that they cannot transmit the agent to a vector for transmission to another host. For example, the amount of virus circulating in the blood is known as *viremia*. When people are incidentally infected with viral encephalitis by a mosquito that had previously fed on a sick horse or chicken, they will not develop viremia sufficient to serve as a source of infection for another mosquito imbibing a little of their blood. Encephalitis cannot, therefore, be spread among people by mosquitoes. Even when clusters of cases occur, they result from exposure to the zoonotic chain.

The final chain in the diagram represents the main focus of this chapter, anthropo-zoonoses, which can be transmitted between animals or between people by vectors and which sometimes cross between the two systems. The first disease for which this complicated system was discovered was yellow fever.

Yellow Fever

Until the 20th century, yellow fever was one of the most feared diseases in the Western Hemisphere. The step-by-step discovery of its transmission chain, with the subsequent recognition of the implications for many other diseases, ranks as one of the milestones of medical science (Figure 3-2).

Pandemics (international epidemics) of yellow fever swept the Americas in the 18th century. The pandemic of 1793–1804, for example, started in Grenada and spread through the Lesser Antilles and Venezuela, finally striking ports from New Orleans to Boston. During this Napoleonic period, the French suffered losses of more than 30,000 men in their garrisons in Guadaloupe, Martinique, and Santo Domingo. By the 19th century yellow fever was endemic from Brazil to Yucatán and throughout the Caribbean; it was seasonally epidemic along the Atlantic seaboard. Then the United States Yellow Fever Commission determined that the infectious agent (a virus, now known to be a group B togavirus) was transmitted by the mosquito *Aedes aegypti*. Using this first proof of arthropod transmission of human disease agents, General Gorgas attacked the mosquito in Cuba, thereby suppressing the endemic focus. Soon after, suppression of yellow fever allowed construction of the Panama Canal. After the 1905 pandemic was stopped in New Orleans by anti-*Ae. aegypti* operations, the newly established Rockefeller Foundation undertook the task

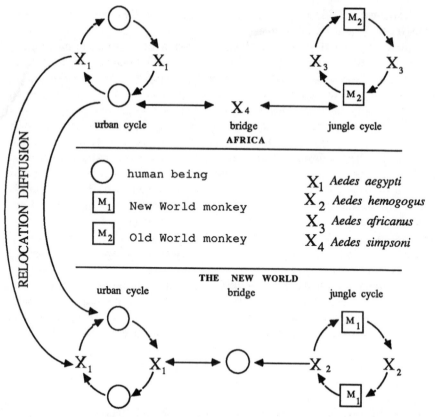

Figure 3-2. Yellow fever transmission chain.

of eradicating *Ae. aegypti* throughout the Caribbean. Partly because the mosquito was an introduced species that was only superficially established, the operation was a success. Yellow fever appeared to have been eliminated.

Then came isolated reports of yellow fever—not epidemics—and investigation discovered that the agent was transmitted among monkeys by mosquitoes, especially *Ae. hemogogus* (see Figure 3-2). This astonishing finding was dubbed the "jungle cycle." Woodcutters and settlers, exposed by their behavior to the monkey-biting jungle mosquito, occasionally contracted the disease; but so long as *Ae. aegypti* was absent there was no urban cycle and no epidemic.

In West Africa, whites had been dying of yellow fever in "the white man's graveyard" for centuries. It was thought, however, that Africans did not get the disease, which was epidemic only in coastal European settlements. Further study revealed that Africans had greater resistance (lower case mortality) to the disease but that in many parts of Africa the infection was almost universal. That is, studies of blood (serology) showed that antibodies to the virus were very common. The mosquito *Ae. aegypti*, which evolved in Africa, was demonstrated to be transmitting the virus among people in the African towns and villages. Further search demonstrated that there was also a jungle cycle: *Ae. africanus* and related mosquitoes were transmitting the virus among monkeys. That yellow fever originated in Africa is indicated by the resistance of human population, the ubiquity of *Ae. aegypti* there, and by a further complication to the system. There is a mosquito, *Ae. simpsoni*, which breeds in the curled leaves of banana plants and other vegetation on the fringe of human settlements, biting people in the villages and monkeys in the forest. Because *Ae. aegypti* is indigenous and exists in many forms fully adapted to various niches, eradication of it seems impossible. Because of *Ae. simpsoni*, which does not have an equivalent in the New World system, the infection that is enzootic in the monkeys (which serve as a reservoir for the disease) can be easily reintroduced to the villages and towns even when the urban transmission chain is broken. The jungle cycle is intimately linked to the village cycle, and yellow fever continues to occur.

The major defense in Africa is the potent vaccine for yellow fever. In tropical America, however, vaccine is used mainly for those occupationally exposed to the jungle cycle. Efforts to prevent the disease are concentrated on preventing *Ae. aegypti* from becoming reestablished in the cities. Since yellow fever does not occur today in the United States, there is little incentive for this country to spend billions of dollars and massive efforts to eradicate *Ae. aegypti*, which is widespread and deeply ensconced. As a consequence, the mosquito is repeatedly exported, via shipped goods, from the United States to Venezuela and other places in Latin America, where it has been repeatedly eradicated.

Once the intricacies of this vectored disease were understood, the puzzles of many other vectored diseases became solvable. Scientists looked for other arthropods capable of transmitting agents and for other animal reservoirs.

Vector Ecology

Most disease agents require specific hosts in which they live and are even more restrictive about the cold-blooded arthropods in which they reproduce. Because the agent usually goes through changes in life stages and multiplies within the vector, most arthropods cannot be biological vectors. An inappropriate arthropod finds its throat blocked, its abdomen too heavy for flight, and its life expectancy so shortened that death occurs to the arthropod and agent before transmission. A vector poorly adapted to an agent implies that the relationship is of recent origin. For example, part of the evidence that typhus is a relatively recent disease in humans is that the human louse itself dies from the infection. Transmission comes through a bite or other abrasion contaminated with louse feces or crushed-louse fluids, instead of from repeated blood meals, which would occur in a well-adapted system.

A full range of arthropods is involved in disease transmission almost everywhere, but one may identify broad patterns that generally coincide with biomes (see Vignette 3-1). At high altitudes and latitudes, the tick is an especially important vector (Table 3-1). This is because it is capable of transmitting rickettsias and viruses transovarially (into its eggs), so that all of the thousands of ticks that hatch from an infected tick's eggs are equally infected, through generations. In this way the disease agent is transmitted despite freezing temperatures, which stop transmission by killing adult mosquitoes and the agent they carry.

Mosquito-transmitted diseases are most important in warm, humid lands where the insect can live long and reproduce frequently (Table 3-2). Reproduction in mosquitoes is temperature dependent and occurs more quickly when

Table 3-1. Some Diseases Vectored by Ticks and Mites

Disease	Agent	Major Vector	Endemic Locale	Reservoir/ Comments
Viral				
Encephalitis, Central European	Togavirus, group B	Tick: *Ixodes* spp.	Europe	R: rodents
Encephalitis, Russian Spring–Summer	Togavirus, group B	Tick	Europe, North Asia	R: rodent: tick transovarial transmission
Louping ill	Togavirus, group B	Tick	Great Britain	R: sheep
Rickettsial				
Rocky Mountain spotted fever (similar to boutonneuse fever, South African tick typhus, India tick typhus, etc.)	*Rickettsia rickettsii*	Tick: *Ixodes* spp.	North America, especially southeast Piedmont	R: small mammals; tick transovarial transmission
Scrub typhus (tsutsugamushi disease)	*Rickettsia tsutsugamushi*	Larval trombiculid mites: *Leptotrombidum akamushi* and *deliens*	Asia: Siberia to Indonesia	R: rodents; mite transovarial transmission
Bacterial				
Tularemia	*Francisella tularensis*	Ticks: dog, rabbit, Lone Star	North America, continental Europe	Zoonosis of rabbits and rodents
Relapsing fever (endemic)	*Borrelia recurrentis*	Ticks: *Ornithrodoros* spp.	Tropical Africa, Mediterranean, Asia Minor, Americas	R: rodents; tick transovarial transmission
Protozoan				
Babesiosis	*Babesia microti*	Tick	Nantucket Island, United States	R: rodents; rare in humans

Table 3-2. Some Diseases Vectored by Mosquitoes

Disease	Agent	Major Vector	Endemic Locale	Reservoir/ Comments
Viral				
Chikungunya	Togavirus, group A	*Aedes* spp.	Africa, Southeast Asia	R: monkeys
Encephalitis, eastern equine	Togavirus, group A	*Aedes spp.*	Americas	R: birds
Encephalitis, western equine	Togavirus, group A	*Culax tarsalis*	Americas	R: small mammals, reptiles?
Encephalitis, Japanese B	Togavirus, group B	*Culex tritaenio-rhynchus, C. gelidus*	East Asia, Pacific Islands	Pig amplifying host
Encephalitis, St. Louis	Togavirus, group B	*Culex* spp.	Americas	R: birds
Encephalitis, Murray Valley	Togavirus group B	*Culex annulirostris*	Australia, New Guinea	R: birds
Encephalitis, La Crosse	Bunyavirus group C	*Anopheles triseriatus*	Central United States	Transovarial transmission
Dengue fever	Togavirus group B	*Aedes aegypti*	World tropics; hemorrhagic form, Southeast Asia	R: monkeys, Southeast Asia
Yellow fever	Togavirus group B	*Aedes aegypti*	Tropical Africa, tropical forest South America	R: monkeys
Protozoan				
Malaria	Plasmodium: vivax (tertian) falciparum (malignant) malariae (quartan) ovale (W. African)	*Anopheles* spp.	Tropics and subtropics	Humans are source
Helminthic				
Filariasis (elephantiasis)	Nematode: *Wuchereria bancrofti*	*Culex fatigans, Aedes* spp.	Tropics and subtropics	No reservoir: infective mircofilaria, night or day
Filariasis (elephantiasis), Malayan	Nematode: *Brugia malayi*	*Mansonia* spp.	Southeast Asia, India, China	periodic, depending on local vector; zoonosis of wild and domestic animals in Malaysia

the temperature is warm. The female mosquito takes blood to get protein to lay its eggs (male mosquitoes do not bite people). A mosquito transmitting malaria in Georgia, if it were to become infected from its first blood meal after emerging as an adult, might infect 11 to 13 people in its life. The same species in New York would be unlikely to infect more than four or five. As temperatures get colder, the metabolism of mosquitoes slows down. Below freezing it stops entirely, and they die. A few mosquitoes, such as the vector of La Crosse encephalitis, sometimes pass the infection transovarially, but this is rare (see Table 3-2). Thus the disease cycle is interrupted annually in cold climates, and a disease like yellow fever can be introduced only as an epidemic in middle latitudes and does not become endemic there, as it was in the Caribbean. In the wet–dry tropics, some mosquitoes have the ability to *estivate*, or to hide and dehydrate during the dry season, and survive. The adult, infected mosquitoes can renew the transmission chain when the rains come. There is growing appreciation of yet another biological strategy. Many viruses transmitted by mosquitoes, including several kinds of encephalitis found in the United States, are actually zoonoses of birds. They are annually reintroduced along the great bird flyways during avian migration and arrive just in time for the new spring mosquitoes to become infected. The virus simply winters in warmer climes.

Flies inhabit a wide range of biomes (see Table 3-3). In general, they are most important in arid places. The flies of Australia, of the wet–dry tropics, and of arid grazing lands are infamous. A few types of these flies are blood-sucking vectors of human disease. The most notorious of these is the tsetse fly, which transmits trypanosomiasis (African sleeping sickness). This disease mainly affects animals. In one form, *Trypanosoma rhodesiense*, it is an anthropo-zoonosis that has a reservoir in African wildlife and is virulent in humans. In another form, *Tr. gambiense*, it is milder, is slower to kill, and has its reservoir mainly among people, with animals being incidental. There are still other forms that affect only animals. The disease has had a major impact on human population distribution and nutrition by making the raising of animals for protein and draft power virtually impossible across large areas of Africa. Even this fly, however, needs water to breed and becomes less active when humidity is very low. The biting blackfly, *Simulium damnosum*, which is the major vector of onchocerciasis (river blindness), lays its eggs in oxygenated, fast-flowing water by anchoring them to underwater stones. Exposure to the disease cycle, as the name suggests, is associated with use of the fertile soils near the streams and rivers or with use of the rivers for laundry, fording, or fishing. Hunter (1980) has described the complexity of the disease system and the multisided, integrated attack that is needed to control it. The various forms of leishmaniasis transmitted by sandflies are characteristic of rather arid terrains, but the fly requires some humidity and is often found only at specific altitudes where relative humidity is high enough. It finds shelter from the sun and low humidity in the plaster and adobe walls of human dwellings and the burrows of rodents.

Table 3-3. Some Diseases Vectored by Flies and Reduviid Bugs

Disease	Agent	Major Vector	Endemic Locale	Reservoir/ Comments
Viral				
Sandfly fever	Bunyavirus, group C	Midge: sand fly *Phlebotomus papatasii*	Mediterranean climate areas, Asia Minor, Central and South America	Human–fly complex crucial
Bacterial				
Tularemia	*Francisella tularensis*	Deer fly	North America, Europe, Japan	Zoonosis of rabbits and rodents
Bartonellosis (Oroya fever, Carrion's disease)	*Bartonella bacilliformis*	Sand fly: *Phlebotomus* spp.	Andean valleys in Peru, Ecuador, Columbia	People are source
Protozoan				
African trypanosomiasis (sleeping sickness)	*Trypanosoma gambiense, Trypanosoma rhodesiense*	Tsetse fly: *Glossina* spp.	Tropical Africa	Humans are source for *gambiense* R: wild game and cattle for *rhodesiense*
American trypanosomiasis	*Trypanosoma cruzi*	Reduviid bugs	Tropical America	R: wild and domestic animals
Leishmaniasis, visceral (Kala-azar)	*Leishmania donovani*	Sand fly: *Phlebotomus* spp.	Tropics and subtropics, Mediterranean	R: dogs, cats, rodents in different places Fatal untreated
Leishmaniasis, cutaneous (Oriental sore espundia, uta)	*Leishmania tropica, L. brasiliensis, L. mexicana*	Sand fly: *Phlebotomus* spp.	Arid margins of Asia, Africa, Mediterranean, South and Central America	R: dogs, rodents
Helminthic				
Onchocerciasis	Nematode: *Onchocerca volvulus*	Blackfly: *Simulium* spp.	Tropical Africa, Central America to Amazon	Humans are source
Loaiasis	Nematode: *Loa loa*	Mangrove fly: *Chrysops* spp.	West and Central Africa	Humans are source

Analyzing the efficacy of a vector requires detailed knowledge of its particular biology. The following characteristics, for example, affect mosquitoes' usefulness as vectors: flight range, altitude range, sex ratio, breeding habits, preferred breeding sites, life expectancy, alternative (nonblood) food, activity time, host preference, resting habits, biting habits, and tolerance for specific virus, microfilaria, protozoan, and other agents.

Some mosquitoes require fresh water with no pollution; others need organic pollution such as sewage. Some breed in containers, tree holes, or leaf tendrils; others need open water. A few like brackish conditions. Some mate for life when they emerge as adults from the water; others hover around blood sources, hoping to find a mate. Some are weak fliers, others are easily carried long distances by wind, and others are strong fliers, even upwind. Most mosquitoes prefer to bite other animals, including snakes, frogs, and birds, rather than humans, but a few species prefer people, and others tolerate them when hungry enough. Some mosquitoes readily enter human dwellings and after feeding rest on the walls; others never enter dwellings and bite only outside. Some are active during morning or evening, others are active only at night. The detailed ecological requirements for specific mosquitoes are inherent in the landscape, and the distribution of relevant species determines where various disease agents can be spread.

Aedes aegypti is an especially dangerous vector, as is evident from its ecology. It prefers to bite people; it is active during times of day when people are active; it breeds in containers and has adapted well to opportunities offered by human dwellings, such as rain gutters, flowerpots, jars of bathwater, toilet bowls, and backyard trash containers. It has readily spread around the world with human help, much as the rat has. This is due to its habit of laying eggs just above the water line in containers. The eggs can thus sail across the ocean and hatch when the water jars are filled in the new port. Aside from the human louse, *Ae. aegypti* is probably our most commensal (domiciliated) arthropod. It transmits several of the most dangerous viruses (see Table 3-2).

THE LANDSCAPE EPIDEMIOLOGY APPROACH

Landscape epidemiology is a geographic delimitation of the territory of a transmitted disease in order to identify cultural pathways for disease control. In Europe several holistic approaches to the study of arthropod-borne or naturally occurring diseases have been developed. In Germany, for example, a holistic ecological approach was developed under the leadership of Helmut J. Jusatz. He produced the first world atlas of epidemic diseases (Rodenwaldt & Jusatz, 1952–1961) which not only portrayed disease distribution but also analyzed its association with climate, topography, hydrology, and flora and fauna. This work was expanded into a series of monographs on the medical geography of such countries as Kenya, Kuwait, and Korea (Jusatz, 1968–1980). This holistic approach and focus on ecological associations, *landschaftökologische*, includes settlement and cultural patterns. In United States medical geography, by contrast, the study of the natural ecology of disease has been weak. Instead, the cultural ecology approach of Jacques May has been very influential. May's disease-mapping project with the American Geographical Society first introduced disease as a geographical subject in

North America. His influence was established through this project and decades of writing and editing books about the ecology of disease and nutrition. The physical and cultural processes elucidated by May and Jusatz need to be integrated into an understanding of transmissible diseases in varying cultures and environments. The modeling approach of Pavlovsky provides a framework to do this.

Landscape epidemiology was developed by a Russian geographer, Eugene N. Pavlovsky. One of the leading parasitologists of the Soviet Union, he was director of the Zoological Institute of the Academy of Sciences, and president of the Geographical and Entomological Societies of the USSR. Because of the importance of the landscape epidemiology approach, the Institute of Medical Geography became a separate entity in the Soviet Academy of Sciences. Pavlovsky (1966) developed the *doctrine of natural nidality*, or the natural focus of disease. In a natural nidus, infection is maintained among wild animals and arthropod vectors. These zoonoses, which Pavlovsky determined and established by fieldwork, occur in particular kinds of terrain. There is a biogeography to the life cycles of the arthropods and various animals involved. Their food sources, soil, climate, slope, exposure, and other ecological parameters determine local distribution and possible occurrence of the disease cycle. By knowing the conditions necessary for specific diseases, scientists can use the landscape to identify disease hazards. Landscape modification can create, or be used to prevent, the establishment of disease cycles. Perhaps the easiest way to understand the concepts of landscape epidemiology is to consider one of Pavlovsky's best known examples, bubonic plague.

Bubonic Plague

The bacteria that are the agents of bubonic plague have been cycling among rodents in the grasslands of Central Asia since prehistory. Plague is a rodent zoonosis transmitted by fleas (Table 3-4). The rodents in the Central Asian grasslands show no symptoms; that is, they are quite resistant to their ancient infection.

Consider for a moment what makes up the living community in which the agent of bubonic plague, the bacterium *Yersinia pestis*, circulates continually (Figure 3-3). The rodents live in burrows deep enough to provide insulation from the severe heat and cold of Central Asia. Under the grass, the horizon of organic soil is deep. The soil is drained well enough that the burrows are unlikely to be flooded, even when the snow melts in spring. The degree of flooding or drainage depends on local slope, as do the amount of rainfall and the solar radiation received. In the burrow are nesting materials of organic nature, grass roots, and a variety of flies, roaches, and other arthropods that take shelter there. Among the worms and arthropods, which might be predators, flea eggs hatch into larvae, which live off the organic matter. Since rats

Table 3-4. Some Diseases Vectored by Fleas and Lice

Disease	Agent	Major Vector	Endemic Locale	Reservoir/ Comments
Rickettsial				
Typhus fever, epidemic	*Rickettsia prowazekii*	Fluids of body louse, *Pediculus humanus*	Cold climes, especially mountains South America, Himalayas, Balkans	Source is people
Typhus fever, murine, endemic	*Rickettsia typhi*	Flea, especially *Xenopsylla cheospis*	Worldwide	Rats
Backterial				
Plague, bubonic	*Yersinia pestis*	Flea, especially *Xenopsylla cheopis*	Widespread natural foci; western United States, Vietnam, South Africa, South America, Central Asia	R: rodents
Relapsing fever (epidemic)	*Borrelia recurrentis*	Louse, *Pediculus humanus*	Local areas in Asia Minor, North and East Africa, South America	Source is people; in outbreaks from nidal tick-borne

breed frequently and have large litters, there are susceptible baby rats for adult flea vectors to infect when they eat. Therefore actively infected young rats are often present to feed the rat fleas when they first emerge from the larval stage. In vacated burrows, adult fleas can maintain infection for months while waiting for new hosts.

The landscape must also have the right kind of vegetation for the rats to eat and a certain mix of predators and competitors to regulate the population. There are also micrometeorological limits to this vegetation and fauna. Sometimes fire serves to scatter the rodents and introduce infected fleas to new burrows. At the margins of the territory, where winter or summer droughts are so severe that the rat population is unstable, it is difficult for the bacteria to continually circulate to new hosts. Experts in the USSR have become quite precise in identifying the slope, exposure, soil type, acid/base balance, vegetation associations, and so forth that mark areas where the transmission chain is most likely to be established. They concentrate on fumigating burrows in that landscape and have managed to reduce greatly the natural focus of bubonic plague in Central Asia.

Occasionally nomads and hunters must have camped in this natural nidus of the disease, and individuals must have been bitten by fleas and died, perhaps scores of miles away. When infected fleas were carried by caravan in bedding

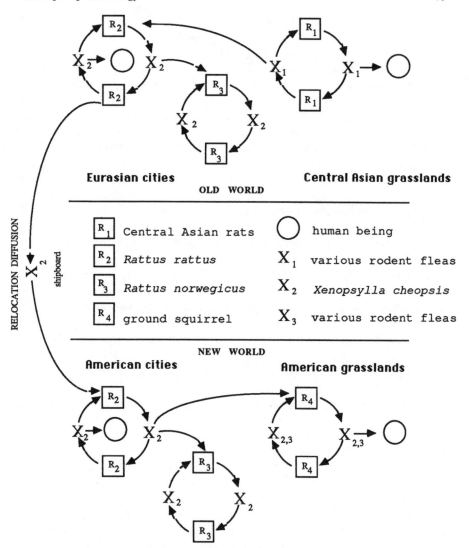

Figure 3-3. Bubonic plague transmission.

or on bodies from this natural nidus to a city like Constantinople (today's Istanbul), a transference, or relocation diffusion, of disease cycle occurred. The hungry fleas bit urban rats, which were quite susceptible to this new infection and died in droves. As the urban rats died and turned cold, their fleas sought new hosts and spread the epizootic. Urban rat populations were, and are, enormous. As the rodents died in the millions, many millions of fleas looking for hosts encountered human beings. One type of flea, *Xenopsylla cheopis*, made an especially dangerous vector because of its willingness to bite humans, who are avoided by most flea species.

The cultural ecology of European medieval cities was ideal for epidemics of plague. Housing was very crowded; water was of limited availability, especially in the quarters of the poor; and hygiene was terrible. Soap was a luxury; hot water was virtually unknown; and food was stored in and near dwellings, which rats knew. In the 13th and 14th centuries areas of agricultural development were pushed to the limits of the transportation system's ability to supply foodstuffs to the growing cities. There was inflation, common malnutrition, and a generally weakened, unhealthy population which was very susceptible to the disease. As people fled the epidemic, or as goods were transported with their stowaway fleas and rodents, the disease agent reached new towns and new rat populations. Bubonic plague swept from east to west across Europe, and then returned from west to east through a new generation of susceptible children. Some people developed a terminal pneumonia that enabled the bacteria to spread by direct means also (through respiratory contagion). The first wave of the Black Death killed a third of the population of Europe. In some places more than half died. Each successive wave, however, had lower case mortality rates as the population began to develop resistance.

As the black rat, *Rattus norwegicus*, spread into Europe in succeeding centuries and outclimbed, outfought, and displaced the brown rat, *R. rattus*, the ecology of the disease changed. The black rat was not as sociable. It preferred to live in sewers and other places where it could avoid people and some of its own kind. The parameters of contact between dead rats, fleas, and people changed. Many of the housing and nutritional conditions changed, too, under new population and economic conditions. The great pandemics abated. Bubonic plague has continued, however, to break out every time war or other severe disturbance has destroyed the ecological balance.

Meanwhile, plague had been carried by ship to the cities of the New World. In the late 19th century, for example, an epidemic in San Francisco led white citizens there to burn out crowded Chinatown. From the urban rat epizootics, wild rodents must have become infected. Today, bubonic plague is enzootic in the rodents of the Southwest. In the grasslands of the southwestern United States, prairie dogs and ground squirrels live in burrows in which most of the nidal conditions of Central Asia are duplicated.

A few people die every year in the southwestern United States from bubonic plague. Frequently, later investigation determines that a cat brought home an infected ground squirrel or that some such similar event occurred. From time to time an epizootic will become evident among ground squirrels in a national park, and the park will have to be closed. A public health threat exists that campers or other travelers in the southwestern arid grasslands will pick up some infected fleas in their baggage (like Asian travelers before them) and transport them to a city where there is a large rat population that has become quite tolerant of our worst rat poisons. Rats infected with the plague have been caught in monitoring traps in Golden Gate Park in San Francisco. Most United States geographers, however, understand little about the biogeography of the natural nidus of bubonic plague in their country and, unlike their Soviet counterparts, have not been involved in any effort at control or eradication.

Plague outbreaks have occurred in most of the world's great ports, and plague has become enzootic among the gerbils of South Africa, field rats of India and of Java, and marmots of Manchuria. Worldwide, more than 220 species of rodents have been shown to harbor plague, although the incidence of plague has been declining. The war in Vietnam showed, however, that when the cultural–biological interface is disturbed, plague can flare into epidemics quickly. Tens of thousands of cases of bubonic plague occurred in refugee-swollen Saigon and the disturbed countryside of South Vietnam. Outbreaks of the deadly pneumonic (contagious) form followed but were contained in the cities by antibiotic therapy and a short-lived vaccine.

Natural nidality involved defining the micrometeorological, geomorphological, and biogeographical limits of a biological community. May (1958a) described these concepts in a different way. He wrote of multifactor diseases: each factor had its own spatial distribution, and all factors had to coincide to create the disease conditions. In the example of plague, the bacterium (agent) is one factor. The rat host (reservoir) is a second. The flea vector is a third. The human host is a fourth. May called a place where the three factors of agent, vector, and reservoir coincide but disease is not known because no humans are present, a "silent zone" of disease. When people, like pastoralists in the grasslands or woodcutters in the forests, penetrate into the silent zone, disease can result. In the USSR, Pavlovsky's hypothesis that the potential for disease could be identified from the simple existence of certain environmental conditions was important in developing mining or industrial towns in Siberia. Planners working with landscape epidemiology could design housing, protective clothing, or work scheduling to shield the human population from exposure to the zoonosis. In the United States, the silent zone concept is being extended to nonbiotic environmental hazards. The mobility of a population of known health status can also bring to light previously unknown silent zones for specific diseases.

Other Applications: Tick-Borne Diseases

Ticks vector an array of viral and rickettsial diseases, from Russian spring–summer encephalitis to boutonneuse fever in India and South Africa (see Table 3-1). Each has its own characteristic landscape ecology. Two brief examples are discussed here.

Central European encephalitis (CEE) is caused by a togavirus vectored mainly by the common tick, *Ixodes ricinus*. The larval, nymph, and adult stages all need a blood meal and are dependent on a variety of animals of different sizes, such as mice, foxes, and deer. Wellmer and Jusatz (1981) studied the landscape epidemiology of the virus and delimited the temperature, humidity, precipitation cycles, vegetation, animal associations, and other environmental conditions needed for its maintenance. They found that limiting factors that create disease-free areas include an annual isotherm of 46° Fahrenheit (8° Celsius), mountain altitudes, and homogenous vegetation such as occurs in pine forest biomes. Lower elevations with dense and diversifed vegetation are required by the disease system. Light soils with humus occur in such areas and may be a condition for sufficient ground moisture. Wellmer and Jusatz determined that the foci are limited to a few square kilometers each, and that a density of only one virus-infected tick per thousand ticks was required to maintain a natural nidus, which the German geographers called a *standortraum* (multifactor location space). They were unable to determine the method of spread. However, they mapped and analyzed the geo-ecological conditions for the repeated infection of specific regions by infected ticks and the recurrent infections of people who entered the natural nidus. (See Vignette 3-3, on field mapping.)

A similar disease system in the United States is Rocky Mountain spotted fever (RMSF), or tick-borne typhus. This disease was first identified in Montana and Idaho (hence its name), but it is most common today in the piedmont region of the southeastern United States. It is caused by the agent *Rickettsia rickettsii*, transmitted by various species of ticks. The dog tick, *Dermacentor variabilis*, is one of the most incriminated. The hosts are mainly small rodents. RMSF has been isolated from the chipmunk, meadow vole, pine vole, white-footed mouse, cotton rat, cottontail rabbit, opossum, and snowshoe hare, and serological (antibody) evidence for infection has been found in many other mammals (Burgdorfer, 1980).

Better publicity has enabled RMSF to be diagnosed earlier and treated with antibiotics, sometimes even before diagnosis. While this complicates statistical reporting, it has brought the case mortality down considerably. The onset of the disease involves fever, chills, aches, nausea, and a rash that spreads from the wrists and ankles. Many of these symptoms are common to other diseases and are easily misdiagnosed without antibody identification. Many cases are so mild they are asymptomatic. Of those who do develop disease symptoms and are not promptly treated, however, 6% to 15% may die.

RMSF has been increasing in the United States, especially in the Southeast. It is hard to determine how much of the increase is due to greater incidence and how much to better reporting. The increase and focus of the disease is attributed by public health entomologists to the rapid extension of suburbs into the wooded, open-field habitat of the tick. The Piedmont is one of the most rapidly growing regions within the southeastern sunbelt. The new suburban houses, parks, jogging paths, and family dog sojourns are in the natural nidus of RMSF. It is normally a zoonosis transmitted among small mammals, but when human residential land use and recreation are extended into the woods it can become established through ticks in houses and lots. Geographers and others do not know, however, how large the foci are, exactly where they occur, or how they change when the landscape is modified.

THE CULTURAL DIMENSION OF WATER-DEPENDENT DISEASE TRANSMISSION

The importance of the human role in habitat modification, of cultural buffering systems, and of human mobility for disease systems was discussed in Chapter 2. These factors are illustrated here with three diseases: malaria, dengue hemorrhagic fever, and schistosomiasis. These three diseases are chosen not only because of their importance, but also because they are water-based in very different ways. They also serve as examples of how activities that promote economic development, especially in the tropics, can modify nidal conditions and affect disease patterns.

Malaria

Malaria is by far the most serious transmissible disease in the world. Malaria was known to kill millions of people each year, but even so, the health improvement in the 1960s in countries where malaria was almost eradicated was eye-opening. Infant mortality rates plummeted. Fertility rates increased as those miscarriages and cases of infertility that were due to malaria disappeared. The terrible pressure malaria has exerted on the human population over the ages is demonstrated by the presence of high-mortality genetic diseases, such as sickle-cell anemia, that were selected by evolutionary processes because they offered protection against the greater mortality costs of malaria. Yet the case mortality rate for malaria is low compared to most serious diseases. Malaria is generally, depending on type, a chronic, debilitating disease. Despite decades in which scientific armament has been marshaled against malaria, an estimated 800 million people still suffer from it. The mosquito vectors are resistant to several insecticides, the agent is resistant to major drugs, and the disease is resurgent in many countries where only a decade ago it was virtually eradicated.

The disease cycle consists of direct transmission of the agent between humans by various species of *Anopheles* mosquitoes. There is no animal reservoir, although monkeys, birds, and other animals have their own forms of malaria. The agent is a protozoan, of a form called a plasmodium, occurring in four types. *Plasmodium vivax* causes benign tertian malaria, historically prevalent in Europe. *P. malariae* causes quartan malaria. *P. ovale*, the least common malaria, occurs only in West Africa. The most dangerous form of malaria is caused by *P. falciparum*. It has more than a 10% case mortality rate and often has unusual symptoms of disorientation, coma, shock, and renal failure. All the forms of malaria have a complicated life cycle within the body. The protozoa pass through the liver, then enter, multiply within, and break out of red blood cells in a synchronized manner (hence the alternation of fever and chills), produce sexual forms, and sexually reproduce within the mosquito after she imbibes them in a blood meal.

May (1958a) described a village in the highlands of Vietnam which had low-grade, endemic malaria. It was a drain on the population's energy, but it was not the worst of these highland villagers' problems. The people lived in houses on stilts and tied their water buffalo and cattle under their houses. They cooked in the houses. These customs were quite different from those of the lowland Vietnamese who moved into the highland area one year. The low-landers built their houses on the ground and kept their animals in proper barns. They also had cooking sheds separate from the houses. They were rapidly driven out of the highlands by epidemic malaria. The local species of *Anopheles* that transmitted the agent was not a strong flier. The height of a house on stilts deterred her. She preferred to bite animals and seldom went past the tethered animals under the house. She, like all *Anopheles*, was active at night and especially in the evening, when the house was full of smoke. The cultural practices of the upland people thus constituted a rather successful protective buffer. Changing nothing in the system except the cultural behavior of the people was enough to render the land uninhabitable.

The experience of the United States illustrates some of the complex forms of cultural interaction with the biotic disease system. Originally Malaria was an Old World disease. The British brought *P. vivax* to North America, and the slaves they imported brought *P. falciparum* from Africa. The early, glowing reports of how healthy the colonies were faded as malaria spread. The nuclear settlements of the plantations were especially good foci of disease transmission. The coastal plantations became so deadly that the owners spent their summers in resorts (the foundation of several beach and mountain resort towns) and cities like Charleston. The greater resistance of Africans to the disease became a common explanation of why the institution of slavery was "necessary."

People did not know the cause of the disease, but they were keen observers. They knew that as the forest around Savannah was removed and land was converted into rice fields, malaria increased. They thought it was the miasma ("mal aria" means bad air) that the trees had protected them from. In what was probably the first environmental legislation for the public health, the city demanded that all land within a mile of the city be used only for dry agriculture. It compensated farmers out of tax revenues for the loss they incurred by not being able to grow rice or indigo. Malaria decreased in Savannah for a few decades, until the disturbances of the Civil War.

Malaria was *the* United States disease of the late 19th century. The confluence of the Ohio and Mississippi was so deadly that many would travel there only in winter. The census of 1890 undertook a survey of cause of death

(Figure 3-4). Although not up to today's statistical standards of reporting, the rates it recorded for malaria, over 7,000 deaths per 100,000 people across the South and more than 1,000 in such states as Michigan, Illinois, and California, are impressive. An average case mortality rate of 5% means that almost the entire population was infected over large areas. Yet by 1930 malaria had disappeared from the North and West, and caused fewer than 25 deaths per 100,000 people in the South except for a few counties. What had happened?

The United States has one important malaria vector, *Anopheles quadrimaculatus*. It breeds in water where it can anchor its eggs on vegetation that intersects the surface in open sunlight and where the currents are not too strong. When the forest was removed across the country, sunlight was let in to marshes and to the poorly drained glaciated land of northern states. Rice plantations planted lots of intersecting vegetation, and much occurred naturally elsewhere. Slow travel by rafts and barges turned canals and rivers into linear transmission channels. Susceptible migrants picked up the infection along the waterways, and travelers who were infected spread it to new places. Houses were of poor quality, and their glassless windows and slatted walls offered little obstruction to mosquitoes. People commonly had poor nutrition and had few animals to divert the mosquitoes. At the turn of the century, the poorly drained land of the glacial till country was deliberately drained for agriculture. Houses were more often constructed of brick or other good material and had glass windows and screens (mainly to keep flies out). Transportation shifted from the rivers and canals to the railroads, and settlements as well

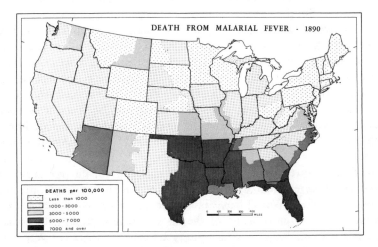

Figure 3-4. Malarial deaths in the United States, 1890. Adapted from *1890 Census of Population* by the U. S. Census Bureau, 1890 (map 17). Washington, DC: U. S. Government Printing Office.

as travelers shifted to the dry upland. The numbers of animals increased. The nutrition of the people improved. Quinine, an antimalarial drug, was available at a more reasonable price when the tariff against its importation (from Java, mainly) was finally dropped. The slower breeding of the mosquito in the cold climate of the North led to the disappearance of northern malaria. In the South the incidence was greatly reduced as the population became more settled (and therefore less mobile), but some counties near sinkholes or the oxbow lakes of the Mississippi remained highly malarious.

Eradication Campaigns

In the late 19th century Laveran discovered the agent of malaria, and Ross proved its transmission by mosquito. A public health official named Carter (1919) was amazed at how little use had been made of this knowledge to prevent malaria. The disease was so common in the South as to be taken for granted, and he had to demonstrate to state and other officials the tremendous economic costs of lost workdays and debilitated farmers. The Rockefeller Foundation became interested. It supported the training and staffing of ento-mologists and other experts and the creation of public health departments at the county level. Carter himself determined the flight range of the United States vector (one mile), its water preferences, and the effects of the water impound-ment for mills. Commonly mills were closed on Sunday, and on Monday the elevated mill ponds were drawn down. He determined that this was good, because it stranded the flotage that anchored larvae.

Carter's findings became relevant when the Tennessee Valley Authority (TVA) proposed to construct dams that would impound large bodies of water in the malarious South. The economic value of developing cheap electricity was obvious, but what was the cost to the public health if malaria again became epidemic? The Army Corps of Engineers worked closely with health officials. All people near the proposed reservoirs were checked for malaria and treated, and all dwellings within a mile of the reservoir were screened. All vegetation was cleared within 10 feet above maximum and below minimum water levels to prevent intersecting vegetation. The water level had to be fluctuated, in spite of navigational needs, at fixed times of the year in order to strand flotage. It all worked, and malaria almost disappeared from the TVA area, instead of becom-ing epidemic.

Similar measures were adopted elsewhere, and levels of malaria were reduced. When threatened by a resurgence of malaria brought by soldiers returning after World War II from fighting in malarious countries, the United States mobilized for eradication of the disease. The military office that had tried to eradicate malaria around military bases where soldiers were training was located in Georgia. It was transformed to head the mobilization against malaria. The weapon was the new insecticide, DDT, which could be sprayed on houses and would for months afterward kill the mosquito vector when it rested on the walls after taking its blood meal. In a massive effort workers sprayed millions of houses across the South and detected and treated thousands of cases. Malaria was eradicated in the United States. The institution that had coordinated the attack became the Center for Disease Control, located in Atlanta, Georgia.

When soldiers returning decades later from the war in Vietnam brought in relapsing malaria cases by the thousands, public health concerns were again raised. The plasmodia, however, found the environment so modified that they could no longer get established in the United States. Few people now sleep where mosquitoes can fly into their bedrooms and bite them, and detection is swift and treatment intensive. The control of malaria in the United States has been so complete that many people forget why reservoir landscapes have exposed banks and fluctuating water levels. Uses for recreation, livestock, and fish cultivation are coming into conflict with water-management laws that were enacted in every state in the South for control of malaria.

Beginning in the late 1950s the World Health Organization attempted to eradicate malaria worldwide. The effort achieved notable successes, especially in Southeast Asia and Central America. By the 1970s, however, health officials talked about control instead of eradication, and in the 1980s they fear even for control. In most places the eradication campaign had made little progress at all, especially in tropical Africa. Many reasons for the failure of eradication have been offered by authorities who gathered in an international forum to learn from the history of the antimalaria effort (Farid, 1980). In some cases maintenance and surveillance failed when, for budgetary reasons, the special malaria teams were combined with multipurpose health programs. The international availability and costs of insecticides have changed. Some (Farid, 1980) believe that when the reduction in malaria led to faster population growth, the capitalist countries cut back their financial support; some even suggest that support was intentionally withdrawn from the eradication effort because of a desire to keep Third World countries poor, sick, and dependent. Whatever faults there were in administration and in financing, however, the original hopes were probably unrealistic because they failed to appreciate the complexities of tropical ecosystems.

In temperate countries, such as the United States, there are few vector species. Sheets of open, fresh water with intersecting vegetation were the breeding area. Spraying walls of houses with a residual insecticide interrupted transmission because the vector bit at night in the houses and then rested on the wall. Figure 3-5 illustrates the contrasting, very complex vector ecology in tropical Malaysia. *Anopheles sundaicus* breeds in brackish water, such as mangrove swamps along the coast. *A. campestris* breeds in rice fields, while *A. letifer* likes coconut, rubber, and swamp trees at low elevations. *A. umbrosus* is the vector of the forest. *A. balabacensis* will breed in the shaded streams of the natural forest or of the rubber tree replacement forests. All of these vectors must compete with many other species of mosquito. Each of the vector species

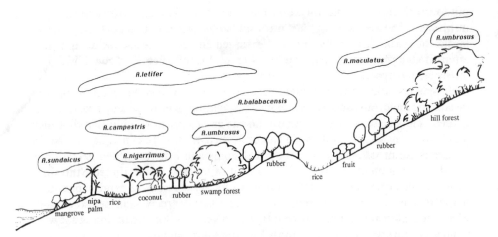

Figure 3-5. Malaria vectors in Malaya. The size of the "clouds" indicates the relative importance in each habitat, or land-use zone, of various species of *Anopheles* mosquitoes in peninsular Malaysia. Adapted from "Malaria in Rural Malaya" by A. A. Sandosham, 1970. *Medical Journal of Malaya, 24.* Copyright 1970 by Medical Association of Malaya. Reprinted by permission.

occurs in relatively small numbers and prefers to bite animals. Some species bite people only outdoors in the early evening and never enter houses. Others enter houses but fly out the window afterwards and rest under vegetation. Most of the *Anopheles* species in Figure 3-5 are normally the vectors of endemic, chronic malaria at low levels. *A. maculatus*, however, breeds in streams exposed to sunlight and readily bites humans. It rapidly increases to very large numbers when given the simple, open habitat it prefers. Thus, when the forest is cut down for land development schemes or plantations and the forest streams are exposed to sunlight, *A. maculatus* rapidly increases in numbers and as a vector of epidemic malaria is one of the most dangerous in the world.

The objective of the eradication campaigns was to interrupt transmission, using some of the known intervention points (Figure 3-6). Residual insecticides sprayed on houses were used to kill the adults after they had taken a blood meal. Humans were examined, and often chloroquine was given as a general prophylaxis to the entire population, to kill the plasmodia in their blood. It was recognized that the mosquitoes could not be suppressed forever. The hope was that when they came back the agent of the disease would no longer exist in that place. This strategy had worked in temperate lands, but the complex ecology of the tropics was frustrating. Although some vectors bit people in their bedrooms at night, rested on the wall, and died, others bit people only outdoors or flew out the window after biting people and never touched the insecticide. Broader attempts to interrupt breeding did not work. Oiling water bodies or stocking them with tilapia fish, which eat mosquite larvae, did not affect mosquitoes breeding in tree holes or hoof prints. People cutting firewood in the mangrove swamps continued to be bitten by *A. sundaicus*. For many such reasons, when the wall-resting, common vector became resistant to the insecticides and rebounded, some malaria was still present in the community, and the transmission chain resumed. Often the population had lost any immunity, however, and epidemics, higher infant mortality, and intensified health problems resulted.

Human mobility also played an important part in eradication failure. Laborers clearing the forest for planting would catch malaria from the *A. maculatus*-borne epidemics and return to their villages. People from areas in which the plasmodia had become resistant to the chloroquine prophylaxis would migrate to other areas and introduce the resistant strain. Prothero (1961) had earlier demonstrated in Africa the importance of labor-shed (a vast area within which people migrate to work) and pastoral mobility patterns for reintroducing an infection to cleared areas. There was international failure to analyze the problem and coordinate efforts at the appropriate scale.

Malaria is spreading in parts of India, Thailand, Central America, and other places where it had almost been eliminated. There is a new generation of drugs coming onto the market, and research has already begun on the next

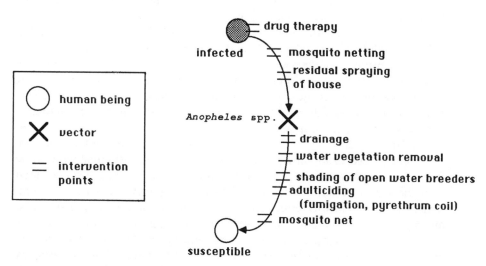

Figure 3-6. Malaria intervention. Malaria is vectored by various species of *Anopheles* mosquitoes, but human malaria only infects humans.

generation of drugs for when resistance develops in the agent. There has been progress toward the production of a vaccine. One of the difficulties with a vaccine is that each stage of the plasmodium's life cycle produces its own proteins. Thus a stage of the plasmodium whose proteins are recognized by human antibodies through vaccination might not prevent disease transmission or even symptoms. People continually infected with malaria do develop resistance—that is, they are not sick with fever and chills all the time—but they do not have a true immunity that prevents the agent from surviving in their bodies.

In some places there is recurrent interest in permanently altering the environment, through drainage and housing, for example, to defeat the ecology of the local disease system rather than relying on the magic bullets of chemistry. No one now even dreams of eradication of this ancient scourge.

Dengue Hemorrhagic Fever

In the 1950s a dangerous new disease began spreading in Southeast Asia. It was first noticed in Manila, but within a few years it was occurring in Thailand. At first it primarily struck Asian children and had a case mortality of over 10%. Investigators considered many factors, including diet, racial susceptibility, and complications from other diseases, without finding the disease agent. Virologists finally isolated the agent, which is identical to the virus that causes dengue fever. Like many arboviruses, it is classed as a togavirus (wearing a covering). Like the virus which causes yellow fever, it is also antigenically identified as a Group B virus (by tests of hemaglutination). It occurs in four types.

Dengue fever had been notorious for the aches (its alias is breakbone fever) and rash it causes, but it had never been a killing disease. The new hemorrhagic form usually puts infected people in the hospital when abdominal pain and high fever occur, 2 or 3 days after infection. There is bleeding into the mucous membranes of the abdomen and into the skin, as well as vomiting. This stage may be followed by a shock syndrome, which was responsible for most of the early deaths. Today the patient is treated for shock. Refinements in therapy have brought the death rate down to 3%. Intense research has focused on dengue fever over the last two decades, and some aspects of the mystery are now clear.

Dengue fever is a monkey zoonosis transmitted by *Aedes albopictus* in the forests of Southeast Asia. This vector will bite people and does breed in containers around human habitation, but it likes vegetation and generally prefers other animals. Dengue fever has been a ubiquitous childhood disease among the rural population of Southeast Asia. With urbanization during the last century, the disease at first disappeared from the urban population, because the vector was not common there. With the arrival of *Ae. aegypti* from Africa, however, dengue became epidemic in the cities, where people had not developed immunities. *Ae. aegypti* is a super vector: in Southeast Asia it breeds in clear water in containers ranging from empty tin cans to large urns that store bath water for the dry season. Dengue became endemic in the urban population, affecting mainly children. It was in this milieu that dengue hemorrhagic fever first emerged.

There are several hypotheses about the etiology of dengue hemorrhagic fever. One is that while thousands of rhesus monkeys were caged on the docks of Singapore, en route to United States laboratories for development of the poliomyelitis vaccine, jungle strains were introduced to the city. Another hypothesis was that the virus had mutated. All four types of the virus were isolated from dying patients, however, and it seemed unlikely that all four types would mutate at the same time. The modified hypothesis is the *Ae. aegypti* is such a super vector that all types of dengue have been able to become more virulent; usually a more virulent virus would face the prospect of endangering its vector. The new disease is indubitably associated with *Ae. aegypti*. As the mosquito has spread among coastal towns and along the railroads and roads, hemorrhagic dengue fever has spread with it. In the urban systems of Southeast Asia, squatter slums and people living on boats have greatly increased the super vector's breeding habitat. As rain gutters, storage jars for rain from roof runoff, ant traps around table legs, flowerpots, and garbage heaps have proliferated, so has the mosquito.

The most widely supported hypothesis today is that dengue hemorrhagic fever is caused by the immune system overresponding to repetitive infection with dengue. As people have migrated to the city from different regions, they have brought all four types of dengue fever into simultaneous circulation. Population mobility and the super vector have combined to increase the chance that a child will be infected with two different types of dengue virus within a relatively short span of time. Some researchers (Halstead, 1980) believe that the immune system of some children reacts so strongly to a second infection by a related, but different, dengue virus that the hemorrhagic-shock syndrome results.

Dengue hemorrhagic fever continues to increase and to spread. The initial biennial frequency of the disease has largely disappeared. It occurs throughout the year, but increases during the peak mosquito-bite periods associated with the monsoon. Better treatment has brought down the death rate, but tens of thousands of children continue to be hospitalized each year.

Dengue fever is endemic in the Caribbean and sometimes epidemic in Florida and the United States. *Ae. aegypti* is common in the United States. Although isolated hemorrhagic-shock cases have occurred in the Pacific and in the Caribbean, epidemics have not arisen. This may be because there are not enough strains in simultaneous circulation on the islands. It is unknown what the critical thresholds of time are between infections, what intensity of vector breeding or of population mobility creates an epidemic, or whether a mutant, virulent virus is involved that has not yet spread from Southeast Asia. Under what conditions will the lethal dengue hemorrhagic fever spread out of Southeast Asia as dengue fever has in past centuries?

Schistosomiasis

Schistosomiasis, endemic in at least 73 countries, is the most rapidly spreading serious infectious disease in the world. It is not, strictly speaking, a vectored disease. Its spread and control, however, can be modeled as a vectored disease.

The parasite is a schistosome, a kind of blood fluke that lives in the veins (see Table 3-5). Although there are five species, three account for most human disease. *Schistosoma japonicum* and *S. mansoni* inhabit the veins around the intestine; *S. hematobium* inhabits the veins around the bladder. *S. mansoni* occurs in Africa, the Arabian peninsula, and Latin America from the Caribbean to northeastern Brazil. *S. hematobium* occurs in Africa, the Middle East, and small foci in India. There is no significant reservoir for these transmissible diseases except people. The rarer *S. intercalatum* is limited to parts of West Africa where it mainly affects sheep, goats, and other nonhuman mammals. Schistosomiasis has been found in several pockets along the Mekong River in mainland Southeast Asia and has recently been associated with a new species, *S. mekongi*. *S. japonicum* occurs in East Asia and limited areas in the Philippines and Sulawesi. In addition to humans, dogs, cats, pigs, water buffalo, field mice, and rats host the agent and comprise a persistent reservoir.

Schistosomiasis (bilharzia in Africa) is not ordinarily fatal. It is a chronic disease that, in cases of intense infection, can be disabling. It weakens people and causes them to die from other things. In places where it occurs, such as the Nile valley, it frequently infects more than half of the population. The symptoms of the disease result from the chronic infection and passage of enormous numbers of eggs. Common symptoms include fever, headache, pain, and bleeding into the urinary tract or intestine. Eggs may block blood vessels. Worms wander from the veins they normally inhabit and go astray into other organs. In cases of heavy infection, the passage of the infesting parasite through the liver can injure the tissue. Individual worms produce eggs for a few years and have been known to do so for more than 20 years.

From the veins around the intestines, eggs are passed in the feces, from those around the bladder they are passed in the urine. Most human waste is deposited into or washed into water: when the schistosome eggs enter water, a form called miracidium hatches and swims downward searching for a snail of the proper genus. The presence of suitable snails has allowed schistosomes to spread to Central and South America; they failed to find suitable snails in the United States. In China, the usual amphibious nature of the intermediate host snails, which support *S. japonica*, allowed a type of intervention (burying and drowning snails) not possible for the aquatic snails in Africa.

After the parasite multiplies in the snail, it breaks out in a form called cercaria and swims to the surface looking for people. When it encounters bathers, it penetrates their skin and follows a complicated internal path to its eventual home. The temperature of the water, the speed of the current, and the hours of sunshine all become critical factors in how long the cercaria survives.

Table 3-5. Some Semi-Vectored Diseases with Intermediate Host

Disease	Agent	Main Intermediate Host(s)	Endemic Locale	Comments
Viral				
Rabies	Rhabdovirus	*Canidae* (foxes, dogs, coyotes, jackals)	Worldwide, except some islands	Also bites of skunks, bats, raccoon: wildlife rabies increasing in United States
Helminthic				
Clonorchiasis (liver fluke)	Fluke: *Clonorchis sinensis*	Freshwater fish, snails: *Amnicolidae*	East and Southeast Asia	Human infection from larvae in cysts in raw fish, eggs passed in feces
Hydatidosis	*Echinococcus granulosus* (dog tapeworm)	Definitive hosts dogs, wolves; intermediate hosts herbivores, dog–sheep–dog cycle most important; also dog–kangaroo–dog, wolf–moose–wolf, dog–cattle–dog	Where dogs are used for herding: Middle East, Australia, Argentia; also Kenya	Human infection from ingestion of eggs from dog fecal contamination
Schistosomiasis (bilharzia)	Fluke: *Schistosoma mansoni, S. haematobium, S. japonicum*	Snails *Biomphalaria* spp. for mansonian, *Bulinus* for other types	*S. japonica*-in East and Southeast Asia; *S. mansoni* in Africa, Northeast South America and Caribbean: *S. haematobium* in Africa, Middle East to India; *S. intercalatum* West Africa	Humans are main source; domestic animals involved with *S. japonicum*; infection by skin contact with water
Trichinosis	Intestinal roundworm, *Trichinella spiralis*	Swine; bears, dog, cats	Variable worldwide, depending on pork habits	Human infection from eating raw or undercooked meat

The cultural side of the equation is the water-contact behavior of the population. To contract schistosomiasis, people do not need to eat or drink anything contaminated. They merely need to walk, swim, bathe, or do their laundry. In semiarid lands or in dry times of the year, human activity often centers on water sources. If the appropriate snail is present and its habitat is favorable, a traveler or immigrant can introduce the schistosome eggs and begin a cycle. The more clustered the population is, the more intense the cycle of infection can become. As the floods of the great rivers are controlled, as irrigation is extended into arid lands or converted from seasonal basin irrigation systems to perennial systems, and as fertilizer and fish ponds enrich the water, so the habitat of the snails that are intermediate hosts is increased, and the disease spreads in Africa and in South America.

The age structure and activity patterns of the population affect the disease's epidemiology. Farmers in the field do not use latrines. Children cannot be kept from swimming in the canals. Women must do laundry. The time of day and the place of contact in the river or canal expose people to different activity levels of the cercariae and to different levels of snail and waste contamination foci. Deep wells with hand pumps to provide water for laundry and bathing have often reduced incidence in women and children. The expansion of irrigated agriculture has increased occupational exposure. Incidental foci of exposure may be things such as the water basins used for ablutions by Moslems before prayers at a mosque.

Figure 3-7 illustrates intervention points in the life cycle of schistosomiasis. Chemoprophylaxis of the population can kill the adult worms. The drugs are powerful, however, and have many side effects. Usually hospitalization is necessary and few peasants can afford the time and lost work. The few United States tourists who get infected can be successfully treated and cured at home, but where a large local population is infected and subject to reinfection, there is no feasible treatment. If all urine and feces were disposed of in a safe manner, contamination of the water could not occur; but given the nature of agricultural work and the activities of rural children, complete sanitary disposal is often not feasible. In many places, such as lower Egypt, the water table is too high to avoid. Fast water, drainage, and concrete linings of ditches that are regularly cleaned can decimate a snail population. Molluscicides have been used to protect certain settlements or urban populations. They are expensive, as they continually wash away downstream, where they may accumulate and result in fish kills and other harmful effects. There is considerable interest in biological controls. Many animals have forms of schistosomiasis that are not infectious for humans (often resulting in "swimmer's itch" in the United States) but that may infect suitable snails and outcompete the human parasite. Some success with biological competition has been achieved in the laboratory, but dominance seems to be temperature dependent and transfers poorly to the field. Clean, safe water can be supplied for domestic use. The water contact behavior of a population can be modified to some extent, if enough is known about the disease hazards. For example, the time of day when certain contact activities occur, such as washing the laundry, might be altered so that water contact occurs when cercariae are less active.

The spread and increase in schistosomiasis are so serious that many kinds of international research are focused on it. Scientists are working on vaccines, on new drug therapies, and on biological competition and predation. Health educators and social scientists are working on modification of population

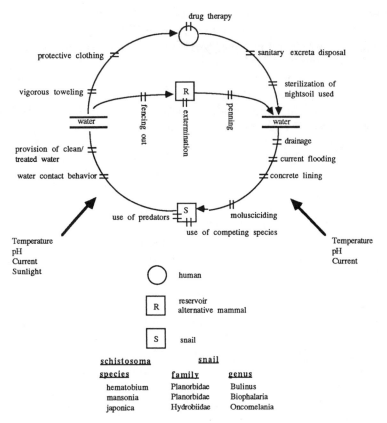

Figure 3-7. Schistosomiasis intervention. Each kind of schistosomiasis has a different genus of snail for intermediate host. Environmental characteristics of bodies of water are important at two points in the schistosome's life.

behavior and the design of protective systems. The complex ecology of the disease differs by type of schistosome, by local snail ecology, by local economy, and by local customs. Population mobilization and a labor-intensive war on snails have decreased and somewhat controlled the disease in China, but it is spreading in the different ecologies of the coffee plantations and new settlements of Brazil and the new irrigation works of East and South Africa. The microgeography of the disease that can guide local intervention efforts still needs to be developed.

Other Transmissible Diseases

It is not possible in one chapter to discuss many of even the most serious transmissible diseases. Onchocerciasis, or river blindness, continues to deny some of the most fertile alluvial land to agricultural settlement in West Africa and parts of Central America. Trypanosomiasis, or African sleeping sickness, affects all efforts of hydroelectric, agricultural, and transportation development in Africa, where it is confined by its uniquely occurring vector, the tsetse fly. Filariasis occurs throughout tropical Asia and Africa. Chagas' disease, or American trypanosomiasis, becomes more serious in South America with every settlement that pushes into the Amazon. Innumerable arboviruses occur whose implications for human health are not even known. The diseases identified by the tables in this chapter are all suitable subjects for analysis using landscape epidemiology.

REGIONALIZATION

In the beginning of this chapter we stated that landscape epidemiology involved four types of regions whose dynamics and locations imparted pattern to the distribution of disease. Let us review some aspects of nidal disease in these terms.

Despite the transportation and mobility transformations of the last few centuries, many nidal diseases are still limited to their realm of evolution. Trypanosomiasis occurs only in Africa because the tsetse fly occurs only there. Schistosomiasis and onchocerciasis must have been introduced to the United States thousands of times, but because the necessary species of snails and blackflies are absent, the diseases cannot establish themselves (although suitable vector substitutes were found in tropical America). Plague, on the other hand, has found enough ecological parallelism in various species of burrowing rodents to have become thoroughly established (by relocation diffusion) in all the grassland biomes.

Biomes, encompassing as they do biotic expression of climate, topography, and soils, broadly categorize the conditions under which a wide variety of nidal diseases flourish, survive, or occur periodically. Mosquito-borne diseases are much more complexly established and intensively transmitted in rain forest and subtropical rice lands than in deciduous forest and cannot be endemic in coniferous regions unless bird migrations or other mediums connect biomes. The disappearance of a disease at a certain elevation in a country's highlands is often the first clue that the disease is transmitted by mosquitoes. Trypanosomiasis disappears as one journeys from the wet savanna to the dry. Mountains and high latitude lands are the favored biome for flea-, louse-, and tick-borne diseases.

The culture realm determines many of the ways the habitat is modified and the population is exposed to or protected from disease. In May's example, (1958a) the lowland Vietnamese who built their houses on the ground were part of the Chinese culture realm, not the Southeast Asian realm in which houses are built on stilts. The dispersed settlement form helped prevent malarial transmission in the United States, an advantage the nuclear European villages did not have during many centuries of endemic malaria. Type of clothes and construction are cultural characteristics that have profoundly affected the importance of disease transmitted by sand flies, mosquitoes, and lice.

Realm of evolution, biome, and culture realm overlap to delimit broad regions of disease occurrence, but it is at the microscale that one can determine the limiting factors of disease transmission and specify intervention points. A transmissible disease does not occur uniformly across the earth's surface. The dynamics of its occurrence from season to season and year to year are affected foot by foot by the immediate weather, soils, topography, land use, and flora and fauna. People can destroy a nidus of disease by exterminating the reservoir or plowing up the vector's habitat and not know what they are doing. They can also bring in water or clear vegetation and create conditions conducive to intense disease transmission without planning for those consequences. Mapping the exact location and conditions where disease transmission is marginal or most intense is of great significance for breaking transmission paths or protecting the population.

CONCLUSION

North American geographers are not very familiar with the approach of landscape epidemiology. They are more used to analyzing the patterns within a city, using socioeconomic, ethnic, and demographic factors, to explain the occurrence of phenomena such as homicide, tuberculosis, or heart attacks and, by regionalizing, to determine the most efficient locations for treatment given the incidence patterns. Asking many of the same geographic questions but using a different methodology, medical geographers can analyze the biogeography of nidal disease systems to specify location and system interaction points that will ease control or eradication. More can be done to minimize transmission of nidal diseases and maximize health education, vaccination, and regulatory protection of targeted groups.

REFERENCES

Abler, R., Adams, J., & Gould, P. (1971). *Spatial organization.* Englewood Cliffs, NJ: Prentice-Hall.
Ackerknecht, E. H. (1945). *Malaria in the upper Mississippi Valley 1760-1900.* Baltimore: Johns Hopkins Press.
Benenson, A. S. (Ed.). (1981). *Control of communicable diseases in man* (13th ed.). Washington, D.C.: American Public Health Association.
Burgdorfer, W. (1980). Spotted fever-group diseases. In J. H. Steele (Ed.), *Section A: Bacterial, rickettsial, and mycotic disease* (pp. 279-302, Vol. II). C.R.C. handbook series in zoonoses, Boca Raton, FL: CRC Press.
Calhoun, J. B. (1979). *The ecology of the Norway rat.* Bethesda, MD: U.S. Department of Health, Education, and Welfare.
Carter, H. R. (1919). The malaria problem of the South. *Public Health Reports, 34,* 1927-1935.

Childs, S. J. R. (1949). *Malaria and colonization in the Carolina low country, 1526–1696.* Johns Hopkins University Studies in Historical and Political Science Series 58, No. 1. Baltimore: Johns Hopkins Press.

Cook, S. F. & Borah, W. (1971–1973). *Essays in population history: Mexico and the Caribbean* (Vols. 1–2). Berkeley: University of California Press.

Crosby, A. W., Jr. (1972). *The Columbian exchange.* Westport, CT: Greenwood Press.

Drake, D. (1854). *A systematic treatise, historical, etiological and practical, on the principal diseases of the interior valley of North America. . . .* New York: Lenox Hill.

Farid, M.A. (1980). Round table: The malaria programme—from euphoria to anarchy. *World Health Forum, 1,* 8–33.

Halstead, S. B. (1980). Dengue haemorrhagic fever: A public health problem and a field of research. *Bulletin of the World Health Organization, 58,* 1–21.

Hunter, J. M. (1980). Strategies for the control of river blindness. In M. S. Meade (Ed.), *Conceptual and methodological issues in medical geography* (pp. 38–76). Chapel Hill, NC: University of North Carolina, Department of Geography.

Jackson, W. A. D. (1985). *The shaping of our world.* New York: Wiley.

Jusatz, H. J. (Ed.). (1968–1980). *Medizinische Landerkunde* (Vols. 1–6). Geomedical Monograph Series. Berlin: Springer-Verlag.

Knight, G. (1971). The ecology of African sleeping sickness. *Annals of the Association of American Geography, 61,* 23–44.

Kucheruk, V. V. (1965). Problems of palaeogenesis of natural pest foci with reference to the history of the rodent fauna. In A. N. Formasov (Ed.), *Fauna and ecology of the rodents* (p. 98). Moscow: Moscow Society of Naturalists.

Markovin, A. P. (1962). Historical sketch of the development of Soviet medical geography. *Soviet Geography: Review and Translation, 3,* 3–19.

May, J. M. (1958a). *The ecology of human disease.* New York: MD Publications.

May, J. M. (1958b). *Studies in disease ecology.* New York: Hafner.

McNeill, W. H. (1977). *Plagues and peoples.* Garden City, NY: Anchor Press.

Meade, M. S. (1976a). Land development and human health in West Malaysia. *Annals of the Association of American Geographers, 66,* 428–439.

Meade, M. S. (1976b). A new disease in Southeast Asia: Man's creation of dengue haemorrrhagic fever. *Pacific Viewpoint, 17,* 133–146.

Meade, M. S. (1977). Medical geography as human ecology: The dimension of population movement. *Geographical Review, 67,* 379–393.

Meade, M. S. (1980). The rise and demise of malaria: Some reflections on southern landscape. *Southeastern geographer, 20,* 77–99.

Moulton, F. R. (Ed.). (1941). *A symposium on human malaria.* Washington, DC: American Association for the Advancement of Science.

Pavlovsky, E. N. (1966). *The natural nidality of transmissible disease* (N. D. Levine, Ed.). Urbana, IL: University of Illinois Press.

Pollitzer, R. (1954). *Plague* (Monograph Series 22). Geneva: World Health Organization.

Prothero, R. M. (1961). Population movements and problems of malaria eradication in Africa. *Bulletin of the World Health Organization, 24,* 405–425.

Prothero, R. M. (1963). Population mobility and trypanosomiasis in Africa. *Bulletin of the World Health Organization, 28,* 615–626.

Rodenwaldt, E., & Jusatz, H. J. (Eds.). (1952–1961). *Welt-Seuchen-Atlas.* Hamburg: Falk.

Roundy, R. W. (1980). The influence of vegetation changes on disease patterns. In M. S. Meade (Ed.), *Conceptual and methodological issues in medical geography* (pp. 16–37). Chapel Hill, NC: University of North Carolina, Department of Geography.

Sandosham, A. A. (1970). Malaria in rural Malaya. *Medical Journal of Malaya, 24,* 221–226.

Sauer, C. O. (1925). The morphology of landscape. *University of California Publications in Geography, 2,* 19–53.

Stevens, J. (1977). American mobilization for the conquest of malaria in the United States. *Journal of the National Malaria Society, 3*, 7–10.

Tuan, Y. F. (1974). *Topophilia*. Englewood Cliffs, NJ: Prentice-Hall.

U.S. Census Bureau. (1890). *1890 Census of Population*. Washington, DC: U. S. Government Printing Office.

Weil, C. & Kvale, K. M. (1985). Current research on geographical aspects of schistosomiasis. *Geographical Review, 75*, 186–216.

Wellmer, H. (1983). Some reflections on the ecology of dengue haemorrhagic fever in Thailand. In N. D. McGlashan & J. R. Blunden (Eds.), *Geographical aspects of Health*. London: Academic Press.

Wellmer, H., & Jusatz, H. J. (1981). Geoecological analysis of the spread of tick-borne encephalitis in Central Europe. *Social Science and Medicine, 5D*, 159–162.

Further Readings

Fonaroff, L. S. (1968). Man and malaria in Trinidad: Ecological perspectives on a changing health hazard. *Annals of the Association of American Geographers, 58*, 526–556.

Gottfried, R. S. (1983). *The black death*. New York: The Free Press.

Haddock, K. C. (1981). Control of schistosomiasis: The Puerto Rican experience. *Social Science and Medicine, 15D*, 501–514.

Harrison, G. (1978). *Mosquitoes, malaria and man: A history of the hostilities since 1880*. New York: E. P. Dutton.

Harwood, R. F., & James, M. T. (1983). *Entomology in human and animal health* (7th ed.). New York: Macmillan.

Learmonth, A. T. A. (1977). Malaria. In G. M. Howe (Ed.), *A world geography of human diseases* (pp. 61–108). London: Academic Press.

Pyle, G. F. & Cook, R. M. (1978). Environmental risk factors of California encephalitis in man. *Geographical Review, 68*, 157–170.

Rosicky, B., & Heyberger, K. (Eds.). (1965). *Theoretical questions of natural foci of disease: Proceedings of a symposium*. Prague: Czechoslovak Academy of Science.

White, G. F., Bradley, D. J. & White, A. (1972). *Drawers of water: Domestic water in East Africa*. Chicago: University of Chicago Press.

Vignette 3-1

PHYSICAL ZONATION OF CLIMATES AND BIOMES

The diurnal and annual variations in intensity of solar radiation and the amount and timing of precipitation, temperature, and humidity have been described, classified, and analyzed in dozens of ways. The variations are complicated by the tilt of the earth's axis, the physiography of mountains and depressions, the contrasts of land and sea, and the size of continents. The broad zones of vegetation, or biomes, that result are also zones of agricultural cropping patterns and of arthropod habitat. The schema in Vignette Figure 3-1 of a prototype continent is simplified but provides a general framework for considering the study of health through landscape epidemiology and biometeorology.

Seasons occur as the tilt of the earth's axis alternately exposes the Northern and Southern Hemispheres to more direct rays from the sun. The vertical rays are limited to the latitudes between the Tropic of Cancer and the Tropic of Capricorn, with the highest annual amounts being received at the equator and the least at the poles. Because hot air rises and cold air sinks, the equatorial area is a zone of low pressure as the air rises, and the poles are zones of cold, subsiding, dry air (see Vignette Figure 3-1). Because cold air can contain less water vapor than an equal volume of warm air, water condenses as the air rises at the equator and cools at higher elevations, producing continually high levels of rainfall over the year. Similarly, the subsiding air at the poles is warmed as it approaches the earth's surface, and because its capacity to hold water vapor is increased, there is little precipitation. Due to processes of atmospheric circulation, air also subsides at around 30 degrees north and south latitude, forming a zone of little precipitation. Because of the lack of cloud cover, this zone receives more radiation at the surface than does the equatorial zone. As subsiding air at 30 degrees latitude and subsiding air at the poles move over the earth's surface and converge, air rises along a broad frontal zone in the middle latitudes and creates a wet, but cooler zone. There is less precipitation in this midlatitude zone than in the equatorial, but because there is also less evaporation and less transpiration from plants, there is high availability of water.

Different vegetational zones, known as biomes, extend over large areas. The biomes are adapted to differences in temperature and precipitation, as well as to seasonal variation. The complex and luxuriant rain forest needs continually wet conditions and high solar radiation, whereas lichen and other tundra vegetation can grow in cold, dry conditions with a short season of solar radiation. As Vignette Figure 3-1 indicates, moving from the poles to the equator, there is a transition of forest type from needleleaf (coniferous, taiga) to temperate deciduous forest, which loses its leaves in the cold season, to subtropical forests that include evergreen plants such as magnolias and palms,

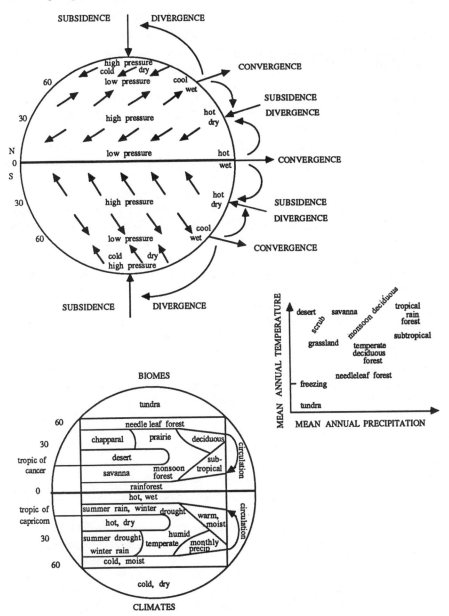

Vignette Figure 3-1. Physical zonation of climates and biomes.

to the tropical rain forest. Desert may be hot or cold, but it is dry; types of desert vegetation such as creosote bush and cactus merge into types of grass-land or evergreen scrub as water increases. The savanna biome is grassland, with scattered monsoon deciduous trees, that has adapted to conditions of summer rain and winter drought. At its dry margins the trees are sparse and the grass short, but as it extends toward the equator the trees increase in density until it merges into forest. Chaparral is the United States name for the biome of evergreen brush that has adapted to the severe conditions of summer drought and winter rain. Grassland, known variously as prairie, steppe, pampas, and veld, extends from short, bunched grass on the arid margins to tall, dense grass on the woodland border.

Broad climatic patterns determine the general pattern of biome location. As can be seem from the prototype continent, (Vignette Figure 3-1) the hot, wet equatorial zone coincides with the tropical rain forest biome, and the cold, dry polar zone with tundra. The subsiding zone of air at 30 degrees latitude creates two broad bands of desert, including the Sahara, Gobi (displaced northwards by Tibetan mountains), and North American deserts in the Northern Hemi-sphere and the Kalahari, Australian, and Atacama deserts in the Southern. The rising-air wet zone is characterized by needleleaf vegetation at its colder latitudes and temperate deciduous at its warmer latitudes. As the climatic zones shift north and south with the sun's rays, they bring wet-dry seasonality to much of the tropics. The area between the equatorial wet and the 30 degrees dry zone supports a savanna or monsoon deciduous vegetation. The area between the 30 degree subsiding zone and the midlatitude wet zone is charac-terized by chaparral. Finally, because of the rotation of the earth, air masses revolve and winds blow in ways that bring humid air from the sea into land on the eastern side of continents but bring only overland, dry air to the western side. On the eastern side of continents, there is a transition from tundra to needleleaf to deciduous to subtropical to rain forest, while on the western side of continents the transition is from tundra to needleleaf to chaparral to desert to savanna to rain forest. Grassy biomes cover the transitional areas.

A last complication is introduced by altitude. The atmosphere cools with increasing altitude. Mountains, furthermore, form barriers so that surface air must go up on the windward side, causing precipitation, and come down on the leeward, forming a "rain shadow," or dry area. An altitudinal zonation of vegetation results that is very similar to the latitudinal zonation that has been presented above. A tall mountain on the equator (such as occurs in Kenya and Ecuador) has a vegetational zonation that proceeds from rain forest at its base, to subtropical, deciduous, needleleaf, tundra, and ice on its windward side and rain forest to savanna, thornbush scrub, needleleaf, and tundra on its leeward side. No matter in what climatic zone people are living, therefore, a great variety of vegetational, agricultural, and living conditions exist for human settlements.

The distribution of diseases transmitted by arthropods—ticks, mosquitoes, flies, and so forth—in different parts of the world can be partly understood from biome distribution. Certain nutritional deficiencies are associated with specific staple crops (see Chapter 2), and cultivation of these crops often follows environmental constraints of biome distribution. Conditions of hazard from air pollution (see Chapter 6) or solar radiation (see Chapter 5) also largely coincide with biome distribution patterns. Understanding the location of major earth biomes can provide a useful framework for understanding the spatial patterns of the distribution of many health hazards.

Vignette 3-2

CULTURAL GEOGRAPHY

Culture is here defined as the accumulation of learned experience transmitted from one generation to another and, potentially, from one group to another across space. Language is both the key element of that culture and its usual mode of transmission. The world's population can be divided into *cultural groups* with common sets of cultural elements that separate them from other groups. Simply put, *cultural geography* is the application of the idea of culture to geographic problems (Jackson, 1985, p. 37).

Our culture can be thought of as the guardian and the marker of our human existence. We are a part of our culture, and that association gives us identity and separates us from other peoples and other places. Culture is inherently conservative. Generally, the more central and important a cultural component, and again language is the obvious example, the more resistant it is to change. Thus cultures sometimes respond slowly to great and very rapid changes in their environment, resulting in periods of confusion and stress. During the last several centuries non-Western cultures have been particularly pressured by the worldwide spread of European influences, the general dissemination of cultural elements through modern transportation and communication, and the sometimes massive movement of people from one physical environment to another.

Change does of course take place within cultures. *Innovation* occurs when a new idea or item comes into existence and is accepted by a society. Innovation is very much a part of a time and place, for it represents a response to a particular situation and depends upon the existing culture to allow it to develop. Similar innovations may develop in response to common conditions or needs in different cultures or times, in *independent invention*. More normally, however, cultures adopt new ideas or items that were developed elsewhere and then transferred in a spatial process called *innovation diffusion* (see Chapter 8).

The cultural geographer may be interested in spatial aspects of almost any part of human culture. Such varied topics as religion, stock-car racing, place names, food preferences, and bluegrass music festivals have all been grist for the cultural geographer's research mill. Perhaps most interesting, however, is the *cultural landscape*, which is the physical imprint of a culture's occupation of territory. That landscape is described and defined by cultural elements such as house types, agricultural field patterns, roads, recreational facilities, and important plant species, alone and in combination. Landscape has been called the geographic expression of human decision (Abler, Adams, & Gould, 1971) and thus of cultural identity.

That cultural landscape is, however, not just the imprint of culture, for the underlying stage, the physical environment, cannot simply be molded and used at will, with its own special features ignored. The natural environment plays an important role in the emergence of a cultural landscape. Should I grow paddy rice, or is wheat the more sensible choice? Should I build my home on stilts, or on the ground? The range of available foodstuffs, particularly in the less developed countries, must to some extent be defined by climate. Our technology and science, our culture, have allowed us to modify and mold that natural landscape. The great expansion *in technology* over the past several centuries has greatly enhanced our impact on the natural landscape, sometimes in totally unexpected or even undesirable ways. No one wants acid rain, nuclear power plant incidents, or the compaction of agricultural soil resulting from the use of heavy farm equipment, but they happen anyway. The landscape that interests so many cultural geographers, then, really represents the interplay between the physical environment and the human imprint on it.

A people often establishes a very special bond with its cultural landscape. This has been called *topophilia*, the love of place, the bond between people and place. We learn to see and understand the world, that is, to *perceive* it, through the filter of our culture. To the extent that we have common experiences or goals, we will perceive similar landscapes in similar (but certainly not identical) ways. Right and wrong or good and bad in the landscape are thus often culturally defined and generally known within the group. Again, this can contribute to *cultural lag* in the landscape as we strive to keep what we know and like.

Finally, despite conservatism and general resistance, change is ever present in culture and on the cultural landscape. A growing population, the decline in soil quality resulting from misuse, the development of the internal combustion machine, the first fast hamburger stand, or the movement of a group into a new physical landscape—these and countless other new ideas, innovations, and changing conditions contribute to the dynamics of culture.

Vignette 3-3

FIELD MAPPING FOR LANDSCAPE EPIDEMIOLOGY

Broad patterns of the occurrence of transmissible diseases may be generalized from biome distribution, as depicted in Vignette 3-1 and discussed in this chapter. Only the largest scale topographic maps, however, which show the smallest areas, have the necessary detail of slope, vegetation, and water occurrence and flow to be useful in landscape epidemiology. It is one thing to establish that malaria, onchocerciasis, or California encephalitis is endemic in a place. It is another to specify exactly where people are exposed or where factors in the natural nidus are most susceptible to intervention or avoidance.

Field mapping is a valuable technique for such analysis. Often a base map can be constructed from air photos. Elevation (slope) can be determined using stereoscopic analysis of air photographs, and detailed contour maps can be constructed for the study area. Sometimes a topographic map of sufficiently large scale is already available. Occasionally the geographer engaged in fieldwork must triangulate between flagged sticks with a proper compass and construct a base map from scratch. Analysis of air photography and field techniques are taught in most geography curriculums.

Vignette Figure 3-3a illustrates the importance of such detailed habitat mapping for understanding the transmission of disease. When the water level rose one year, the foundations of houses became islands of higher land. Rats left their preferred sedge nesting areas and moved into close proximity with humans. Thus a change in river level resulted in greatly increased opportunity for the transmission of flea-, mite-, and tick-borne diseases to the human population.

Meade's investigation of health consequences from the resettlement of people in the land development progam of Malaysia included a study of biotic changes and increase or decrease of transmissible disease risk (Meade, 1976a). A field map for a land settlement scheme and a traditional village was constructed using a Bronson compass and branches, stuck in cans of cement, by which to triangulate (Vignette Figure 3-3b). In that way paths, orchards and vegetable gardens, and areas marshy after storms could be mapped as well as houses and particular buildings. Rat traps were set on a spatial sampling frame for fixed periods of time. After the rats were chloroformed, their ears were examined for the species of mite that transmits scrub typhus. Thus, although laboratory studies could not be done to identify antibodies or to culture the rickettsias, the presence of all other elements necessary for the disease's transmission could be ascertained. The micromobility of the population could then be studied in relation to dangerous habitat.

The land scheme represented a reversal from the traditional village habitat danger. In the village, most of the *lalang*, a tall, coarse, invasive grass that was

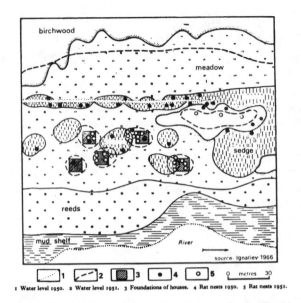

Vignette Figure 3-3a. Microscale mapping of landscape. From "Problems of Palaeo-genesis of Natural Pest Foci with Reference to the History of the Rodent Fauna" by V. V. Kucheruk, in A. N. Formasov, (Ed.), *Fauna and Ecology of the Rodents* (p. 98), Proceedings of the Study of Flora and Fauna of the USSR, Moscow: Moscow Society of Naturalists.

preferred rat habitat, occurred in open space associated with the rubber trees where the men tapping latex were occupationally exposed to rat mites. Villag-ers walked on the road in this linear settlement and had little exposure. The land scheme, however, had created a nuclear settlement containing consider-able vacant space that was occupied by lalang. The houses had quarter-acre lots that often extended into the ravines, where settlers planted papaya trees and tapioca that the field rats also consumed. Because of the form of the settlement, women and children going to the school, clinic, or store walked on paths through the ravines and fields of grass and were exposed to mites. The area near the school and latex collection station was especially heavily popu-lated by rats and their mites. The men tapping the rubber trees, in contrast, had been disciplined to maintain clean cultivation and control the grass and weeds. Few rats were trapped in the rubber cropland. Much of the vacant land within the settlement had been planned for future development. The lalang always invades such vacant, sunny lots. Leaving the undeveloped land under trees, or planting rubber trees there until it was developed, might have reduced the hazard of scrub typhus.

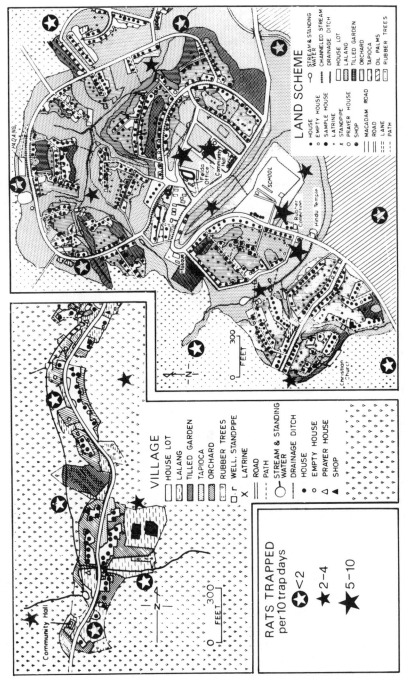

Vignette Figure 3-3b. Land-use habitat for scrub typhus and the changes resulting from land development.

4

Developmental Change and Human Health

The human population of earth, now 5 billion, has doubled since World War II. Demographers project that it will at least double again before stabilizing late in the 21st century. After growing at an increasing rate for centuries, population growth peaked at a little over 2% a year in the late 1960s and began to slow down. Although the present growth rate of 1.7% is projected to decrease to 1.5% by the end of this century, these lower rates are occurring to a greatly enlarged base, resulting in the absolutely largest population increases in history. The human population will grow by more than 800 million per decade for the next 40 years. Its age structure will change during this time. The major causes of death will change. The human population will undergo the profound alteration of becoming primarily urban.

The purpose of this chapter is to convey the complexity and dynamics of population–habitat–behavior interactions that comprise disease systems and the consequent need for careful evaluation of health consequences in developmental impact analysis. The concepts developed in the chapters on cultural ecology and landscape epidemiology are applied here to show how social and economic change affect disease occurrence. Since population growth and urbanization are among the most powerful forces in today's world, models of population transition will be presented first. Then some ways in which ecological relationships can complicate the transition are discussed. Changes in disease prevalence resulting from development in Africa, Latin America, and Southeast Asia are reviewed, and the impact of urbanization is discussed.

THE DEMOGRAPHIC TRANSITIONS

There have been two waves of population growth. Beginning in the 1700s, European population grew for 200 years, sometimes reaching growth rates of a little over 1% a year. Europeans migrated to the Americas, Australia, and elsewhere so that the increased population was redistributed on earth. Today Europeans and their descendents elsewhere have, or are approaching, fertility below replacement reproduction.

The second wave began to build mainly after World War II, when many Third World countries achieved independence. Population growth first in Latin America, then Asia, and now Africa has crested at more than 3% (in a few countries even 4%) a year. There are no "new worlds" to populate, but pressure for redistribution has become evident. Some of the most rapid growth is occurring in places with the most meager environmental and economic resources. A large part of the world's income and wealth has become concentrated in some countries, and population and growth in others. The World Bank estimates that the "industrial market" countries, with 18% of the world's population, have a per capita gross domestic product of $10,444. That is more than 41 times the $250 of low-income less developed countries, with 54% of the population.

What we understand of population growth and change is modeled on the European experience, which is not entirely applicable to current circumstances. The following three models of transition—demographic, mortality, and mobility—are interconnected as shown in Table 4-1.

The Demographic Transition

Mortality and fertility have both been high, fluctuating around each other, for most of human existence. The progression from this state of affairs to the present conditions of low mortality and fertility in the most economically developed countries is known as the demographic transition.

The Swedish demographic experience as shown in Figure 4-1 illustrates the model. Mortality declines, as shall be discussed shortly. For a while fertility continues at its historically high level. During this early stage of the transition *natural increase*, the difference between births and deaths, accelerates. Then, under conditions usually associated with urbanization and economic demands for a more educated workforce, the birth rate begins to fall. In the classic model, the change to a "small-family norm" happens because, although children add to the wealth of a family in a rural agrarian society, they drain a family's wealth in an urban manufacturing society. Research in historical demography, however, has demonstrated that in some parts of rural, agrarian France fertility declined sooner than in urban, industrial England. More attention has been paid recently to cultural attitudes about childbearing, to the status and especially the education of women, and to diffusion of the *idea* of birth control. The transition is completed when birth rates fall below death rates, as has happened in several European countries.

The circumstances of Third World countries today are different in many ways. The magnitude of population increase is unprecedented, with rates of growth more than triple the European experience. The demands that such rapid growth places on an economy, social institutions, and the environment may result in increasing death rates. The decrease in both infant mortality and

Table 4-1. Multiple Transitions

Stage	Demographic	Mortality	Mobility
Historic stable	Birthrates and death rates fluctuating at over 35 per 1,000; population increase imperceptible; >45% population aged under 15, <5% over 65	Life expectancy under 35 years; infant mortality >180 per 1,000; death results mainly from infectious diseases	Individual migration alost unknown except for marriage; circulation local for basic agricultural, commercial and religious needs
Early transition	Death rates fall; birthrates remain high; population grows rapidly by natural increase	Infectious diseases are controlled by sanitation, vaccination, suppression of vectors, and medical treatment; young population has low death rate	Migration from rural areas to cities, agricultural frontiers, and foreign countries; circulation increases socially and for labor
Late transition	Birthrates fall as death rates stabilize; population growth slows and population ages	Degenerative diseases become major causes of death; life expectancy exceeds 65 years	Migration to cities and agricultural frontiers slackens; emigration virtually ceases; circulation increases in structural complexity and intensity
Future stable	Birthrates and death rates fluctuating at less than 20 per 1,000; population increase imperceptible or population declining; <20% population aged under 15, >15% over 65	Life expectancy over 70 years; infant mortality <10 per 1,000; death results mainly from degenerative diseases	Multipurpose and vigorous circulation; migration mainly between cities and for retirement; immigration of labor

fertility appears to have stalled in several countries at levels considerably above those of the European model. The initial conditions are different. In many developing countries the fertility rates were higher at the start of the transition, mainly because age at marriage was higher in Europe. Modern economic activities are no longer so labor intensive as they were, and employment opportunities are limited. Demographers have identified *delayed models*, as for Mexico (Figure 4-1), in which birth rates remain elevated longer than expected for the levels of urbanization and economic development; and *accelerated models*, as for China, in which concerted government programs affecting education, old-age security, employment, infant mortality, and contraceptive means and knowledge have resulted in lowered fertility even under agrarian conditions. With today's means of contraception and sterilization it is

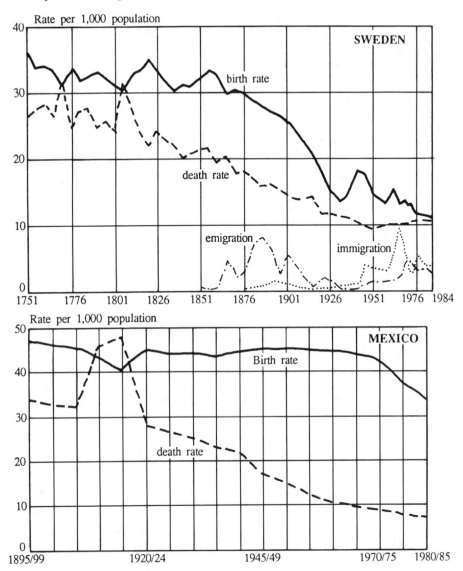

Figure 4-1. The demographic transition. Two different periods of transition are reflected in the experiences of Sweden and Mexico. From "World Population in Transition" by T. W. Merrick, 1986. *Population Reference Bureau Bulletin*, 41(2), p. 9. Copyright 1986 by Population Reference Bureau. Reprinted by permission.

possible for a population to lower its fertility very rapidly, although the *population momentum* for growth that results from a large proportion of children continues for decades.

The Mortality Transition

One does not expect people in London to be dying by the thousands from malaria or cholera, although they have in the past. Nor would one expect cancer to be a leading cause of death in India. In any population the major causes of death are related to the levels of economic and institutional development. The changes in health problems that come with economic and social development are often called the mortality, or epidemiological, transition. As is the case with the demographic transition, much of what we know about the mortality transition comes from European and United States experience. There are serious questions about its relevance to conditions in the Third World, as well as about the next stage for the First and Second Worlds. Developmental changes have brought many unanticipated consequences.

In the European and United States, or Western, experience, preindustrial populations had high birth rates and death rates. One in five babies commonly died before it was a year old. The death rate for toddlers and small children was almost as high. Life expectancy at birth was 30 years, give or take 5. Most people, especially in childhood, died from infectious diseases such as smallpox, malaria, pneumonia, typhoid, or whooping cough (pertussis). These diseases were nearly always present in the population. People also died from great epidemics of cholera, yellow fever, and bubonic plague. Some deaths were caused by vectored diseases, but many more deaths resulted from bacteria or viruses that multiplied in filth and were often transmitted through water or milk. As the rural poor, displaced by enclosure of the commons and other agricultural changes, flooded the industrializing cities, the combination of crowding in abominable housing conditions, overwork, and malnutrition caused the death rates to soar even higher. For a century the life expectancy of people in urban areas compared unfavorably with the more salubrious countryside. Most cities would have experienced absolute population loss without constant in-migration.

As the negative consequences of industrialization became recognized, perceptual changes led to social movements to improve hygiene and aesthetics. Sewage and garbage were attacked, wells and aqueducts were built, trees were planted, and other improvements were made. The advent and diffusion of germ theory in the late 19th century (see Chapter 9) was followed not only by improved treatment of diseases with new drugs and sterilization, but by a public health revolution. Milk was pasteurized, water was chlorinated, and sewage and garbage were removed. The newly recognized importance of sunlight and fresh air led to new housing codes and types of construction. Transportation changes allowed the population to spread out and commute to the central city. Death rates declined, population grew, and people who could not be accommodated even by the prosperous, growing industrial economies of Europe poured out to the Americas, Australia, and other parts of the world. After more than a century, birth rates declined. In several European countries mortality is greater than natality, a condition usually considered to mark the end of the demographic transition: mortality and natality begin to fluctuate around a new, low level of stabilization. Other countries with advanced economies, such as the United States, Canada, and Japan, have almost reached the end of demographic transition.

Under conditions of low birth rates and death rates, death results mainly from degenerative and chronic diseases. Because few children die from infectious diseases, life expectancy at birth is usually over 60 and often over 70 years (see Figure 4-2). Heart disease, stroke, and cancer of various kinds become the major causes of death, with kidney disease, diabetes, cirrhosis of the liver, vehicular accidents, and suicide also being very important (Figure 4-3). Only influenza and pneumonia have continued to be mortal infectious diseases of consequence for the population living in the more developed countries. Because technological innovations, industrialization, and economic development diffuse from a center, one would expect a spatial pattern in the spread of the mortality transition. In North Carolina, for example, the Piedmont Region was the first to urbanize and industrialize. The coastal plain lagged behind, remaining agricultural and poor. The mortality transition from infectious to degenerative causes of death arrived later in the coastal region (Figure 4-4).

After World War II death rates began to fall rapidly in many newly independent, low-income countries. Infant mortality rates that had been more than 200 per 1,000 fell to perhaps 90, and life expectancy rose to 55 years in many countries. Several factors account for the rather precipitous decline in mortality and consequent growth in population. Antibiotics and other drugs became widely available for the first time. Although biomedical health services are still not available to many rural poor, they are generally accessible in the urban areas. Many countries organized and established programs to vaccinate babies for smallpox, diptheria, and whooping cough. DDT and other insecticides were used to attack malaria, yellow fever, and many vectored diseases, with results that were often dramatic. World Health Organization campaigns to eradicate smallpox and to obliterate yaws (which, having a bacterial agent, responded to a single dose of penicillin) were sensationally successful. Most people concerned with health or development came to assume that the Third World would follow rather closely the Western model and that the demographic problem was controlling births. In the last few years, however, it has been realized that the death rates have generally stopped declining, far short of the levels expected on the basis of the Western experience. Development programs that have been successful in increasing incomes have sometimes not reduced mortality.

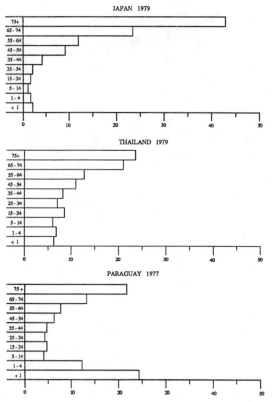

Figure 4-2. Percent of deaths by age. Paraguay, Thailand, and Japan represent three stages of the mortality transition. Adapted from data in *World Health Statistics Annual 1981* by the World Health Organization, 1981, Geneva: Author.

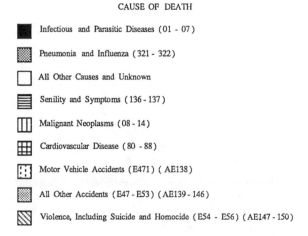

PERCENT OF DEATHS BY CAUSE

JAPAN   1979

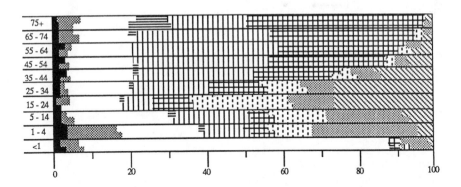

THAILAND   1979

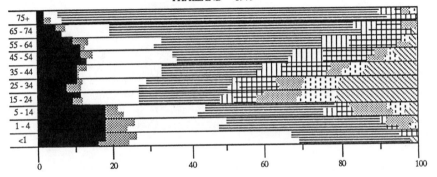

PARAGUAY   1977

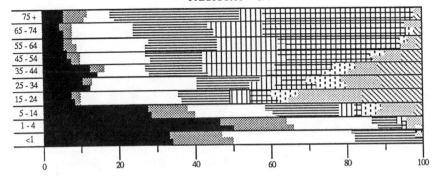

Figure 4-3. The percents of deaths due to different categories of cause as populations pass through the mortality transition, stages of which are represented here by Paraguay, Thailand, and Japan. (Key is on facing page.) Adapted from data in *World Health Statistics Annual 1981* by the World Health Organization, 1981, Geneva: Author.

Rates/100,000

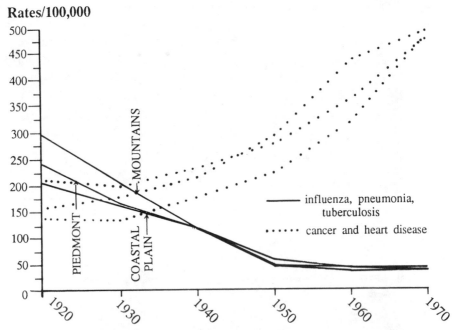

Figure 4-4. Mortality transition in North Carolina. The year where the infectious and degenerative diseases lines cross indicates a region's transition. From *Changing Mortality Patterns in North Carolina, 1920–1972: A Regional Analysis* by G. H. Rice, 1983, unpublished doctoral dissertation, University of North Carolina, Chapel Hill. Reprinted by permission.

Intervention technology from the developed world has had a major impact on the treatment and prevention of many infectious diseases. Only in a few places, however, have there been more pervasive changes in sanitation, housing, and water for drinking and cleaning. Many of the deliberate changes carried out for economic development, moreover, have had unanticipated adverse impacts on health.

The Mobility Transition

Population movement exists at many scales in space and time and is strongly affected by developmental changes. Prothero (1961; 1965) was the first to conceptualize the importance of population mobility in the control of malaria in East Africa; maps of worksheds, animal-herding territories, pilgrimage routes, and other forms of culture-specific population movement are vital to this international effort.

Gould and Prothero (1975) developed a typology of mobility. In studying human mobility, one must separate permanent moves (migration), or those intended to be so, from temporary and recurring ones (circulation). *Migration* is usually measured when the move crosses political boundaries. *Circulation* may involve the daily herding of animals or commuting to an office, the periodic pilgrimage to Mecca or a shopping trip, the seasonal movement to harvest crops or go to college, and the long-term movement for apprenticeship, job transfer, or labor opportunities in the city. By their nature, circulation movements have reverse flows about equal to their forward flows and so have greater potential for disease circulation than does unidirectional migration.

Zelinsky (1971) suggested a mobility transition that adds the dynamic relationship of development to this typology of mobility. In a premodern society, with high birth rates and death rates, there is little residential mobility except for that following marriage (see Table 4-1). Circulation is limited to travel for customary religious observances, agricultural needs, local commerce, and warfare. As the demographic transition begins and death rates fall, a massive movement begins, from the countryside to cities, to new frontiers, and to foreign countries. Social and labor circulation increases. Later in the transition fertility falls, movement to cities and frontiers slackens, and emigration virtually ceases; however, circulation still increases in intensity and in structural complexity. In an advanced society, fertility and mortality have stabilized at low levels. Migration is primarily between cities, although there is significant in-migration of unskilled workers. Circulation is vigorous and accelerating, and includes pleasure-oriented trips as well as moves for social and economic purposes. Zelinsky suggested that in the future the technology of communications and delivery systems may decrease migration and some forms of circulation.

The effects of these changes in mobility, associated with economic development, on patterns of disease diffusion are important subjects for future study. Understanding the types of mobility and anticipating their transformations provide a basis for conceptualizing disease diffusion systems at any scale.

MOBILITY AND EXPOSURE

The importance of population movement for the spread of disease has been known since the time of Hippocrates. Recently, the continuing increase in the numbers of international air passengers has been of great concern to health officials. Air travel defies quarantine and other precautions because the trip is accomplished before symptoms appear.

When a carrier (person with an infectious agent) moves, he or she can infect people along the way and at the destination. Conversely, if a traveler is

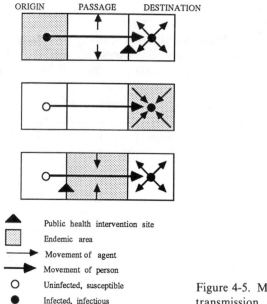

▲   Public health intervention site
□   Endemic area
——▶   Movement of agent
——▶   Movement of person
○   Uninfected, susceptible
●   Infected, infectious

Figure 4-5. Mobility and disease agent transmission.

susceptible to an infection, he or she will probably become sick if the infection is endemic in the destination area. The most likely transmission patterns are illustrated in Figure 4-5. There is always some reverse flow of disillusioned emigrants or from visits and other purposes. In time the schematized flows may be reversed. Because travelers are usually under some psychological and biometeorological stress (see Chapter 5), they are all the more likely to succumb to infections such as "traveler's diarrhea." An infected person traveling on mass transportation comes into contact with many susceptibles. There used to be a fear, before smallpox was eradicated, that an international visitor would introduce the disease to New York City. By the time symptoms developed and were identified, the traveler could have been in contact with thousands of people in subways, buses, and public places. The public health quarantine was established centuries ago to prevent the infected traveler from entering the destination area. More recently, vaccination has been used to prevent susceptible travelers from succumbing to infection at their destination. Vaccination is useful for only some diseases, and quarantine has largely become outmoded by the speed of travel. Neither has proved very appropriate for rural–urban or other within-country movements.

Perhaps the best way to understand the importance of mobility for disease ecology is to turn to the microscale. Within a district, valley, village, yard, house, or room, areal differences in health hazards can be identified. The child standing in the kitchen next to boiling water is exposed to different hazards than the one across the room tending the cat litter; the man driving his car on a busy street is exposed to different hazards than the farmer driving his tractor through the dust. One can visualize surfaces of different kinds of hazards, try to define hazard regions, and then study population mobility in order to understand exposure to them. Vignette 4-1 discusses one method of collecting and analyzing mobility data.

Roundy (1976, 1978) considered microlevel mobility and exposure to different habitats in Ethiopia. Figure 4-6 shows some of his ways of summarizing exposure. The following are some of the hazards he identified for the areal cells:

- *individual*: trachoma, scabies, ringworm
- *household*: tuberculosis, ringworm, tapeworm, ascariasis, salmonellosis, hydatidosis, cutaneous leishmaniasis
- *compound*: ascariasis, tapeworm, poliomyelitis, hookworm, tetanus
- *settlement*: tuberculosis, hepatitis, amebiasis, salmonellosis, measles, whooping cough
- *production area*: tapeworm, amebiasis, schistosomiasis, bovine tuberculosis, rabies, malaria
- *further-ranging area of contact*: tuberculosis, syphilis, schistosomiasis, malaria, diphtheria, yellow fever, river blindness, yaws, filariasis, visceral leishmaniasis

These and other hazards can be identified with particular habitats. Parts B, C, and D of the figure summarize population exposure to habitats by age/sex role, purpose of mobility, and time of day.

Development changes cause changes in disease hazards and exposures. The Ethiopian government built a road to develop a new regional market town, which it located in a valley to facilitate the road-building. The valley was the habitat of malaria, river blindness, filariasis, and other hazards to which the population previously was only occasionally exposed. In the future, some children might be taken from animal-herding areas and sent to school. Other plans might include planting eucalyptus for firewood, drilling a deep well, digging latrines, establishing a bus route, and increasing the density of grazing animals. All such changes would alter the disease hazards of the various habitats or alter population exposure to them.

ECOLOGICAL COMPLICATIONS

Odum (1978) and others have described the characteristics (listed in Table 4-2) of mature and youthful ecosystems. These characteristics require some elaboration.

The most mature ecosystem, a tropical rain forest, has an enormous variety of species competing. Each species tends to be specialized, to be limited

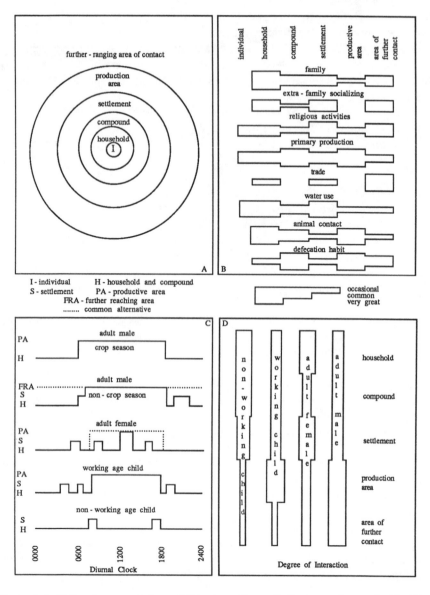

Figure 4-6. Cells of exposure in an Ethiopian village. A diagrams the areal cells. The width of the bars in B signifies the relative exposure of various population groups to different habitats. The line graph in C shows time of day of exposure of population subgroups to different habitats. The width of the bars in D signifies degree of interaction. From *Hazards of Communicable Disease Transmission Resulting from Cultural Behavior in Ethiopian Rural Highland-Dwelling Populations: A Cultural-Medical Geographic Study* by R. W. Roundy, 1977. Copyright 1977 by R. Roundy. Reprinted by permission.

Table 4-2. Some Generalized Characteristics of Ecocystems

	Youthful	Mature
Number of species	Few	Many
Number of species individuals	Many	Few
Niche	Generalists	Very specialized
Growth	Rapid	Slow
Reproduction	Frequent, prolific	Infrequent, limited
Dissemination	Wide, fast	Local, slow
Size tendency	Small	Large
Information, feedback	Low	High
Food chains	Linear	Complex
Nutrient conservation	Low	High
Entropy	High	Low
Overall stability	Low	High
Prototype animals	Mice, sparrows	Elephant, chimpanzee
Prototype plants	Dandelion, pine	Mahogany tree

Note. This information was, for the most part, derived from *Ecology: The Link Between the Natural and the Social Sciences.* (2nd ed., by E. P. Odum, 1978, New York: Holt, Rinehart, & Winston.)

in the number of individuals who can live in its specialized niches, and to grow slowly, reproduce slowly, and disseminate slowly. The different life forms are regulated through feedback stemming from competition, predation, symbiosis, and commensalism. This feedback creates a resilient system that dampens fluctuations while maximizing the use of energy in life support. The system took millions of years to develop its complexities. It is highly vulnerable to human intervention. Once large areas of rain forest are destroyed, as by bulldozers, reestablishment of the intricate relationships by the slow-reproducing, slow-growing, and poorly disseminating life forms will take a humanly incomprehensible length of time. In contrast, a young or simple ecosystem such as tundra or a recently colonized landslide has a few generalized species that are able to withstand extremes of weather and soil conditions. They must reproduce and grow quickly and disseminate easily. Although there are few species, each has enormous numbers of individuals. The ecosystems are subject to great fluctuations, as there are few sources of feedback to stabilize them. A flock of birds, a herd of caribou, or a snake passing through can be devastating for plants and rodents in a young ecosystem. Only the sheer numbers, rapid reproduction, and dissemination of species maintain such simple systems.

One of the most profound human influences on the earth has been *ecological simplification.* People create the simplest possible ecosystem for agriculture. When farmers plant a crop, they do not want competition from any bird, insect, or animal; they want it all themselves. They want the crop to grow rapidly, to produce lots of seed, and to be devastated by harvest. Because there are so few of the balances that exist in complex ecosystems, rats, finches, or the wrong kind of rain can also be devastating.

The simple, young ecosystems that people create with their houses, road clearings, farms, meadows, and herds of domesticated animals are ideal for epidemics. When arthropods or disease agents inimical to human health are present in a simple ecosystem, they exist in large, explosively growing numbers with few controls. In a rain forest there may be several species of *Anopheles* mosquitoes capable of transmitting malaria, but they are competing with hundreds of other species of mosquito. Each of the vector species, one of many species that might bite a person passing by, exists in relatively small numbers. Malaria occasionally gets transmitted, but more often it does not, and no focus is developed. When trees are felled, exposing streams to sunlight, conditions similar to a young ecosystem are created. The species of *Anopheles* that breed in those conditions have large numbers of individuals, reproduce rapidly and so take blood meals frequently, have few controls, and create epidemic malaria.

An anthropologist has demonstrated the perils of ecological simplification. Forest-dwelling aborigines in Malaysia have many kinds of intestinal parasites: roundworm, tapeworm, whipworm, fluke, and hookworm. People have only a light infection of any one parasite, however, because they move frequently and foci do not build up. Under these circumstances the parasitic infections seldom result in disease. When people move to the fringes of towns and become sedentary, however, the ecological conditions are simplified, and only a few species of parasitic worms can exist. An intense focus of transmission is formed, which results in high levels of infection involving many individuals of each species. The helminthic infections become diseases (Dunn, 1972).

Humans have created intense transmission of and intimate human exposure to animal diseases by domesticating and herding animals. Animal intestinal worms and fleas, vectored viral diseases, and salmonella infections have become part of the human environment. Besides intentional domestications, humans have also "domiciliated" many animals and arthropods. Audy (1965) uses the term domiciliated for creatures that have become closely adapted to living with people and depend on people for their great numbers and success, but that were not deliberately domesticated and are not kept by people for any purpose. Domiciliated animals include the house mouse and brown and black rats. Domiciliated arthropods include bedbugs and *Aedes aegypti*. The human louse is so domiciliated that it would become extinct without humans. The intimacy with humans is so great, and the house or barn ecosystem so simple, that when a disease that can be vectored or passed by herded animals or domiciliated arthropods is introduced, the intensity of the foci can produce incidence rates like those of the black death.

Audy described the influences that the processes of creating a mosaic vegetation, herding animals, and domiciliating vectors and reservoir animals has had on disease habitat. In Figure 4-7, a virgin forest is cleared by a farmer

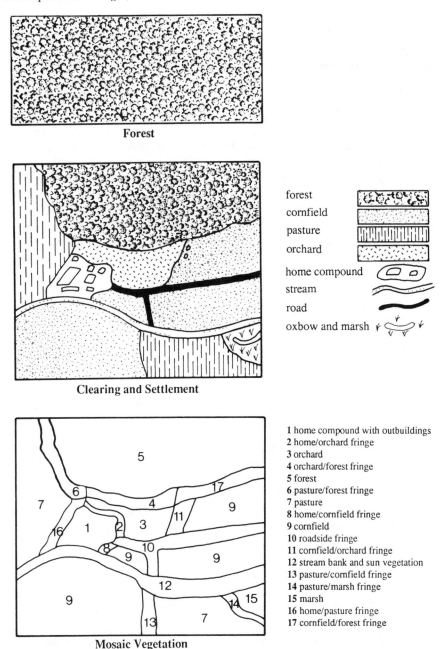

Figure 4-7. Ecosystem simplification and multiplication. Epidemic capacity is increased when a complex ecosystem, characterized by diversity of species with few individuals, is replaced by a mosaic of habitats, each characterized by a few species with many individuals.

for cultivation and pasture. What had been a relatively complex ecosystem is replaced by a *series* of simple ecosystems. The risk of epidemic inherent in a simple ecosystem is multiplied by the number of such systems located in this small area. One additional complication has been introduced. The fringe between two habitats is itself a habitat conducive to passing infection between ecosystems. For example, birds and rodents might shelter in the edge of the woods and venture out to exploit the cornfield. Birds exposed to forest mosquitoes could pass the encephalitis virus to field mosquitos, which in turn infect field hands. In the case of African yellow fever (Chapter 3), *Aedes simpsoni* breed in the tendrils of the banana trees that were planted in the forest fringe. Because they bite both forest monkeys and village humans, they are able to pass the virus repeatedly to the settlement-dwelling *Ae. aegypti.*

Development obviously changes the ecosystem in ways that affect disease transmission. The potential for disease introduction and human exposure is further altered by changes in human mobility. Although the following section discusses ecological simplification and mobility changes and some of their consequences for land development and population resettlement in the Third World, the processes are just as relevant for the most developed lands. Hazards there might include dangerous road curves and toxic chemicals in the groundwater instead of insect bites and worm eggs, yet it is microlevel mobility and exposures to precise environments for different durations at different times that result in the pattern of morbidity.

LAND DEVELOPMENT AND HEALTH CONSEQUENCES

Throughout the Third World the increase in population has led to attempts to increase food production by expanding agricultural land and to industrialize by harnessing hydroelectric power from local rivers. The efforts to exploit the resources of Siberia and the coal fields of the western United States have previously led to both planned and spontaneous resettlement on undeveloped lands. Landscape epidemiology was developed in the Soviet Union largely as a response to the problems of settling Siberia. Let us review some of the cultural and ecological changes that affect the exposure of the population to disease hazards.

As the transition in birth rates and death rates proceeds, there is an increase in migration to agricultural frontiers and cities and a continual increase in circulation. The diffusion of different strains of old and new disease agents takes on new dynamics as susceptibles, infectives, and immunes mix on a new scale. The density of human-host foci increases as villages and cities grow. Often there is a change in house type. Not only does shelter in the city slums tend to utilize material different from that in farming villages, but available material for frontier housing is often limited. On the other hand, as affluence increases, screens and window glass become more common, letting light into previously dark interiors. Throughout the Third World corrugated iron ("zinc") roofs have been replacing thatch because they last longer. A secondary consequence is that the nesting sites of many arthropods and rats, and the snakes that control them, are being destroyed. Not only the settlement size and building materials, but often the settlement pattern is changed. Villages that traditionally were nuclear may become elongated, linear villages along roads. People who lived in nuclear villages may move to a dispersed pattern on the frontier. Such changes alter the exposure to arthropod-borne disease and to fecal contamination of water. The provision of water itself is often profoundly altered. Deep tube wells may make safe, abundant water available for bathing, laundry, and drinking to people who previously had access only to scummed, hand-dug wells during the dry season. In cities people will have flush toilets, sewerage systems, and chlorinated water. Even if a government expands such systems as fast as the budget will allow, however, a large portion of the city's population may consist of in-migrants living in squatter areas without any amenities for hygiene. The circulation of population in a city and the roles of peddler and pedicab driver, frequently assumed by migrants, assure that the disease focus of such squatter slums is a problem for the whole city. Improved and extended transportation systems contribute to growing population circulation. They assure that motor vehicle accidents will increase, that disease agents will spread more widely more quickly, and that tsetse flies, schistosomes, and other components of disease systems will travel long distances for relocation. Transportation systems also assure that food from outside sources will be available to alleviate local crop failures or other disasters and that existing health service facilities will become more accessible.

Agricultural development affects the cultural–ecological process of animal herding. There is increased penning, control of breeding, and higher quality feed. The nature of barns and enclosures often changes. Animals do less scavenging. They are harvested, or culled, more regularly so that the age structure of the herd becomes younger, and the proportion of immune animals to susceptible ones changes. The socioeconomic role of contact with animals may change. It may come to involve more men than boys and become an occupational specialty rather than a source of general population exposure.

Some appreciation of the effects of cultural–ecological changes on disease hazards can be gained by considering the case of India. The Green Revolution (the development of high-yielding varieties of food crops and the production and socioeconomic consequences of the technology that is needed to produce these varieties) has often been studied for its effects on the food supply, foreign exchange, banking system, infrastructure, social and political systems, and so forth, in poor countries. Its effect on animal husbandry has rarely been considered, even though any change in contact between humans and animals may affect disease transmission (see Table 4-3).

The Green Revolution has had a significant impact on rice and wheat production in India, especially in the Punjab. Machines are replacing the

Table 4-3. Transmission of Some Anthropozoonotic Diseases in India

Transmission	Disease	Agent	Definitive Host	Reservoir/ Other Hosts
CONTACT				
Skin contact	Ringworm	Fungus: *Mycosis canis, M.* spp.	Cats	Dog, cattle
Skin–water contact	Leptospirosis (Weil's disease)	Bacteria: *Leptospira*	Rats	Cattle, dogs, swine
Skin–water contact	Strongyloidiasis	Helminth: *Strongyloides*	Dog, cat	Monkeys, pigs
Skin–soil contact	Cutaneous larva migrans	Helminth: *Ancylostoma braziliense*	Cat, dog	
Carcass contact	Anthrax	Bacteria: *Bacillus anthracis*	Livestock	Spores in soil
INHALATION				
Aerosols	Bovine tuberculosis	Bacteria: *Mycobacterium bovis*	Cattle	Buffalo
Aerosols	Q fever	Rickettsia: *Coxiella burnetii*	Cattle, sheep, goats	Tick vector
BITES				
	Rabies	Virus	Dog, cat	Mongoose, bats
	Pasteurellosis	Bacteria: *Pasteurella multocida*	Domestic animals	Wild animals
VECTORED				
Ticks	Boutonneuse fever	*Rickettsia conorii*	Dogs	Rabbits, rodents
Fleas	Murine typhus	*Rickettsia typhi*	Rats	
Mosquitoes (*Culex*)	Filariasis	Nematode, lymphatic	Dog, cattle, buffalo, pig, cat in South Asia	
Mosquitoes (*Culex*)	Japanese encephalitis	Virus	Pigs	Birds, horses
Sandflies	Leishmaniasis (kala-azar)	Protozoa: *Leishmania tropica*	Dogs?	

bullocks in the fields. Water buffalo are being raised in large herds in nonrice areas, fed surplus grain, and used to develop a dairy industry. An improved transportation system supplies Delhi and other major urban centers with dairy products from the Punjab. Strong cultural resistance prevents the slaughtering of animals and culling of herds to improve breeding stock or control disease among asymptomatic animals. Any infections occurring within these dairies will encounter a changing control and distribution system as marketing and regulations develop. Mass penning and stabling of village animals has not been customary. Most have been sheltered in a shed within the family compound

Table 4-3. (*continued*)

Transmission	Disease	Agent	Definitive Host	Reservoir/Other Hosts
INGESTED				
Water	Salmonellosis	Bacteria: *Salmonella*	Domestic animals	Rodents
	Yersiniosis	bacteria: *Yersinia pseudotuberculosis*	Domestic animals	Birds
Dust and dirt	Ascariasis	Roundworm: *Ascaris* spp.	Pig	Sheep, goats, cattle
	Visceral larva migrans	Hookworm: *Toxocara canis, T. cati*	Dog, cat	Pig
	Beef tapeworm	*Taenia saginata*	Cattle	Human feces
	Hydatidosis	Tapeworm: *Echinococcus granulosus*	Dog	Sheep, cattle
	Capillaria	Liver worm: *Capillaria hepatica*	Dog, cat	Rodents
	Toxoplasmosis	Protozoa: *Toxoplasma gondii*	Cat	Transplacental
Meat	Beef tapeworm	*Taenia saginata*	Cattle, buffalo	Yak
	Pork tapeworm	*Taenia solium*	Pigs	
	Toxoplasmosis	Protozoa: *Toxoplasma gondii*	Goat, swine, cattle, sheep, dogs	
	Anthrax	Bacteria: *Bacillus anthracis*	Cattle, goat, sheep	
	Gnathostomiasis	Helminth: *Gnathostoma spinigerum*	Pig, cat, dog	Tigers
Milk	Brucellosis (undulant fever)	Bacteria: *Brucella* spp.	Cows, buffalo	Sheep, goats
	Tuberculosis	Bacteria: *Mycobacterium bovis*	Cattle	Buffalo

courtyard or a wing of the house. In these places the dung was accessible, and small children played in contaminated yards. Women traditionally plastered the house with cow dung and made it into fuel cakes for their ovens. Now electricity is becoming available in most of the villages, so the manure is increasingly available for use as fertilizer. Farmers are being urged to feed pigs in pens rather than let them scavenge. Increased herding of dairy buffalo, goats, and pigs leads to greater ecological simplification, to concentration instead of dispersal, and hence to fewer types of infection at greater intensities. The proportion and component of the population exposed to the animals and their products will be altered. Many mosquitos may find animals no longer accessible and bite people instead.

India has almost the entire inventory of domesticated animals: elephants, yaks, donkeys, mules, ponies, horses, dogs, cats, and pigs. It is first in the world in numbers of cattle,water buffalo, camels, and goats, and sixth in sheep. For cultural as well as economic reasons, people have an unusually intimate connection with animals. As developmental changes in India affect the quality and resistance of animal and human hosts, increase the intensity of animal herding, displace animals, change occupational connections, modify water and food pathways, and alter habitat for arthropod vectors, the significance and distribution of various diseases transmitted from animals to humans will also be affected. What these effects will be, what can be planned, what needs to be studied, and what kinds of intervention can be made remain unanswered.

Irrigation and Water Impoundment

Nothing except the use of fire can compare with the ecological consequences of the human ability to control water. The health hazards resulting from impoundment of water and irrigation of crops have long been recognized, although not often evaluated by engineers and economists. Much of the present appreciation of their importance stems from the construction of the Aswan High Dam in Egypt and the spread of schistosomiasis.

Basin irrigation is ancient practice in upper Egypt. The basins constructed to store the floodwater of the Nile were eventually drained to supply irrigated fields, subsequently planted to crops, and eventually dried up completely before again being flooded. At the height of Ethiopia's rainy season, the flooding Nile brought vital silt and water to the lower river valley. It also washed out the snails and their eggs that had survived the dry season, and physically damaged the snails' gills, thus, serving to control the snail population. With the coming of the Aswan High Dam and perennial irrigation, the snail population expanded greatly. Because the nutrient-rich irrigation water flowed slowly through the agricultural fields, was cleared of silt by reservoir sedimentation, and often supported aquatic vegetation, it was an excellent snail habitat. As predicted by health professionals, schistosomiasis, which had been endemic at low levels in the region from the time of the pharaohs, exploded to become a universal scourge. The same situation has been repeated monotonously, from the Gezira irrigation scheme in the Sudan to Lake Volta in Ghana. Africa is a dry continent. Irrigation is absolutely crucial to countries' hopes of feeding their growing populations, but the price is high. In Kenya, in a district on the shores of Lake Victoria, the rate of schistosomiasis infection for children entering elementary school went from virtually nothing to 100% in less than 10 years.

Kloos (1985) has shown how complex and dynamic the ecology of schistosomiasis can be and also the important part human behavior plays (see Figure

3-7). In Ethiopia's Awash Valley, water resources development is increasing *Schistosoma mansoni* in irrigation schemes in the upper valley by improving the habitat for its *Bulinus* snail intermediate host; these snails, however, remain territorially limited by temperature, silt, and other habitat conditions. The same development, by decreasing flooding, is decreasing *Biomphalaria* snails that host *S. haematobium* in the river lowlands. Endangering behavior, such as using the canals for laundry, bathing, and swimming, remains unchanged even by the provision of safe well water. At the same time, alterations in mobility are increasing the prevalence of *S. mansoni* among pastoralists, who come to the irrigated farmland partly to get health care, and of *S. haematobium* among migrant laborers.

In Brazil, too, expanding irrigation means expanding the territory of schistosomiasis. In Southeast Asia there are only a few foci, including a small but critical one in the Mekong River. If the plans for damming and controlling that great river ever materialize, one of the feared consequences is the spread of schistosomiasis throughout the river basin.

Water impoundment also has important implications for malaria. As discussed in Chapter 3, malaria was an important consideration in building the Tennessee Valley Authority dams in the southeastern United States. The significance of irrigation for encephalitis is less often recognized. One of its best-known consequences was the relocation-diffusion and intensification of encephalitis in Southern California as the *Culex* mosquitoes, vectors of the virus, bred in the organically polluted irrigation water (Reeves & Hammond, 1962).

There are also disease systems that are adversely affected by impoundments and irrigation. Around many of the reservoir impoundments in West Africa, the blackfly *Simulium damnosum* had decreased as its breeding sites in rapids and fast-flowing streams have been eliminated by flooding reservoirs and altering river gradients. Consequently, river blindness has often decreased. Irrigation water can be important for hygiene, supplying abundant water for bathing and laundry in areas where previously water was too precious for such uses during the dry season. Their universal role as sewage collector, however, can make irrigation ditches linear transmission sites. Even large impoundments can serve as concentration basins for parasitic diseases, as was demonstrated by the increase in clonorchiasis (see Table 3-5) around water impoundment resettlements in Thailand in the 1970s.

Population Resettlement

Countries in Southeast Asia and South America are vigorously promoting population resettlement in wilderness areas as a way of developing their national resources and relieving population pressure on declining land resources. The Federal Land Development Authority (FELDA) in Malaysia has resettled hundreds of thousands of farmers who needed land on development "schemes," or projects, carved out of rain forest to grow rubber trees and oil palms. Although not without political and economic problems, the land development schemes have become models for other countries. They have played an important role in making Malaysia internationally dominant in palm oil production and have created decent livings for poor farmers who otherwise might have gone to the urban squatter slums. They have been much studied by political scientists and economists for their costs and benefits and for their organizational successes and failures in increasing human capital. Interested in the alterations being wrought worldwide by land development programs, Meade (1976, 1977, 1978) studied the FELDA schemes' impact on health.

People who had applied for land were admitted from all over Peninsular Malaysia. They spoke different dialects, and in those schemes to which Chinese and Indian Malaysians were admitted along with Malays, they had different religions, diets, and other cultural practices. Most were far from their home villages for the first time. For the first few years, while the tree crops matured, they faced raw land that had to be cultivated, fertilized, drained, and planted. Some settlers could not speak to their neighbors and in any case were not used to mixing with strangers. It took 2–3 years for communities to form. Occasionally, before that time some lonely settlers, unable to get to town or visit home villages left the schemes. Women, accustomed to giving birth to their first child in their mother's house, were especially stressed. The men were required to report to work early in the morning even in poor weather, a kind of regimentation previously unknown to many of them. Malay women did not traditionally hoe plantation fields or work on roads, but they did these and other jobs on the land schemes. Sometimes older children had to stay home from school to take care of small children while the mother worked, the children eating plain cold rice for lunch.

Settlers' houses had a bedroom, a living room area, and a kitchen. They were much smaller than substantial village houses, and early settler families, selected because they had large numbers of children, found the dwellings too small. Later, smaller families were selected, and settlers were aided in expanding their houses. The corrugated iron roofs were long-lasting and vermin-free, but since the whole area had been clear-felled and was exposed to tropical sun the houses frequently heated to over 110° Fahrenheit (43° Celsius) during the afternoon. Each house had land around it for gardening. It takes years for fruit trees to mature, however, and it was difficult for government agents to get the Malay settlers to grow vegetables (which are usually provided by Chinese in the highlands). Almost all of the adults used the latrine by the house, although most of them were not previously used to it, but since toddlers are not diapered, some house compounds were fecally contaminated. A few children's play groups maintained helminthic infections until school age, and the school became a focus for infecting the children who had grown up in more sanitary homes. Every few houses shared a standpipe that brought chlorinated water. Although the system often broke down because of leakage and siltation problems, the enteric infections and fevers in one scheme, studied in detail, were clearly reduced compared to the nearby traditional village.

There were five kinds of increased hazards. The worst was malaria. Many of the schemes constructed in rain forest areas repeated the experience of early rubber plantations: the ecosystem was simplified. Trees were cut down and the vegetation was burned, exposing the streams to sunshine and creating breeding habitat for *Anopheles maculatus*. Houses were sprayed with insecticide twice a year as part of Malaysia's rather successful campaign against malaria, but the success of chloroquine and insecticides had led to neglect of traditional concern for drainage, shading, oiling, and other techniques that had orginally been developed in Malaysia. The mosquitoes were becoming resistant to the insecticides and the malaria plasmodia resistant to chloroquine. The clinics serving the areas of low-density population where remote schemes were established were not enlarged before the settlers came: their capacity was adequate for the pre-settler population. Settlers sick with malaria were therefore driven in landrovers for hours over rutted and flooded land to clinics with no capacity to treat them, were given chloroquine pills, and sent home. The resulting underdosage, combined with new strains, led to increased prevalence of chloroquine-resistant malaria and its unusual expressions. In at least one scheme, rare cerebral malaria led to the perception of spirit possession and consequent abandonment of the land. As the schemes matured and crop trees grew to shade the water, the epidemic receded.

In the center of a land scheme, vacant land was often left for future post offices, bus stations, and shops. This land was quickly invaded by a coarse grass. Located amidst growing trees, tapioca plants, and human waste material, the grass made an excellent habitat for rats. As discussed in Vignette 3-3, a reversal of traditional population exposure to the risk of scrub typhus occurred in the land scheme. Women and children were exposed instead of men, who had been occupationally exposed in the rubber forest.

The settlers who stayed until their trees were mature and producing generally achieved a higher income than their relatives in the villages. Because of their more remote location, however, everything cost more, and nutritional circumstances had changed. The FELDA store stocked dried and canned food only. The settlers did little fishing or hunting, and fresh fruit and vegetables were limited. In the study scheme, settlers had almost given up trying to raise chickens because, having been raised to scavenge and range freely, the chickens were constantly stricken by epizootics and repeatedly wiped out. The settlers had come from small villages of a few hundred people to a scheme of thousands, but the consequences of urbanization (herding of humans and animals) for contagion was reflected in the chicken population.

Remote location and higher income also meant proliferation of motorcycles. Roads were winding, hilly, and frequently wet; overloaded lumber trucks were frequently uncontrollable; and many cycle drivers were unlicensed and helmetless. Motorcycle accidents and mortality became a definite factor in the health of the scheme population.

Finally, changes in mobility lay behind not only vehicular accidents, but also diffusion of infectious disease. In the traditional village, almost everyone walked or rode bicycles to visit and go marketing on a daily basis. On the scheme, usually the male head of household rode a motorcycle into town on Saturday to do shopping and go to the mosque, sometimes taking small children with him. Most other people stayed on the scheme for years at a time. When people did leave, it was not for local mobility but to take an interstate bus back to their home villages. At the time of the study there was a cholera epidemic on the coast. Despite the poor wells of the village and the safe water supply of the scheme, it was probably the scheme that was most at danger of having cholera introduced, perhaps because its settlers were more likely to take trips into infected areas, such as when they brought back seafood as gifts.

One of the advantages of studying land development in Malaysia is that subtle changes can be identified. Elsewhere, in Africa especially, trypanosomiasis, malaria, and schistosomiasis can overwhelm everything else.

Development of the Amazon

The agricultural frontiers in South America attract settlers mainly to the Amazon basin from the Andes Mountains and the Brazilian coast. This basin land was long protected from settlement by the presence of yellow fever and malaria. Gade (1979) suggests that diseases such as leishmaniasis prevented the Incas from extending their hegemony into the tropical lowlands. Leishmaniasis certainly is one of the curses of settlers today, for malaria and yellow fever have been controlled sufficiently to make settlement attractive for the first time since the Colombian exchange.

Again, mobility patterns are critically important. Much of the movement into the Amazon basin is really circulation. People move seasonally to take advantage of different altitudinal zones in the Andes or to work on plantations or road construction. Weil (1981) points out that little work has been done to define the labor and settlement areas of periodic population movements and disease introduction. There is special concern that onchocerciasis may be spread from endemic areas along the northern border into the basin, where three species of *Simulium* exists. Potentially large areas of the Amazon could be denied to settlement by river blindness. Schistosomiasis is also spreading with the new agricultural developments.

There are special health problems for population mobility in this region. People migrating from the Andes are adapted to living at high elevations. Respiratory and circulatory problems are common for these settlers in the wet lowlands. Water impoundment, deforestation, and other ecological simplifications have increased some disease hazards. Helminthic infection levels have increased, and fecal contamination of water has spread enteritis. Game has been depleted, and diets are often poor in protein. As Weil points out, however, many of the health problems of the settlers are simply those of protracted poverty. Many settlers are faced within a few years with nutrient depletion of the soil, laterization, weed invasion, and other losses. They sell out to cattle ranchers and move to a new frontier. Little health care is available. Those involved in capital-intensive agribusiness have money to buy food from afar, to get vaccinations and prophylactic drugs, and to have protective housing, deeper wells, less dangerous work, and access to health care when they do get sick.

African Development and Unintended Changes

The medical geography of developmental changes in Africa has been the most investigated of any region. Multiple infections with different types of helminths and protozoa are common. Much of the rural African population has close contact with vectored disease. The three diseases that can render land totally uninhabitable—malaria, river blindness, and sleeping sickness—seem to have evolved in the African realm. They still deny expanses of good agricultural land to the growing and sometimes malnourished population. Much of the pastoral mobility and labor movement to plantations and mines is international in scope, to a degree reached on no other contininent. While African peoples go through the demographic transition, the circulation that characterizes the mobility transition continues to grow and to have a profound impact on diffusion of disease agents and relocation of nidal systems. It is here that the spread of schistosomiasis made politicians and development specialists concerned about ecological and health consequences.

The comprehensive review by Hughes and Hunter (1970) of the literature on development and health in Africa sensitized a generation of scholars and experts. Hughes and Hunter described, for example, how the evacuation of tribes from their agricultural and grazing land was carried out by colonial governments because of outbreaks of rinderpest or smallpox or in order to stop tribal warfare. Such evacuations allowed brush and tsetse to claim the abandoned territory, so that returning herdsmen or farmers years later found the land denied them by trypanosomiasis. They described how roads had become linear transmission sites for disease. Tsetse, blackflies, mosquitoes, and disease agents of many kinds become concentrated where roads cross streams and rivers and people stop to eat, bathe, and urinate. New reservoirs and roads mean new trading paths and new exposures and transmissions. They described some of the nutritional and social–psychological consequences of development.

Hunter and Hughes explained the difficulty of intervening in the complex disease systems and multiple levels of infection. They emphasized most of all, however, the need for a comprehensive, ecologically informed approach to the developmental projects designed to break the synergism between poverty and disease.

In Africa especially, examples of the inadvertent consequences of developmental changes have multiplied. Demonstration projects in parts of Kenya replaced bullocks with tractors and other machinery on the rather large farms. Arboviral diseases and malaria increased, as the mosquitoes now had only people to provide their blood meals. In Thailand perennial irrigation of rice fields and intensified husbandry of pigs led to intensified Japanese B encephalitis, because the pigs are hosts that carry high levels of viremia and amplify the transmission. In Central America increased cultivation of cotton for export to Japan has led to a greatly increased use of insecticides. The contaminated environment is producing vectors of malaria and arboviruses resistant to control. In the Pacific Islands, some fish become poisonous with ciguatoxin when they consume a type of dinoflagellate that grows in simple ecosystems, such as the disturbed habitats created by construction of ports and airstrips. The consequences for export earnings and tourism are serious.

It should be evident that the old assumption of positive association between economic development and improved health is not necessarily true. By far the most common inadvertent consequences result from changes in nutrition.

Nutritional Changes

Many people assume that most of the malnutrition in the world is a consequence of poverty, and that economic development that increases people's income will improve their nutritional status. Certainly for those suffering from absolute deficiency of calories and protein, the economic ability to "demand" food can mean the difference between life and death. For malnutrition, however, the evidence is that increased prosperity can have an adverse effect, at least initially.

Food preference plays a crucial role in nutrition. The excess calories and fat of United States diet and the consequent obesity and serum cholesterol are often considered to be a new kind of malnutrition. Except for a few people who are vegetarians for religious reasons, however, the evidence is that those whose diets are based on vegetable protein and grain are happy to increase their meat consumption at the first economic opportunity. Japan is the most obvious example, but the preference for meat when accessible has become obvious from the Soviet Union to India. Other changes in preference are more problematic. Often the first purchase made with increased income is Coca Cola or alcohol. Sugar consumption increases, from soft drinks especially, and consequently cavities increase in places where dental care belongs only to the rich. Colonialism and the continuing power of the Western world have led many people to give bread a status higher than the locally produced staple. New members of the technocratic class and residents of cities will buy bread when they can, even when the indigenous millets and other grains are more nutritious. Highly polished white rice is prized. The association of beriberi and the spread of rice-milling technology is described in Chapter 2.

The changes in diet consequent upon developmental alterations take many forms. When the Tonga people were moved in order to construct the Kariba Dam in Africa, they were resettled in land where their traditional crops did not grow. They resisted the strange new seeds the government gave them, and they no longer caught fish and rodents along the river to balance their diet: consequently they endured severe malnutrition. Similarly, people in Southeast Asia who used to gather a few leaves from appropriate trees to garnish their staple have to learn in developing areas to buy or grow vegetables.

There are many ways that technology and development affect nutrition. Roads, bridges, electricity, and refrigeration can have tremendous beneficial results. Food can be delivered to areas where crops fail. Varieties of food can be supplied year-round rather than seasonally. Intestinal diseases can be curbed. Modern processing and storing can deny the rodents and vermin their free meals and save more food for people. Information on weather, farming practices, or new strains of food crops can be disseminated quickly. A few years of education can increase nutritional knowledge and improve child care. New foods and fish cultivation can be introduced, and poultry and livestock diseases can be controlled.

Adverse effects of development are deeply intertwined with societal changes. Women, drawn into economic activities outside the house, have fewer children. Usually they also stop breast-feeding and start bottle-feeding. Other women do so simply because bottle-feeding becomes associated with higher status. For monetary reasons powdered milk is often overdiluted with unsafe water in bottles that cannot be boiled because of energy shortage (of firewood or charcoal). The consequences of these practices can be epidemics of enteritis and high infant mortality.

By far the most profound, baffling, and frustrating deterioration in nutrition often follows the successful development of the cash economy and improvement in local incomes. If a cash crop is introduced and processing, transportation, marketing for it are arranged through great effort, some of the best and most powerful changes of the developmental process are begun. Incomes may increase, accessibility to health care and education may improve, and loyalty to the central government can be cemented. It is still assumed by most planners that if the economic basis of the standard of living is improved for the peasantry, their nutritional status and disease status will improve too.

Repeatedly it has been found that cash cropping and monoculture displaces local cultivation of food. People may devote to the cash crop all their land or all their labor, if they are involved as laborers in agribusiness. Local millets or vegetables are not grown. Chickens are neglected. No one has time to go hunting. Money is available to buy food from the store, and because of roads and development, food can now be available year-round. But people who are not used to working in a money economy tend to spend all their money soon after they earn it. They often spend it on alcohol, consumer goods like radios and cloth, and treats like ice cream and soda pop. The result commonly is that households are soon out of money and reduced to the cheapest, poorest of foods long before the next harvest of their cash crop. Food preference is important: people choose status foods and sweets and no longer have access to the nutritious but coarse homegrown fare. People in villages that have planted remunerative cash crops, such as cocoa in Ghana or coffee in Central America or Kenya, often are wealthier than neighboring villagers but also more apathetic, sick, and malnourished. Plantations and agribusiness are sometimes criticized for demanding that all of their laborers' attention be devoted to the cash crop and for providing no land for vegetables, fruit, or staple foods. As one travels in the Third World, however, one sees many private farmers who devote all their land and time to a remunerative cash crop and who suffer the same consequences.

URBANIZATION

One of the most profound demographic changes, comparable only to population growth, is the urbanization of the world's population. Most Latin American countries are already more than 60% urban. Africa and Asia are less than a quarter urban, but changing rapidly. Where populations are growing 3% a year, capital cities are often growing at 8% and 10% a year. The distribution of the urban population is a problem, relevant to economic and population geography, but the growth of the urban population has major consequences for patterns of disease and health.

The process of urbanization in the Third World intensifies most of the types of change discussed in this chapter. The European model of a city assumes that the population is removed from vectored diseases except, initially, for those vectored by rat fleas and human lice. This is not necessarily true in the tropics. Dengue fever transmitted by the domiciliated *Aedes aegypti* seems to have been transformed into a hemorrhagic form in the intense transmission found in urban areas (Chapter 3). Filariasis has actually increased in some parts of Indian cities where the vector breeds in sewage-contaminated drainage ditches. Yellow fever remains an urban threat. Helminthic infections proliferate in the squatter settlements.

The consequences of the inadequate, crowded, and often windowless housing for the rural migrants pouring into the cities are more congruent with the Western experience. Tuberculosis has become a scourge in the cities of Asia and especially Africa, where the population may be genetically susceptible. Measles is a fatal disease in many populations. Measles, influenza, and other contagious diseases thrive in urbanization.

Malnutrition is a problem in the cities. More urban women bottle-feed their babies. More urban people turn to corn flours and other food sources that can be quickly prepared, and turn their backs on the millets, sesame seeds, groundnuts, and other wholesome foods of the countryside. Mineral and vitamin deficiencies proliferate.

Everywhere, mobility changes are at the heart of health changes. Vehicular accidents increase. Migration into cities and circulation between village and city disseminate tuberculosis, cholera, and venereal disease. Agricultural families suffer at planting and harvest time from the loss of good workers, with nutritional consequences to both rural and urban populations. Families are separated, and young men are sent off alone.

The mental and social stress of such rapid and extreme change, of culture conflict between ethnic groups and generations, of despair, isolation, loneliness, and crushed hopes, lead to mental and social diseases. Alcoholism increases. Venereal diseases increase, and the infertility that can result adds to the social stress. Suicide in many places is epidemic. Mental illness has become a serious problem in many rapidly growing cities and has added substantially to the recognition of the important role that the indigenous health practitioners of various cultures have to play in coping with ill health (see Chapter 11).

Most of the positive health changes that accompany urbanization are related to the demographic transition. Health services become more accessible, infectious diseases more controllable, sewage and water more treatable, the population easier to reach with vaccination. As more children live and the population grows older, the diseases of developed economies come to the fore.

CONCLUSIONS APPLIED TO MORE DEVELOPED COUNTRIES

Although most of the developmental changes discussed in this chapter have been concerned with the Third World, the reader should not think that the process of change is not as consequential for health in the industrialized world. New industries and economic bases bring with them new occupational exposures, waste products, and hazards. Settlement changes resulting from migration to nonmetropolitan areas lead more people into contact with natural foci of vectored disease. Water systems and sewerage systems interlock in new ways and water is reused many more times. Toxic chemicals enter groundwater, offshore fish are poisoned with heavy metals, agricultural runoff carries potent chemicals into municipal drinking water. Social diseases such as alcoholism and homicide, sexually transmitted diseases, and mental diseases continue to evolve. Many of these influences are considered in Chapters 6 and 7, on the pollution syndrome and diseases in developed places.

Medical geographers need to become involved in developing prospective medical geographies of health and disease. Much is known of the etiology of various diseases. Many consequences of specific types of changes are collectively understood among the biological, medical, and social sciences. These include the consequences of changing age structure through fertility behavior and immigration, changing house type, changing contact with natural foci of vectored diseases, changing transportation systems and population circulation, changing occupational hazards, changing exercise and food habits, changing recreational habits, and changing demands on the environment for cleansing human sewage and industrial wastes and providing safer water for consumption and hygiene. The projection of present trends is not enough. Their integration, from a geographic perspective, is crucial everywhere to sound planning and health promotion.

REFERENCES

Audy, J. R. (1965). Types of human influence on natural foci of disease. In B. Rosicky & K. Heyberger (Eds.), *Theoretical questions of natural foci of disease: Proceedings of a Symposium* (pp. 245–253). Prague: Czechoslovak Academy of Science.

Dunn, F. L. (1968). Epidemiological factors: Health and disease in hunter-gatherers. In M. H. Logan & E. E. Hunt, Jr., (Eds.), *Health and the human condition: Perspectives on medical anthropology* (pp. 107–118). North Scituate, MA: Duxbury Press.

Dunn, F. L. (1972). Intestinal parasitism in Malayan aborigines (Orang Asli). *Bulletin of the World Health Organization, 46,* 99–113.

Gade, D. W. (1979). Inca and colonial settlement, coca cultivation and endemic disease in the tropical forest. *Journal of Historical Geography, 5,* 263–279.

Gould, W. T. S., & Prothero, R. M. (1975). Space and time in African population mobility. In L. A. Kosinski, & R. M. Prothero (Eds.), *People on the move.* London: Methuen.

Gwatkin, D. R. (1980). Indications of change in developing country mortality trends: The end of an era? *Population and Development Review, 6,* 615–644.

Harinasuta, C., Jetanasen, S., Impand, P., & Maegraith, B. G. (1970). Health problems and socio-economic development: Investigation on the pattern of endemicity of the diseases occurring following the construction of dams in northeast Thailand. *Southeast Asian Journal of Tropical Medicine and Public Health, 1,* 530–552.

Hubbert, W. T., McCulloch, W. F., & Schnurrenberger, P. R. (Eds.). (1975). *Diseases transmitted from animals to man* (6th ed.). Springfield, IL: Charles C. Thomas.

Hughes, C. C., & Hunter, J. M. (1970). Disease and development in Africa. *Social Science and Medicine, 3,* 443–493.

Kloos, H. (1985). Water resources development and schistosomiasis ecology in the Awash Valley, Ethiopia. *Social Science and Medicine, 20,* 609–625.

Lewis, N. D. (1986). Disease and development: Ciguatera fish poisoning. *Social Science and Medicine, 23,* 983–993.

Mayer, J. D., (1980). Migrant studies and medical geography: Conceptual problems and methodological issues. In M. S. Meade (Ed.), *Conceptual and methodological issues in medical geography* (pp. 136–154). Chapel Hill, NC: University of North Carolina, Department of Geography.

Meade, M. S. (1976). Land development and human health in West Malaysia. *Annals of the Association of American Geographers, 66,* 428–439.

Meade, M. S. (1977). Medical geography as human ecology: The dimension of population movement. *Geographical Review, 67,* 379–393.

Meade, M. S. (1978). Community health and changing hazards in a voluntary agricultural resettlement. *Social Science and Medicine, 12,* 95–102.

Merrick, T. W. (1986). World population in transition. *Population Reference Bureau Bulletin, 41* (2).

Odum, E. P. (1978). *Ecology: The link between the natural and the social sciences* (2nd ed.). New York: Holt, Rinehart, & Winston.

Omran, A. R. (1977). Epidemiologic transition in the U. S.: The health factor in population change. *Population Bulletin, 32* (2).

Palloni, A. (1981). Mortality in Latin America: Emerging patterns. *Population and Development Review, 7,* 623–649.

Prothero, R. M. (1961). Population movements and problems of malaria eradication in Africa. *World Health Organization Bulletin, 24,* 405–425.

Prothero, R. M. (1965). *Migrants and malaria.* London: Longmans.

Reeves, W. C., & Hammon, W. MD. (1962). Epidemiology of the arthropod-borne viral encephalitides in Kern County, California, 1943–1952. Berkeley, CA: University of California Press, (Publications in public health 4).

Rice, G. H. *Changing mortality patterns in North Carolina, 1920–1972: A regional analysis.* Unpublished doctoral dissertation, University of North Carolina, Chapel Hill.

Roundy, R. W. (1976). Altitudinal mobility and disease hazards for Ethiopian populations. *Economic Geography, 52,* 103–115.

Roundy, R. W. (1977). Hazards of communicable disease transmission resulting from cultural behavior in Ethiopian rural highland-dwelling populations: A cultural-medical study. Doctoral dissertation, University of California, Los Angeles. Ann Arbor, MI: University Microfilms.

Roundy, R. W. (1978). A model for combining human behavior and disease ecology to assess disease hazards in a country: Rural Ethiopia as a model. *Social Science and Medicine, 12,* 121–130.

Surtees, G. (1970). Effects of irrigation on mosquito populations and mosquito-borne disease in man, with particular reference to rice field extension. *International Joural of Environmental Studies, 1,* 35–42.

Surtees, G. (1971). Urbanization and the epidemiology of mosquito-borne disease. *Abstracts on Hygiene, 46,* 121–134.

United Nations, Department of International Economic and Social Affairs. (1985). *World population prospectus: Estimates and projections as assessed in 1982.* New York: United Nations.

Weil, C. (1981). Health problems associated with agricultural colonization in Latin America. *Social Science and Medicine, 15 D,* 449–461.

Wolfe, B. L., & Behrman, J. R. (1983). Is income overrated in determining adequate nutrition? *Economic Development and Cultural Change, 31,* 525–549.

World Bank. (1984). *World development report 1984.* New York: Oxford University Press.

World Fertility Survey. (1984). *World fertility survey: Major findings and implications.* London: Author.

World Health Organization. (1981). *World health statistics annual 1981.* Geneva: Author.

Yuill, R. S. (1971). The standard deviation ellipse: An updated tool for spatial description. *Geografiska Annaler, 53 B,* 28–39.

Further Reading

Armstrong, R. W. (1973). Tracing exposure to specific environments in medical geography. *Geographic Analysis, 5,* 122–132.

Desowitz, R. S. (1981). *New Guinea tapeworms and Jewish grandmothers.* New York: W. W. Norton.

Farvar, M. T., & Milton, J. (Eds.). (1972). *The careless technology: Ecology and international development.* New York: Tom Stacey.

Howe, G. M. (1972). *Man, environment and disease in Britain.* New York: Barnes & Noble.

Howe, G. M. (Ed.). (1977). *A world geography of human diseases.* London: Academic Press.

Kvale, K. M. (1981). Schistosomiasis in Brazil: Preliminary results from a case study of a new focus. *Social Science and Medicine, 15D*, 489–500.

Stanley, N. F., & Alpers, M. P. (Eds.). (1975). *Man-made lakes and human health.* New York: Academic Press.

Takemoto, T., Suzuki, T., Kashiwazaki, H., Mori, S., Hirata, F., Taja, O., & Vexina, E. (1981). The human impact of colonization and parasite infestation in subtropical lowlands of Bolivia. *Social Science and Medicine, 15D*, 133–139.

Thompson, K. (1969). Insalubrious California: Perception and reality. *Annals of the Association of American Geographers, 59*, 50–64.

Thompson, K. (1969). Irrigation as a menace to health in California: A nineteenth century view. *Geographical Review, 59*, 195–214.

World Health Organization. (1976). *Water resources development and health: A selected Bibliography* (Document MPD/76.6). Geneva: Author.

Vignette 4-1

MICROSPATIAL EXPOSURE ANALYSIS

Detailed mobility information is essential for study of exposure to specific environments. For example, knowledge of water contact behavior is critical to understanding the incidence of schistosomiasis. Knowing about exposure to orchard mosquitoes in the evening or stream mosquitoes in the morning helps to identify hazards from specific vectored diseases. Often the environment must be carefully described. The home at night, for example, affects health differently depending on whether the house is screened, the beds are netted, the room is smoke-filled, and so on. At a public health level, the appropriate population for education, intervention, or monitoring can often be targeted through knowledge of critical environments and population mobility. Knowledge of microspatial mobility may help identify risk factors for diseases of unknown etiology.

Vignette Figure 4-1 illustrates one means of collecting such microspatial mobility data. Respondents are asked to recall their mobility over a 24-hour period, yesterday. "Yesterday" should not be a holiday or a special day of the week. The day is marked off in equal time segments by three columns of dots (in the figure a 15-minute segment is used). The interviewer talks the respondent through the day. A line is drawn down the first column of dots until the first trip, to the yard or to another place, depending on the scale being analyzed. The shift of the vertical line from column one to column two and back to column one, represent a change in environment. Notice that the line passes on the diagonal to the dot in the next column. A horizontal line would result in counting the same time segment twice. A line to the third column marks the means of travel. When the trip takes 15 minutes (one time segment) or more, the third column dots represent environmental exposure. The distance of the trip can be noted in appropriate measures (in the example, miles and blocks), and the location can be specified so that map coordinates can be assigned later. Coding requires that one specify environments appropriately for the research purpose. The number of dots that the vertical line covers for each environment provide an easy summary. The number of trips, the distances traveled, and the means of travel by purpose are also easily summarized from the form. The scale and environments involved in another interview might specify particular fields, water holes, marketplaces, rooms within a factory, or districts within a city.

Aggregating the responses to mobility interviews permits assessment of differences in exposure to environments and of the extent and means of mobility. Vignette Table 4-1 represents the results obtained from a hypothetical study population with an illness and a healthy control population, matched by age, sex, ethnicity, income level, and other appropriate criteria. Since there are

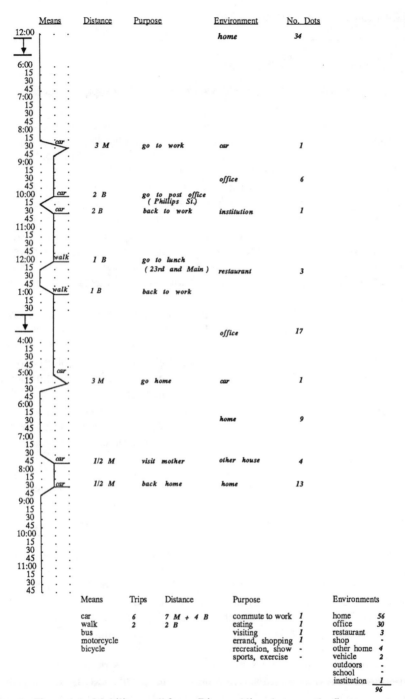

Means	Distance	Purpose	Environment	No. Dots
12:00			home	34

	Means	Distance	Purpose	Environment	No. Dots
8:00–30	car	3 M	go to work	car	1
9:00–30				office	6
10:00	car	2 B	go to post office (Phillips St.)		
10:30	car	2 B	back to work	institution	1
12:00	walk	1 B	go to lunch (23rd and Main)	restaurant	3
1:00	walk	1 B	back to work		
				office	17
5:00–15	car	3 M	go home	car	1
6:00–30				home	9
7:00–45	car	1/2 M	visit mother	other house	4
8:00–30	car	1/2 M	back home	home	13

Means	Trips	Distance	Purpose		Environments	
car	6	7 M + 4 B	commute to work	1	home	56
walk	2	2 B	eating	1	office	30
bus			visiting	1	restaurant	3
motorcycle			errand, shopping	1	shop	-
bicycle			recreation, show	-	other home	4
			sports, exercise	-	vehicle	2
					outdoors	-
					school	-
					institution	1
						96

Vignette Figure 4-1. Mobility recall form. Diagonal lines between the first two columns of dots show change of habitat, while the third column is used to indicate the means of transportation, which may itself become a habitat exposure.

138

Vignette Table 4-1. Risk of Exposure

| Environment | Dots of Exposure of Population[a] | | Control Population Relative Risk |
	Sick	Healthy	
Total home	1,613	1,498	1.08
Home 7 A.M. to 7 P.M.	260	196	1.3
Office	86	490	.18
Factory	518	58	8.9
Restaurant	28	86	.33
Shop	29	55	.53
Other home	86	60	1.43
Vehicles	115	58	1.98
Outdoors	29	144	.20
School	346	403	.86
Institution	5	18	.28
Church	25	10	2.5
Total dots of exposure	2,880	2,880	

[a]Population = 30 sick and 30 healthy individuals.

96 15-minute segments (dots) in a day, each population of 30 has a total of 2,880 dots of exposure to some environment. The sick population has a high relative risk (Vignette 1-2) for exposure to a factory environment and to vehicles during the commute to work. It is much less exposed to the outdoors (perhaps for exercise or recreation). There are also social differences between the two populations, as reflected in more time spent by the sick population in visiting friends, going to church, and staying home, and less time spent in restaurants or institutions such as banks or post offices. The significance of differences between the population can be assessed statistically by using a *t*-test.

Vignette 4-2

STANDARD DEVIATIONAL ELLIPSES

One of the methods used to describe the spread and orientation of sets of points in space is to construct standard deviational ellipses (SDEs). The basic idea derives from the standard deviation of a set of data, which is an indication of variability, dispersion, or spread of a set of data values. The SDE extends this concept of variability to two dimensions. Since the ellipse encloses approximately two thirds of the data points (just as one standard deviation on either side of the mean takes in approximately two thirds of a set of data in one dimension), its size, especially when compared to other ellipses, is a good indicator of how spread out the points are. The major axis of the ellipse is a good indicator of the main direction along which the pattern of points is distributed. Another useful feature of the technique is that each data point can be weighted by the number of people there or any other data item that is concentrated at each point.

The first step in constructing an SDE is to find the mean center of the data points, using the formulas:

$$\bar{x} = \frac{\Sigma x_i \, w_i}{\Sigma \, w_i}$$

and

$$\bar{y} = \frac{\Sigma y_i \, w_i}{\Sigma \, w_i},$$

where $\bar{x}$ and $\bar{y}$ are the coordinates of the mean center, x_i and y_i are the coordinates of the data points, and w_i is the weight at each point.

The second step is to transform the mean center and the original data points by subtracting $\bar{x}$ from the x-coordinates and $\bar{y}$ from the y-coordinates. This has the effect of changing the mean center to the origin point (0,0) and shifting the original data points accordingly.

The next step is to calculate the angles of orientation of the major and minor axes of the SDE using the formula

$$\tan \theta = \frac{- (\Sigma x_i^2 w_i - \Sigma y_i^2 w_i) \pm \sqrt{(\Sigma x_i^2 w_i - \Sigma y_i^2 w_i) + 4\Sigma x_i y_i w_i)^2}}{2 \, \Sigma x_i y_i w_i},$$

where x_i and y_i are the data point coordinates and w_i the weights, as before. Notice that, because of the plus and minus signs in the formula, there are two angles, θ, that will result (those who remember their algebra may recognize the form of this formula as the general solution of a quadratic equation). The two angles will be exactly 90 degrees apart.

Once the two angles are found, one can use them to calculate the length of the major and minor axes using the formula

$$\sigma = \sqrt{\frac{\Sigma(y_i \cos\theta - x_i \sin\theta)^2 w_i}{\Sigma w_i}}.$$

The value calculated using the angle of the major axis is half the length of the minor axis, as it represents one standard deviation away from the major axis. Similarly, the value calculated using the angle of the minor axis is half the length of the major axis.

The formulas shown above can be worked out with the aid of a calculator or programmed into a microcomputer.

Suppose that we wish to know to what extent the activity space of a group of people corresponds to the location of hospitals in an urban area. In Vignette Figure 4-2, the triangle represents the block where the study population resides. The solid dots represent places where the population went to work, shop, and carry out other daily activities. The size of the dots represents the number of trips made to these places during 1 week. The x- and y-coordinates of the dots and the number of trips (weights) were entered into the SDE formulas, and the resulting ellipse was drawn on the graph to represent the study population's activity space. Another SDE was constructed based on the sites of four hospitals (squares on the figure) and their bed count (weights).

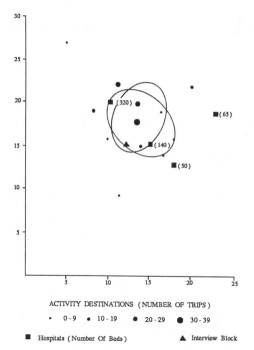

Vignette Figure 4-2. Plotted points and standard deviational ellipses. Standard deviational ellipses are used to compare daily activity spaces of a population and the location and capacity of hospitals in an urban area.

5

The Biometeorology of Health Status

Biometeorology considers how variation and change in the physical and chemical characteristics of the atmosphere affect variation and change in the physico-chemical systems of living organisms. The human mammal is sensitive to a far wider range of atmospheric characteristics than is commonly realized. These include not only temperature, humidity, air movement, solar radiation, sound, and gaseous pollution but also infrasound, magnetism, electrical charge, and atmospheric pressure. In addition, a host of indirect influences derive from the biometeorology of each location on earth. A knowledge of climate is necessary to understand what crops are grown with what reliability or to find pattern in the occurrence of arthropod habitats and vectored diseases. These influences are explained in Vignette 3-1.

Biometeorological conditions are important in establishing the health status of the population. In terms of the triangle of population–habitat–behavior interactions, this chapter is especially concerned with the influences of the physical environment on the internal, physiological status of the population. The physiological, chemico-electrical system of the body forms the habitat for disease agents as well as the mechanism for coping with the social, physical, or other pressures. Each individual is unique in his or her combination of enzymes, hormones, sensitivity thresholds, membrane permeability, and every other feature. For individuals and whole populations, however, physiological status is also influenced by the time of day, and the light, temperature, and altitude of a place. The susceptibility of a person to disease, toxins, and pharmaceutical drugs can be altered by biometeorological changes. This in turn affects the health status of the population vertex of the ecological system. Although there is a tendency in medicine to treat all people as a homogeneous population, in fact human characteristics differ from place to place as well as individual to individual.

This chapter surveys the effects of weather and of electromagnetic radiation on human health. Biological rhythms and acclimatization to temperature and altitude are examined. The physiological basis of climatic influence is described in some detail: sound geographical hypotheses and choice of relevant variables depend on an understanding of the disease process. The influence of elements of weather and their change and passage is described. The effect of biometeorological influences on the seasonal occurrence of death and birth is also considered.

DIRECT BIOMETEOROLOGICAL INFLUENCES

The Radiation Spectrum

The spectrum of electromagnetic radiation is illustrated in Figure 5-1. Very short wavelengths of radiation are known as *ionizing* radiation because they can detach electrons and damage atomic structure. Cosmic rays and x-rays can penetrate into genes and disrupt DNA sequences, which are the body's blueprints. X-rays are principally a human-made health hazard; the atmosphere shields us from their natural sources.

Ulraviolet light is best known for causing sunburn. Because the damage it does depends on the intensity and duration of exposure, it creates some important occupational hazards. Welding, for example, produces such intense ultraviolet radiation that an instant's unshielded exposure can cause severe inflammation of the eye's membrane. Only a small band within the range of ultraviolet radiation can penetrate the atmosphere to any extent. Ultraviolet radiation interacts with oxygen to produce ozone and form what is called the "germicidal band" within the stratosphere. In this zone, air-borne bacteria and fungus spores are killed, and the atmosphere is cleansed.

Ultraviolet radiation also activates a human enzyme to produce vitamin D. The dependence of the human body on ultraviolet radiation to catalyze the body's synthesis of vitamin D posed something of a dilemma for human evolution as people spread over the globe. Too much ultraviolet radiation can cause skin cancer and genetic mutation; not enough can cause a lack of vitamin D, which is necessary to the proper metabolism of calcium. The calcium-

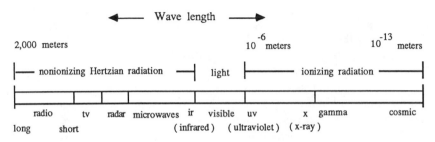

Figure 5-1. The spectrum of electromagnetic transmission. Radiation, from cosmic to radio, is a continuum of wavelength.

deficiency disease, rickets, is characterized not only by bowed legs and mis-shapen backs but also by deformed pelvic bones, which can cause women to die in childbirth. Some radiation must penetrate the skin; too much must be kept out. Yet people live from the equator to the Arctic Circle, with great variation in both annual and seasonal levels of radiation. The evolutionary solution is the deposition of melanin, a dark pigment, in the skin surface of people who live in regions of high radiation and its absence in the skin of people in high latitude areas of low overall radiation. This adaptation to radiation shields people of equatorial lands and exposes people of the high latitudes. Tanning is the solution for people who live in changeable environments with great fluctuations in radiation levels. The ultraviolet radiation mobilizes pigment that is deposited during times of excessive radiation and removed during seasons of low radiation. Behavioral interference with this rather elegant solution, however, is culturally widespread. Europe after the Industrial Revolution built tall buildings and shrouded them in palls of black smoke, with the result that children were exposed to little sunlight. In some Moslem lands, women in purdah are so cloistered indoors and shrouded outdoors that rickets can be a problem even in places with maximum solar radiation. The greatest interference with this evolutionary design is continual human migration. Slavery moved tropical Africans to high latitudes, and northern Europeans colonized Australia, the southern United States, and other sunny places. Negative health consequences were inevitable.

Visible light is biologically very powerful. It stimulates photoreceptive tissue, such as the retina in human eyes. It also changes the electrical charge of protoplasm (the physical base of all living activities) and the viscosity, permeability, and colloidal behavior of proteins. It thereby causes living things to turn toward light and energizes photosynthesis. Light seems to be the most important factor influencing biological rhythms.

Infrared and longer wave radiations are important for heating the earth's atmosphere and surface and cold-blooded plants and animals, either directly or through reradiation by the earth. The health hazards of these long-wave radiations result from their function as penetrating heart sources. Infrared radiation used to be a cause of blindness in such occupations as the manufacture of glass, but careful shielding and new industrial processes have removed most such occupational hazards. Today, microwaves are of some concern as the machinery that produces them proliferates. The waves can penetrate the body at considerable distance and cause internal heating that the body's sensors do not detect. The eyes are at greatest risk because the lens of the eye has no blood circulation to remove the heat. At the end of the spectrum are the longer radio waves. They are all around us but are not known to have ever done anyone harm, not even people intensely exposed because of their occupation.

All of these types of electromagnetic radiation vary over the surface of the earth. The earth's magnetism affects the path and concentration of much radiation. The tilt of the earth results in the seasonal and latitudinal differences described in Vignette 3-1. High elevations that project through thousands of feet of atmosphere experience different amounts and constitution of radiation. Human-made sources of several kinds of radiation are becoming important, and they have a very irregular distribution. Consequences of radiation have been much studied in terms of cell chemistry, microbiology, and physiology, but the spatial distribution of hazards and effects has received little attention, except for skin cancer.

Biological Rhythms

The presence or absence of light is one of the oldest and most universal selective pressures to which all living things have to adapt. An important part of that adaptation has been the development of daily and seasonal rhythms of many body processes. It is clearly beneficial to an organism to be able to anticipate when cold, night, rain, high tide, or spring flowers and grass are going to come so that the slow physiological processes of growing hair, losing leaves, storing food, or coming into rut can be initiated in good time. The precise synchronicity of breeding swarms of many sea creatures and insects, the beaching and egg-laying of turtles and crabs that travel great distances to appear simultaneously at a specific place, and the migration of birds and bats are phenomena that have long fascinated scientists and testified to the existence of biological clocks. The equivalent body rhythms of human beings have been recognized much more recently.

The existence of endogenous "clocks" is now well established for a wide variety of animals, including humans. Even in constant temperature and total darkness some biological processes continue to oscillate rhythmically. The rhythms with a span of about 24 hours (which may stretch to 25 or 26 or shorten to 23) are known as *circadian*. In humans they include sleep and wakefulness; body temperature; cognitive performance; serum hormone levels; urinary cycles and excretion of ketosteroids, chloride, sodium, and urea; and ionization of blood calcium and phosphate, which affects regulating hormones. It is clear that the physiological state of individuals is constantly changing. Ten-fold differences in the susceptibility of mice to bacterial toxins at different times of night have been demonstrated, and it is likely that human circadian rhythms also influence susceptibility to infection and toxins.

The study of circadian and other rhythms can help us understand the ways in which environment directly affects and alters the status of a population. Reception and interpretation of environmental stimuli influence virtually all the body's major organs and endocrine systems (Figure 5-2). The brain's central mediator of the various light and thermoregulation responses is the hypothalamus, comprised of many separate nuclei and nerve connections. It has a direct connection with the retina, outside of the main optic channel to the brain. In mammals the pineal gland, controlled by the hypothalamus, acts as

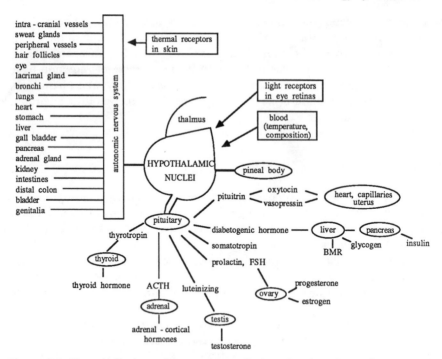

Figure 5-2. Hypothalamic environmental mediation. Sensing the environment, the hypothalamus acclimatizes the body through its direct neural connections to many organs and through the hormones it induces the pituitary to send.

an endocrine transducer (that which converts one form of energy to another). It synthesizes and secretes the hormones melatonin and serotonin, two of the most powerful chemical transmitters in the brain. During the day the pineal synthesizes and stores serotonin to be secreted at night. The filling and emptying of the pineal serotonin "reservoir" seems to be the body's master clock. Besides regulating circadian rhythms of sleep and other bodily functions, the serotonin cycle in the pineal gland is implicated as the clock that controls the development of puberty.

The hypothalamus recognizes light and dark and knows when it is time for another sleep cycle. By controlling the pituitary, the body's master gland, the hypothalamus can adjust most of the body's metabolism, endocrine system, blood vessels, and other organs as appropriate for environmental conditions. As we shall see, this is important for acclimatization.

Most people become aware of circadian rhythms when they travel longitudinally. The physical and mental effects of de-synchronizing the body's rhythms from those of the environment are familiar as jet lag. It is not a matter of fatigue: the effects are the same even if one slept well on the airplane. Rather, body temperature and kidney and hormone cycles are no longer appropriate for the schedule of social activities and daylight. Each biological clock is reset, advancing or slowing an hour or two each cycle, until all cycles are again synchronized with the environment. During the period of jet lag, however, the body may be more susceptible to infectious agents and, through impaired judgment and coordination, to accidents.

The large volume of international travel has made circadian rhythms and their adjustment a matter of concern to business, sports, and tourist industries as well as to diplomats and the military. One topic of interest is the importance of circadian rhythms for work shifts. When people continuously work at night, first the sleep cycle, then the body-temperature cycle, and eventually, it is thought, all the other cycles will be appropriately set to the social activity pattern. In some professions, however, shifts are regularly changed around or even rotated. This means that nurses, policemen, and other 24-hour–service personnel may frequently be involved in critical situations while effectively in a state of jet lag.

Most research on circadian rhythms has been medical. There has been no study of their spatial distribution, their effects on population mobility, or the impact on them from cultural transformation of the environment. Are people who live in the long winter nights of high latitudes, for example, more susceptible to infectious agents than those in equatorial lands? Light is used today to induce chickens to lay more eggs and to bring cattle and sheep back into estrus after a miscarriage or failed fertilization. Does nighttime electric lighting affect human fecundity at all? Do certain schedules of economic activity impact disease susceptibilities? Properly designed studies of these questions would include contrasts between places and cultures and could be very illuminating.

Acclimatization

Acclimatization is the genetically given capacity of humans to adapt, over long periods, their physiological systems to heat, cold, and altitude. People are adapted to living in an extraordinary range of temperatures and other environmental conditions. They live and reproduce in hot deserts and Arctic tundra, as well as in the high Andes and Himalayas.

Behavioral and genetic adaptations are of great importance to acclimatization. Cultural geographers have studied adaptive differences in house-type design and construction, the materials and design of clothing, and the cultural rhythms of daily activity patterns. Genetic adaptations often involve the morphology of the body. In many animal populations, for example, long limbs and slimness serve to dissipate heat efficiently, while squat bodies with a low surface-to-volume ratio serve to conserve heat. Physical anthropologists have identified some such tendencies among human populations, but extensive migration has created as much exception to the rules (tall Norwegians in the Arctic and compact Chinese in tropical Asia) as conformity (compact Eskimos in the Arctic and tall, slender Watutsi in tropical Africa). As important as genetic and behavioral adaptations are, this chapter looks at the physical processes of acclimatization and the implications for health that result from variations in physiologic status of the population.

The physiological processes of thermoregulation are known in some detail. The human body must maintain a temperature of around 98.6° Fahrenheit (37° Celsius) regardless of the environmental temperature. Over short periods, people react by shivering to generate heat from muscle movement and sweating to cool the body by evaporation. People everywhere sweat and shiver at the same skin temperature. Acclimatized people, however, use other physiological means to control body temperature. Their skin temperatures may not reach the threshold values for sweating or shivering.

The hypothalmus learns about environmental temperature from nerve channels to thermoreceptors in the skin, and about body temperature from nerves and the hypothalamic blood vessels. The hypothalamus causes the peripheral blood vessels to dilate (vasodilation) under heat stress, so that more blood can be brought to the surface for cooling, and to constrict (vasoconstriction) in protection against heat loss, so that more blood is held in the core of the body. Over time, the density of the capillary network near the skin will change. Most animals sweat from a few sites (armpits, for example) controlled by the sympathetic nervous system. Humans have evolved an unusual capacity for eccrine (excretory) sweating by glands located all over the body. This involves excreting dilute solutions of water, salts, urea, sugar, and lactic acid for purposes of cooling the surface of the body through evaporation. Under conditions of prolonged heat, the efficiency of this sweating may be greatly increased. The acclimatized person sweats from the body's entire surface; the flux of water can reach several liters in an hour. The consequent drain of blood electrolytes (such as magnesium and sodium) is enormous, even though urinary output is considerably diminished.

By far the most complex adaptations to temperature involve the endocrine system (see Figure 5-2). Through the thyroid gland, the *basal metabolic rate* is altered to generate more or less heat from food. Since different kinds of food, such as fats or carbohydrates, have different capacities for heat generation, dietary needs change. Levels of thyroxin, which regulates carbohydrate metabolism, change, as do levels of iodine required to make thyroxin. Thus, someone living at a cold, high elevation would need more iodine to maintain a high basal metabolic rate and would be more at risk of goiter than someone in a warm place, even if iodine were equally available in the environment. In fact, iodine is *less* available in glaciated mountains, which have soils deficient in iodine and are usually remote from the transportation systems that would facilitate trade and the import of seafood, so goiter is often a serious problem of such areas.

Another endocrine system, the adrenocortical system, stimulates the pancreas to produce more or less insulin and controls the metabolism of the liver. It also affects capillary resistance and the removal and excretion of waste products in the blood. Even the relative amounts of the types of cells that comprise blood change from winter to summer. Blood volume also increases under heat stress and decreases under cold stress. An important adaptation to cold is deposition of a layer of fat under the skin.

The work of the heart, deposition in the blood vessels, blood-clotting time, blood sugar levels, and innumerable other physiological characteristics differ in people acclimatized to different degrees of heat or cold and even in an individual in different seasons.

Acclimatization to high altitude affects far fewer people but in some ways is more dramatic than adaptations to heat or cold. The greatest adaptive stress is from low oxygen tension in the rarified atmosphere: more red blood cells must be produced, and hemoglobin (and iron nutrition) must be increased. The body must change its endocrine balance to prevent acidosis resulting from changed levels of carbon dioxide in the blood. Permeability of cell and capillary walls and the facility with which red blood cells give up oxygen to body tissue must be increased. Lung size, surfaces, and permeability are altered. When fully adjusted to elevations of over 10,000 feet (3,000 meters), the physiology of people alters sufficiently to create much more susceptible, and often hazardous, reactions to therapeutic doses of drugs. Fecundity seems to be lower at high altitudes. When people acclimatized to high altitudes descend to sea level to seek work, they may be susceptible to respiratory infections.

THE INFLUENCES OF THE WEATHER

Weather, as distinct from climate, consists of atmospheric conditions experienced by people on an immediate basis. It changes hour by hour and mile by mile. The frequency and amount that it changes—the variability of weather—is one of its most important characteristics. The extremes of conditions reached are also significant for health.

Air masses, huge chunks of relatively homogeneous air flowing over the earth's surface with a depth of up to a few kilometers, are as familiar to television weather report watchers as the fronts that separate air masses. Although tremendous energy exchanges occur at the fronts, the air masses do not generally mix. They retain their own characteristic temperature, humidity, atmospheric (barometric) pressure, and other properties. The fronts that separate them are accompanied by winds and, especially if associated with squalls and thunderstorms, the generation of ionization, extremely low frequency (ELF) waves, and infrasound (too low in frequency for humans to hear). Let us review these individual elements and their health implications before considering what happens when fronts pass or extreme conditions develop.

Temperature, Humidity, and Air Movement

Temperature, humidity, and movement of air together determine how readily the human body loses its heat. Such heat is generated by metabolic activity, even when the body is at rest. We have considered the major physiological processes by which the body controls the generation, conservation, and dissipation of heat during acclimatization. The ability of the body to radiate heat depends on the surrounding temperature. The effectiveness of evaporative cooling depends on the surrounding humidity and air movement. Temperature–humidity, "comfort," and wind-chill measures portray environmental stress much more accurately than a simple measurement of temperature, which is commonly used by medical scientists.

The most obvious health impact of the weather attends extremes of heat and cold. Death from exposure to cold involves no mysterious mechanisms: when the body's core temperature falls below a certain point, its chemical reactions cease. Heat causes increased peripheral circulation, which can be stressful in its own right, and copious sweating can drain the body of electrolytes. In heatstroke, the sweating mechanism shuts down, allowing body temperatures to rise rapidly. Certain proteins in the brain can permanently change, even if the body is cooled before death results.

Heat waves have long been associated with increased mortality. Persistently high temperatures seem more closely related to mortality than the peak temperatures reached. The most consistent predictors seem to be the average temperature and the number of successive hot days. Increased mortality usually occurs in the first 2 or 3 days following the heat wave's peak. Greenberg (1983) studied the 1980 Texas heat wave and compared it with the heat wave of 1950 and the period from 1970–1979. Temperatures of over 100° Fahrenheit (37.7° Celsius) existed for 61 of 71 days, and 107 deaths were attributed to heat. The risk of death was highest among males, blacks, those engaged in heavy labor, and the elderly. Death rates were higher during heat waves of former years; Greenberg presumes that is because of the increased prevalence of air conditioning today.

There are several predisposing factors for suffering heatstroke. These include the presence of such degenerative diseases as cardiovascular disease, renal (kidney) disease, and hypertension; preexisting acute diarrheal or febrile disease; salt deficiency and dehydration; and the use of drugs that affect the thermoregulatory system—alcohol, amphetamines, and such therapeutic drugs as diuretics and anticholinergics. People who undertake sustained activity in the heat are at risk, as are people whose lack of fitness or acclimatization makes their cardiovascular systems less able to cope with the necessary heat dissipation.

Heat waves have also been associated with increased crime, especially homicide, but the pattern is not systematic or consistent. A plausible explanation is that heat interferes with sleep and therefore with dream cycles. The prolonged disturbance or prevention of dreaming is known to cause psychological distress.

An increase in the number of deaths following heat waves has been repeatedly observed. It is not clear, however, how much of the increase in mortality occurs to those about to die from degenerative disease anyway. Several studies have reported less-than-normal mortality shortly after the peak associated with the heatwave, but the decrease does not always offset the previous increase. There is some concern that people surrounded by air conditioning may be losing their acclimatization to heat so that they are less able to cope with extreme heat stress. Mortality, however, is certainly higher among those socioeconomic groups least likely to enjoy air-conditioned homes.

Air-Borne Life

The air is charged with living things. The transport and survival of bacteria, viruses, fungi, and allergens such as pollen depend on certain conditions of atmospheric temperature, humidity, condensation, and movement. Agents of human disease are injected into the air by coughing and sneezing, by the shedding of hair and dead skin, and by the spray of cooling towers, air conditioners, and irrigation systems. Soil bacteria, fungi, and pollen are picked up by the wind. Dispersal is dependent on atmospheric turbulence. To all such life-forms ultraviolet radiation ultimately is lethal.

Temperature and humidity are the limiting factors for survival in air; they act together with different effects on different organisms. Bacteria often can withstand extremes of temperature but may be seriously affected by humidity. The ubiquitous intestinal bacteria *Escherichia coli* survives for less than 3 hours at 68° Fahrenheit (20° Celsius) and 50% relative humidity, but mycoplasma of human pulmonary origin survived up to 5 hours at 82° Fahrenheit (28° Celsius) and 50% humidity. Viruses can multiply only in living organisms, but they can survive in the air anywhere from a few seconds to a few hours. Polioviruses have been shown to be progressively desiccated in the air, but to have sufficient longevity for dissemination over several miles. Tolerance and preference for environmental conditions are highly specific to the type of organism. In one study, for example, poliovirus survived best at 80% relative humidity, but vaccinia virus survived best at 20% (Hyslopo, 1978).

Air-borne bacteria and viruses have caused epidemics miles from their sources. An outbreak of Q fever in San Francisco, for example, was traced to the fumes from the fat-rendering plant of a slaughter house. The combined effects of temperature and humidity are thought to result in different survival times for air-borne microorganisms in summer and winter and thus to be a critical factor in the seasonal incidence of disease (described later in this chapter).

Both fungi and pollen are suspended in the air. A few fungal infections of the lungs are serious diseases, especially histoplasmosis and coccidioidomycosis. Their distribution is clearly related to temperature, humidity, rainfall, and local winds, which pick up the fungi from disturbed soil. The distribution of human disease may be limited by prevailing winds as well as soil temperature and moisture. Ragweed, the most common pollen source for the hayfever that afflicts more than 6% of the United States population, grows best on cultivated land. Again, rain, wind, and humidity affect dispersal.

Atmospheric Pressure

The weight of the atmosphere presses on the surface of the earth. Subsiding air exerts more pressure than rising air. One atmosphere of pressure, usually defined as 1,013 millibars (mb), is felt as 14.7 pounds per square inch (1,034 grams per square centimeter) at sea level. Atmospheric (barometric) pressure changes are slight in equatorial zones. Over North America normal pressure changes between air masses are on the order of 25 mb. At high latitudes pressure changes may reach 120 mb. The most extreme low and high pressures ever recorded on earth amount to a difference of about 3 pounds per square inch (211 grams per square centimeter). (Altitudinal changes, such as those encountered in balloons or depressurized aircraft, can cause much greater pressure differences).

The major mechanism postulated for why normal changes in atmospheric pressure affect the body's biochemistry is that body volume expands slightly, leading to retention of water and therefore to alteration of electrolyte balance. Eventually, levels of disequilibrium are reached that trigger intervention by higher homeostatic control systems, to restore proper fluid levels. These changes would result in water storage in certain parts of the body—in joints, eyes, and so forth—and could cause joint pain, glaucoma pain, increased blood pressure, blood clotting, and general irritability.

Winds

Winds blow between the different atmospheric pressures that characterize air masses and between zones of subsiding and rising air (Vignette 3-1) as part of the heat balancing of atmospheric circulation. The strength, frequency, and duration of winds vary greatly from place to place.

Air movement is important for cooling the body and for dispersing disease agents. A "red" snow that fell in Sweden in 1969 was caused by high concentrations of bacteria later traced through winds between a low-pressure air mass over Scandinavia and a high-pressure air mass over the Soviet Union, to sandstorms north of the Black Sea. Not only microorganisms ride the winds. Hunter (1980) has shown that the blackflies that are the vectors of onchocerciasis can be disseminated over hundreds of miles by the wind systems associated with the arrival of the wet season in the savanna lands of West Africa. This monsoon effect is a major obstacle to the World Health Organization's attempts to control onchocerciasis over a large area of the Ivory Coast, Mali, Ghana, and Burkina Faso (formerly Upper Volta).

Winds have another effect. The movement of air, especially air with low humidity, promotes the ionization of atmospheric gases. Electrons are stripped from their atoms, producing positive and negative charges. In the extreme form associated with convective build-up of cumulus clouds, segregation of electric charges may result in lightning. The degree of atmospheric electrical charge is one characteristic of an air mass.

Dry winds, such as the Chinook, Santa Ana, foehn, sirocco, and harmattan, are associated across cultures with irritation, bad temper, accidents, and violence. Despite an extraordinary richness of folklore, however, there has been little in the way of controlled studies. The heat itself used to be implicated, but suspicion is now directed at ionization.

Atmospheric Ionization

Ionization in the atmosphere has several sources. It results from radiation impact as well as the friction of movement. The charged molecules are electrical and chemical stimuli with undetermined consequences for human health. Persinger (1980) has reviewed and explained many of the phenomena involved with the "electromagnetic stimuli of the weather matrix." Much of the following discussion draws upon his book.

Atmospheric ions could react with chemical processes in the body by giving or taking electrons, if they had access to biochemical pathways; therefore their potential to be culprits in the etiology of suicide, homicide, aggression, infection, migraines, conjunctivitis, and respiratory congestion needs to be considered seriously. Their physical influence as environmental electrical stimuli is generally dismissed on the grounds that any current the ions might induce in the human body by their presence in the atmosphere would be too small to have effect, even around the brain neurons' electric field. Chemically, however, there is a pathway for possible interaction: by absorption into the blood stream through the lungs, the ions might alter the pH by acting as oxidizing or reducing agents. This could influence a host of chemical pathways and homeostatic sensors and controls.

The serotonin irritation syndrome of migraines, nausea, irritability, edema, conjunctivitis, and respiratory congestion has been correlated with the accumulation of positive ions that occurs as much as a day before the weather changes. The serotonin hypothesis, in brief, is that in the brain a major chemical transmitter whose activities are associated with aggression, sexual activity, pain sensitivity, sleep duration, and other functions is affected by bursts of atmospheric ions. The mechanism is not explained yet, but if it exists, it would be relevant to depression, irritability, suicides, reports of pain, and other cross-cultural correlates of the weather.

Unpleasant effects have usually been correlated with positive ions, whereas negative ions are generally associated with feeling good. Positive ion charges occur naturally before thunderstorms, and they accompany hot, dry winds. Humans generate positive ions by such means as heating systems, in which cold, dry air is heated, making it drier, and blown through metal pipes. Dry throats, mouths, and membranes, which become more permeable to viruses, may result from sleeping and living in heated environments. Therapeutic claims made for machines that generate negative ions for bedrooms have not been proven.

Extremely Low Frequency Waves and Infrasound

A few scientifically sound studies have been done on the effects of geomagnetic storms upon moods or brain function. Most of these have focused on reaction times, which are quantifiable measures. It is possible that the intensity and quantity of ELF waves (3 Hz to 3 kHz or so) generated by such storms could influence the brain since they include the frequencies of the brain's own transmissions. Most ELF energy (generated, for example, by lightning and human-made sources) remains between the earth and the ionosphere and travels very long distances without appreciable attenuation. The major variations in the number and intensity of ELF waves and their sources result from diurnal and seasonal changes.

Infrasound shares some of these ELF characteristics, in that it can travel long distances with little loss of energy. At frequencies too low for humans to hear, the waves have the potential to generate whole-body vibration by resonance. They are generated by severe weather disturbances and, traveling at the speed of sound, could be detectable weather precursors. In laboratory studies of infrasound, 10 Hz at 115 decibels can cause lethargy, euphoria, and loss of time judgment; these are frequencies generated inside closed automobiles traveling at 100 kilometers an hour! At other frequencies, infrasound is associated with nausea and dizziness. The research literature, however, is sparse. There are few weather stations that measure infrasound and fewer correlation studies under controlled conditions.

Passage of Fronts

The passage of a front marks a change of air mass. Especially when the passage is accompanied by squalls, it entails winds, electric charge, and the generation of ELF waves and infrasound. As the air masses change, so do many of the environmental conditions that have been affecting people: temperature, humidity, brightness or cloud cover, suspended living organisms, and atmospheric pressure. The body must cope with all these stimuli and their changes. Accumulated stresses can lower the level of health so that individuals fail to rally (Chapter 2). It is thus not surprising that the passage of fronts has often been related to increased mortality.

There have been numerous studies on the influence of frontal passage and weather changes on morbidity and mortality rates. Weather types have been classified in many ways, and various weather elements have been examined singly or together. The best correlations have been attained when weather is treated as a complex of elements. Those variables most often found to be important include mean temperature, diurnal change, mean dew point, mean visibility, mean barometric pressure, day-to-day change in barometric pressure, mean wind speed, and various comfort indices used to adjust temperature for humidity and wind speed. The most consistent relationships of the rates have been to intensity of change in the weather. The most consistent mortality and morbidity associations with weather conditions have been overall mortality, cardiovascular disease, and asthma; diabetes, respiratory illness, and psychiatric conditions also have some associations with weather.

SEASONALITY OF DEATH AND BIRTH

Death Seasonality

It has been known for at least 2,500 years that some causes of death occur more often in one season than in another. A hundred years ago it was common knowledge in the United States that in summer people died of malaria, yellow fever, cholera, typhoid, gastroenteritis, and tuberculosis. In winter people died of influenza, stroke, and cold-related causes.

The literature on seasonality of mortality has been dominated by one person, Masako Sakamoto-Momiyama of the Meteorological Research Institute in Tokyo. Her work over the last 25 years discovered, explained, and popularized the importance of seasonal patterns and how they are changing.

After analyzing the seasonal pattern of occurrence of scores of diseases at various latitudes, in various climates, and among different age, socioeconomic, race, and ethnic groups, Sakamoto-Momiyama developed general models of variation (Figure 5-3). Mortality patterns in most countries have been bimodal; that is, they have had both winter and summer peaks. With economic and social development, the summer peak has disappeared, and mortality now has its highest incidence in winter. Sakamoto-Momiyama differentiates several types of shift in seasonality. In the *transitory type* (cancer Figure 5-3), the summer peak of incidence gradually shifts into autumn as it decreases, perhaps moving toward a winter peak. The *reversing type* reverses from a summer to a winter peak. It has two subcategories; diseases for which the former summer

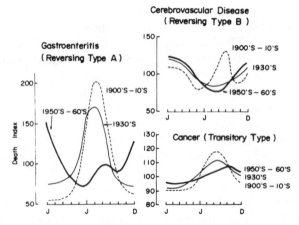

Figure 5-3. Models of seasonal shifts in mortality. When the death index by month is plotted for several time periods, the peak of mortality for some diseases reverses from summer to winter; for others the summer peak gradually disappears, leaving a previous winter peak as the only seasonality; for still others the summer peak decreases and shifts toward fall. From *Seasonality in Human Mortality* (p. 33) by M. Sakamoto-Momiyama, 1977, Tokyo: University of Tokyo Press. Copyright 1977 by University of Tokyo Press. Reprinted by permission.

peak becomes a trough and a new winter peak is created (reversing type A), and diseases that formerly had both a winter and a summer peak but lost the summer one (reversing type B). Gastroenteritis, tuberculosis, and beriberi are type A; many of the degenerative diseases associated with aging are type B.

As Sakamoto-Momiyama relates (1977, p. xiv), while doing reasearch in 1965–1966 in New York she was surprised to find the winter peak disappearing for many causes of death in the United States. Her detailed studies of infant mortality in particular led her to revise her previous idea that technological progress would lead to concentration of mortality in winter, as she had concluded from her experiences in Japan. Instead, she hypothesized that a "deseasonalization" of mortality would occur. Her latest works note that Japan and western Europe are following the United States and Scandinavia into this deseasonalized pattern of mortality. (Sakamoto-Momiyama, 1977)

There can be many reasons why mortality incidence loses seasonality under conditions of economic and technological development. Heavy labor and exposure to the elements decreases. Diet changes, and a variety of nutritious foods becomes available all year. Medical technology advances, and health care becomes more accessible. For some causes of death one can point to specific socio-technological developments; refrigeration and the chlorination of water supplies control summer gastroenteritis, for example; the conquest of polio means that swimming pools no longer start summer epidemics;

changes in house type, screening, occupation, and other factors have led to the eradication of diseases like malaria and yellow fever in some countries. Sakamoto-Momiyama (1977) has advanced the hypothesis that much of the deseasonalization of mortality is due to central heating and air conditioning. She points to the relationships between temperature and cerebrovascular disease (Figure 5-4). Sweden and New York City, which have widespread central heating, show consistent linear relationships without changes of slope at high temperatures, while Western Europe and Honshu in Japan, where many houses are not heated, have seasonal changes of slope. Hokkaido, which is cold in winter, uses heating systems that are turned off when the temperature reaches 54° Fahrenheit (12°–13° Celsius); the winter slope is small until that point and afterwards resembles Honshu's (represented by Tokyo in Figure 5-4).

Much research has been stimulated by these studies: some of the statistical procedures essential to seasonality research are discussed in Vignette 5-1. Although many diseases have been analyzed in detail by epidemiologists, few studies have been done on geographical patterns, and few national studies have been completed by geographers. Kevan and Chapman (1980), for example, confirm the deseasonalization of many diseases in Canada but find great seasonality continuing for bronchitis, pneumonia, influenza, and circulatory diseases (winter) and violence, accidents, and poisoning (summer).

Influenza continues to be a winter disease. The association of cooling winter temperatures with increased influenza mortality, after a certain lag, is indubitable. The association persists over widespread locations and different climates. The reason is unknown. Speculation blames these factors: biometeo-

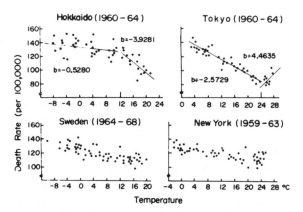

Figure 5-4. Association between cardiovascular mortality and temperature. In Hokkaido (northern Japan), Sweden, and New York, in deseasonalization of mortality seems to be associated with use of central heating. From *Seasonality in Human Mortality* (p. 165) by M. Sakamoto-Momiyama, 1977, Tokyo: University of Tokyo Press. Copyright 1977 by University of Tokyo Press. Reprinted by permission.

rological conditions cause the nasopharynx and trachea to be dry, and membranes become more susceptible to virus penetration; the virus can survive in the air, between hosts, more easily when the air is relatively dry and cold than when it is hot and humid; the lower solar radiation, and hence ultraviolet radiation, promote virus survival; the body's metabolic changes make it more susceptible; people gather indoors in closer proximity for longer periods in winter, schools are in session, and even recreation tends to be indoors—a point emphasized by the importance of room density as a correlate in many influenza epidemic studies. Any explanation of influenza's seasonality must account for its winter incidence in Florida and Hawaii, on the Great Plains and in New England, and in perpetually crowded institutions such as prisons.

Mortality from heart disease is negatively correlated with temperature in several midlatitude countries. Between 30° and 70° Fahrenheit (-1° and 21° Celsius) correlation coefficients of -.95 have been found. A 2.5% increase in mortality has been estimated for each degree Celsius decrease in mean monthly temperature, although the exact relationship differs by climate area. Sometimes mortality from ischemic heart disease but not cerebrovascular disease, and sometimes the reverse, have been correlated with temperature change.

In general, respiratory disease shows the strongest association with winter, and cancer shows the least seasonality—which is hardly surprising, given its long incubation period. The meteorological conditions and physiologic stresses of acclimatization are undoubtedly involved in seasonality, but so are culture rhythms and technology. The etiology of most seasonal patterns of mortality remains puzzling.

Birth Seasonality

Seasonality of birth, found everywhere, has been much less studied than seasonality of death. The seasonal incidence of birth in Japan is quite different from that in countries of Europe, and that of the United States is again distinct. The index of birth (see Vignette 5-1) in Figure 5-5 illustrates the bimodal pattern that has persisted in the United States for decades. The relative peaks differ in magnitude and timing across the country; they have changed over time, but the manner of this change has not yet been studied. There have been a few good descriptions of the seasonal patterns by region, race, and other population characteristics in the United States and Europe. In some studies, the amplitude (height above and below the index line) has been found to be greater for illegitimate births and births to the poor, to the less educated, and to blacks, although there are a few indications that these relationships may be changing. Despite minor differences in the amplitude of seasonality among population subgroups in the United States, however, the pattern of seasonality is the same.

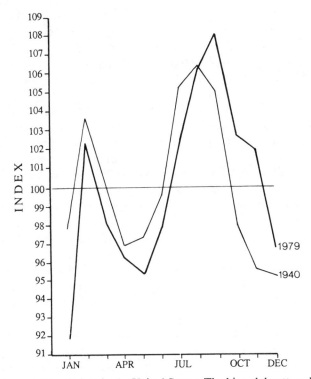

Figure 5-5. Seasonality of birth in the United States. The bimodal pattern has remained remarkably constant for decades.

One of the first studies ever to note and research seasonality of birth was done by a geographer, Ellsworth Huntington (1938). Although his statistical methodology would not pass in an undergraduate term paper today (he did not even adjust for the number of days in a month), he asked some profound questions about the seasonal patterns he documented with data collected from all over the world. He ran transects through the United States, and in the Soviet Union he compared the same ethnic groups in different locales and different ethnic groups in the same locale. He was interested in the effect of climate upon man and was one of the best-known proponents of a school of thought known as environmental determinism (Chapter 1). Probably because of the subsequent discreditation of environmental determinism, geographers have ever since ignored the questions he raised.

One possible reason for birth seasonality is that high temperatures may have an effect upon spermatogenesis (the creation of sperm). High temperatures may also injure sperm. Animal studies have shown that testicular tissue can be injured at very high temperatures. As with other biometeorological effects, threshold and range are presently unknown. How high a temperature over how long is necessary to affect fertility? Is a single episode, such as playing tennis in tight athletic clothing on a hot day, enough exposure? Is birth also becoming deseasonalized?

The seasonal patterns of birth may have nothing to do with temperature. They may be a result of agricultural patterns, school calendars, the timing of holidays and vacations, Christmas–New Year celebrations, and leisure time in the United States, and of equivalent events in other cultures. When people migrate, does their seasonality of birth pattern change relatively suddenly under the new environmental conditions or gradually, over a generation or more, through cultural adaptation? What lags and periods of temperature should be compared, what extremes or averages? In Figure 5-6 birth indices for the mountain, Piedmont, and coastal plain regions of North Carolina have been converted into the first harmonic curves, sine waves showing relative amplitudes and times of peaks and troughs. These are certainly the patterns one would expect if the higher temperatures of the coastal plain and lower temperatures of the mountains made any difference to conception.

There are innumerable consequences for health and disease from birth seasonality. The most obvious is the variable need for health services. Obstetrical wards built for peak periods will have many idle beds during the trough. Birthing classes and maternity leaves from work are affected. More significant effects, however, may lie with congenital birth defects, premature births, and neonatal mortality. There is considerable seasonal variation in neonatal mortality, but its correlates are not clear. The point often made by scientists who study congenital disease is that, if there is seasonality, then the condition is not the result of genetics alone. Something in the environment after conception, whether in the womb, at birth, or after birth, must actually cause the expression of the disease. Since bones, nerves, palates, and endocrine systems form at different times in the development of the fetus, biometeorological influences that might be relevant to congenital disease may not occur at the time of conception at all. This is especially true for the one mental disease that universally shows a seasonality of birth. There is a strong peak of birth for schizophrenics during late winter and early spring. These schizophrenics would have been fetuses in their third month, when the central nervous system is forming, during the late summer heat that is associated (in the United States) with the minimum of conception.

The problem of separating environmental and cultural factors is complex, however, as mothers are likely to be eating different foods, exercising in different ways, and following different work and sleeping habits during summer than they are in winter. It is suspected that for some congenital diseases seasonality may result from the activity of a viral agent. This is an analogy to the congenital effects of rubella (German measles) infections. Most physicians studying a seasonal pattern in a congenital disease will look to changes in food, drink, sickness, and medicines before they will consider the weather.

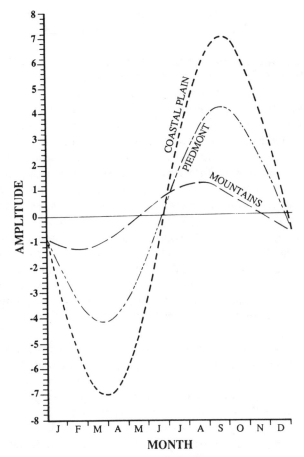

Figure 5-6. Births by region in North Carolina, 1969–1971. The sine waves of the first harmonics show a greater amplitude of seasonality, but similar timing, that might be associated with heat as an etiological factor.

Other Biometeorological Effects

There are many other, more indirect, biometeorological effects on mortality and natality. Chief among these is the effect of climate upon agricultural cycles and cropping patterns, and hence upon many nutritional problems. The "hunger season" recognized in wet–dry climatic zones of Africa, for example, creates nutritional susceptibility to infectious diseases and other stresses. In many places the weather cycle creates periods of high and low employment, of transportation difficulties and poor access to health services, and of stagnant air or atmospheric inversions and intense pollution episodes. Some pollution effects are considered in Chapter 6.

CONCLUSION

There is a vast store of folklore on weather sensitivity. Old war wounds, creaking joints, and restlessness are notorious for being able to predict changes in the weather. Mood swings, migraine headaches, asthma, rheumatism, and heart problems are popularly associated with various weather conditions. As we have seen, it is certainly possible that popular wisdom is not pure myth. Changes in barometric pressure, temperature, ionization, and other weather elements may plausibly affect the biochemistry of the body.

There is remarkably little proof, however, of the effects of weather on health, aside from temperature extremes. The connections between weather conditions and human biochemistry and behavior are very complex matters. Among the difficulties facing the researcher are the classification of morbidity and mortality data, the range and location of weather conditions, and the limitations on doing controlled studies with human beings as subjects. Conditions and data vary with age, season, socioeconomic status, and cultural buffering through the use, for example, of air heating and cooling. Most of the research done so far has been accomplished in Europe with its generally moderate marine climate. The climates of the United States and the seasonality of its latitudes vary in more substantial ways and offer many opportunities for research.

The present status of research is such that little is known of the relevant periodicities. The body's clocks vary from seconds and hours to days, months, and possibly years. Variations in temperature, pressure, or other weather elements and their rates of change occur over seconds, minutes, hours, days, months, and years. Finding the pathways for influences is difficult because of the problem of specifying the relevant periodicity (should one correlate heart attacks with temperature change in a day or over a number of days, with passage of a front or with the number of fronts that passed within a limited time) and our ignorance of threshold or relevant range (is a temperature change from 80° to 60° Fahrenheit the same in effect as one from 40° to 20°; can winter in countries whose mean daily temperatures never get below freezing be compared with that in countries whose temperatures stay below freezing for weeks). The general scientific law of initial value, applied to biometeorological studies, would state that any weather influence is going to differ with both the level from which it started and the level of the physiology it is affecting. For example, if a certain influence raises blood pressure, the health result depends to a large degree on the initial level of the blood pressure. At what point, then, can we detect effects? The tremendous range of individual variation in physiology further complicates study.

In sum, health effects from biometeorological changes are believed to occur, and plausible pathways for many influences exist. There is a clear need for broad-based, multifactorial, methodologically sound research.

REFERENCES

Bovallius, A., Roffey, R., & Henningson, E. (1978). Long range air transmission of bacteria. *Applied and Environmental Microbiology, 35,* 1231–1232.

Calot, G., & Blayo, C. (1982). Recent course of fertility in western Europe. *Population Studies, 36,* 349–372.

Campbell, D. E., & Beets, J. L. (1979). The relationship of climatological variables to selected vital statistics. *International Journal of Biometeorology, 23,* 107–114.

Cech, I., Youngs, K., Smolensky, M. H., & Sargent, F. (1972). Day-to-day and seasonal fluctuations of urban mortality in Houston, Texas. *Biometeorology, 23,* 77–87.

Center for Disease Control. (1980). *Morbidity and Mortality Weekly Report, Annual Summary 1980.* Atlanta: U.S. Department of Health and Human Services.

Cowgill, U. M. (1966). Season of birth in man: Contemporary situation with special reference to Europe and the Southern Hemisphere. *Ecology, 47,* 614–623.

Dalen, P. (1975). *Season of birth: A study of schizophrenia and other mental disorders.* Amsterdam: North Holland.

Driscoll, D. M. (1971). The relationship between weather and mortality in ten major metropolitan areas in the United States, 1962–1965. *International Journal of Biometeorology, 15,* 23–39.

Driscoll, D. M. (1983). Human biometeorology in the 1970's. *International Journal of Environmental Studies, 20,* 137–147.

Fellman, B. (1985). A clockwork gland. *Science 85, 6,* 76–81.

Folk, G. E., Jr. (1974). *Textbook of environmental physiology.* Philadelphia: Lea & Febiger.

Greenberg, J. H., Bromberg, J., Reed, C. M., Gustafson, T. L., & Beauchamp, R. A. (1983). The epidemiology of heat-related deaths in Texas—1950, 1970–79, and 1980. *American Journal of Public Health, 73,* 805–807.

Hansen, J. B. (1970). The relation between barometric pressure and the incidence of peripheral arterial embolism. *International Journal of Biometeorology, 14,* 391–397.

Hollander, J. L. (1963). Environment and musculoskeletal diseases. *Archives of Environment and Health, 6,* 527–36.

Hunter, J. (1980). Strategies for the control of river blindness. In M. S. Meade (Ed.), *Conceptual and methodological issues in medical geography* (pp. 38-76) Chapel Hill, NC: University of North Carolina, Department of Geography.

Huntington, E. (1938). *Season of Birth.* New York: Wiley.

Hyslopo, N. S. G. (1978). Observations on the survival of pathogens in water and air at ambient temperatures and relative humidity. In M. W. Loutit & J. A. R. Miles (Eds.), *Microbial ecology* (pp. 197–205) Berlin: Springer-Verlag.

Keller, D. A., & Nugent, R. P. (1983). Seasonal patterns in perinatal mortality and preterm delivery. *American Journal of Epidemiology, 118,* 689–698.

Kevan, S. M., & Chapman, R. H. (1980). Variations in monthly death rates in Canada: A preliminary investigation. In F. A. Barrett (Ed.), *Canadian studies in medical geography* (pp. 67-77). Downsview, Ontario, Canada: York University, Department of Geography.

Krueger, A. P., & Reed, E. J. (1976). Biological impact of small air ions. *Science, 193,* 209–213.

Persinger, M. A. (1980). *The weather matrix and human behavior.* New York: Praeger.

Rose, G. (1966). Cold weather and ischemic heart disease. *British Journal of Preventative and Social Medicine, 20,* 97–100.

Rosenberg, H. M. (1966). Seasonal variation of births in the United States, 1933–63. In *Vital and Health Statistics* (Series 21, No. 9, pp. 1–42). Washington, DC: U.S. Department of Health, Education, and Welfare.

Sakamoto-Momiyama, M. (1977). *Seasonality in human mortality.* Tokyo: University of Tokyo Press.

Sakamoto-Momiyama, M., & Katayama, K. (1967). A medical-climatological study in the seasonal variations of mortality in the United States of America. *Papers in Meteorology and Geophysics, 18,* 209–232.

Sakamoto-Momiyama, M., & Katayama, K. (1971). Statistical analysis of seasonal variation in mortality. *Journal of the Meteorological Society of Japan, 49,* 494–509.

States, S. J. (1976). Weather and death in Birmingham, Alabama. *Environmental Research, 12,* 340–354.

States, S. J. (1977). Weather and deaths in Pittsburgh, Pennsylvania: A comparison with Birmingham, Alabama. *International Journal of Biometeorology, 21,* 7–15.

Susser, M. (1973). *Causal thinking in the health sciences.* New York: Oxford University Press.

Takahashi, J. S., & Zatz, M. (1978). Regulation of circadian rhythmicity. *Science, 217,* 1104–1110.

West, R. R., & Lowe, C. R. (1976). Mortality for ischaemic heart disease: Inter-town variation and its association with climate in England and Wales. *International Journal of Epidemiology, 5,* 195–201.

Further Reading

Barry, R. G., & Chorley, R. J. (1982). *Atmosphere, weather, and climate* (4th ed.). New York: Holt, Rinehart, & Winston.

Critchfield, H. J. (1983). *General climatology.* Englewood Cliffs, NJ: Prentice-Hall.

Jusatz, H. J. (1966). The importance of biometeorological and geomedical aspects in human ecology. *International Journal of Biometeorology, 10,* 323–334.

Kavaler, L. (1981). *A matter of degree: Heat, life, and death.* New York: Harper & Row.

Rapoport, A. (1969). *House form and culture.* Englewood Cliffs, NJ: Prentice-Hall.

Terjung, W. H. (1966). Physiologic climates of the conterminous United States: A bioclimatic classification based on man. *Annals of the Association of American Geographers, 56,* 141–179.

Tromp, S. W. (1980). *Biometeorology.* London: Wiley/Heyden.

Vignette 5-1

MONTHLY INDEXES

The study of seasonality of events is complex because the pattern of the whole year has to be compared, whether as months or as weeks. Various, often complex, statistical analyses (such as correlation–regression, Box–Jenkins ARIMA modeling, and spectral analysis) have been used to study seasonality. Beware. It is easy to summarize and compare patterns in a careless and invalid manner. Whatever methods of analysis are used, three methods have proved essential to preparing the pattern for analysis or comparison. These are smoothing, adjustment, and indexing.

Smoothing serves to tone down random fluctuation and other "noise," which can be especially important when total numbers are small. A pattern, that was clouded and confused in the raw numbers often shines through smoothed data. There are several methods of smoothing. The simplest and most common is to create a running average. For example, 1980, 1981, and 1982 yield a mean value used for 1981; 1981, 1982, and 1983 yield a mean value for 1982; and so on. Smoothing in this manner is more appropriate for yearly running totals, however, than for months, because of the variation in the number of days in a month. When smoothing monthly data in this manner, one ends up with periods of 92 days for one average and 89 days for another. The irregular pattern of month length increases such period differences rather than smooths them. Another method of smoothing, illustrated in Figure 5-6, involves trigonometric functions such as sine waves (in harmonic analysis) to identify the periodicity and strength of variation.

The most important step is to *adjust* for the number of days in a month. Comparing deaths in February and August will not do. There are different types of adjustment formulas, but they all do essentially the following:

$$I_i = \frac{M_i}{D_i} \times \frac{365}{12}$$

where i = the month, I = the monthly death index, M = the number of deaths, D = number of days.

The adjusted monthly deaths may be used to create monthly adjusted death rates by dividing by the total population. The procedure also works for age categories, to create age-adjusted monthly rates.

Indexing portrays each month's incidence as the percentage above or below what would be, under conditions of even distribution throughout the year, the monthly average after adjustment for the number of days (Figure

Vignette Table 5-1. Monthly Adjustment and Indexing

Month	Number of Days	Reported Morbidity from Mumps, 1980	Smoothed Number	Adjusted Number	Index
January	31	1,095		1,074	125.9
February[a]	29	1,168	1,238	1,225	143.6
March	31	1,451	1,230	1,424	166.8
April	30	1,070	1,202	1,085	127.1
May	31	1,085	935	1,065	124.7
June	30	651	672	660	77.3
July	31	280	382	275	32.2
August	31	215	238	211	24.7
September	30	220	278	223	26.1
October	31	398	323	391	45.8
November	30	352	366	357	41.8
December	31	349		342	40.1
Total		3,534 + 42 unknown			

[a]Leap year

Note. Mumps morbidity from *Morbidity and Mortality Weekly Report, Annual Summary 1980* (p. 3) by the Center for Disease Control, 1980, Atlanta: U.S. Department of Health and Human Services.

5-5). When the above equation is multiplied by $(1,000/TM)$, where TM is the total mortality, the result is an index based on 100.

The results of using these three methods is illustrated in Vignette Table 5-1.

6

The Pollution Syndrome

The economically highly developed countries and their subregions are characterized by a new kind of environment. Since the beginning of the Industrial Revolution, exponentially increasing kinds and amounts of metals, gases, and chemicals have been added to the air, water, and soil of these places. Many of the chemicals have never existed on earth before. Because biological processes to break them down have not evolved yet, some chemicals persist for very long periods. Some of the gases and especially the metals have always been widespread health hazards but have never been so concentrated. The spread of industrialization around the world also brings technologically created hazards to health. The purpose of this chapter is to survey environmental processes and their spatial context as a background for research on diseases in developed countries.

Dubos (1965) points out that humankind is now adapting, genetically and culturally, to the environments that humans have built. We spend most of our time in our homes and workplaces. In industrialized countries most people live in cities, but even in rural areas cultivated land and settlements create the environmental stimuli that surround people. Coping with these new stimuli is largely under the control of culture, because genetically we as a species change very slowly. Dubos notes, for example, that there is no reason to think that our eardrums are any more able to withstand vibration than those of cavemen 200 generations ago. Yet, noise levels have increased dramatically.

Normally audible sound consists of vibrations from 30 to 20,000 cycles per second. Loudness is measured in decibels (dB), a logarithmic scale based on sound pressure levels in which a doubling of the intensity of the sound is represented by an increase of 3 dB. Because the ear is more sensitive to sounds in the 4,000 to 6,000 cycle range, it takes less intensity to injure it in these ranges than at higher or lower frequencies. The "A scale" (dBA) was constructed to assess total noise by measuring four frequencies and adjusting for their different effects. Exposure to levels above 90 dBA can produce deafness. Legally, therefore, the maximum occupational noise exposure for 8 hours in a day has been set at 90 dBA. Hearing loss, however, is thought to occur in susceptible individuals exposed to 75 dBA over an 8-hour daily exposure for a working lifetime and in a considerable portion of those similarly exposed at 85 dBA. The best available information indicates that over 3 million people in the United States are exposed to 90 dBA or above at their workplace, over 4 million additional people are exposed to between 85 and 90 dBA, and almost the entire industrial workforce is exposed to levels exceeding 75 dBA. The United States Environmental Protection Agency (1973) estimated that 6 million people had enough hearing loss from noise exposure for them to be designated handicapped.

Although this chapter focuses on chemicals and radiation, other serious health consequences, such as the effects of noise, are also associated with the environmental changes that accompany economic development. If the chemical and physical insults described in the discussion of Audy's definition of health (Chapter 2) were mapped, the surfaces of the industrialized countries would appear as jagged mountain ranges and stratospheric plateaus.

TOXIC HAZARDS OF NATURAL AND ECONOMIC ORIGINS

Toxic hazards to health and life are nothing new. In fact, they occur naturally in the environment. Poisonous and carcinogenic gases are emitted by swamps and volcanoes. Toxic chemicals are produced by fungi parasitic on rye, wheat, peanuts, and other crops, especially in wet years. One such fungal toxin that causes ergotism, for example, has been blamed by scientists for everything from the medieval Saint Vitus' dance to the Salem witch trials. Food plants, such as soy beans, cabbage, and wheat, produce an array of chemicals, designed to protect themselves from fungi and arthropods, that can damage the liver, destroy red blood cells, block the absorption of protein or iodine, cause allergies, and generally poison livestock and humans. The air can be polluted with the spores of fungi that cause the occasionally fatal blastomycosis (valley fever) and coccidioidomycosis (desert fever). Well and stream water can be radioactive because of source area material. To these and other naturally occurring pollutants, however, humans are adding exponential increases of their cultural creations.

Table 6-1 lists a few of the pollutants and health effects of current concern. Of perhaps greatest concern are the increasing numbers of chlorinated hydrocarbons and organophosphates, many of which are known to be carcinogenic or teratogenic (causing malformation of the fetus). Many substances that are dangerous as pollutants are important economically. Lead is an important stabilizer, and it seems essential in many batteries. Polychlorinated biphenyls (PCBs) are noninflammable, have a high plasticizing ability, and have a high dielectric constant. They therefore are widely used in transformers and capacitors, as heat transfer and hydraulic fluids, as plasticizers in adhesives and sealants, and as anticorrosion coats for electric wires, lumber, and concrete. As PCBs leak, leach and vaporize, however, they become air and water pollutants of great concern and occasional accidental polluters of food and feed. They are stored in fatty tissue, pass along the food chain, and persist for long periods in the environment. The organophosphates (see Table 6-1) have become so useful that the world's food supply relies heavily on them.

Lead offers an excellent example of how our cultural capacity to alter the environment is outpacing our biological ability to adapt. Lead is common in the rocks on the earth's surface and therefore in its waters. It usually occurs at low levels, and in some places it probably has always posed a health hazard. Lead is a systemic poison that interacts with a range of body chemicals. Mainly it interferes in blood formation by retarding the maturation of red blood cells in bone marrow, but it is implicated in chronic nerve disease and brain damage because it affects copper metabolism, among other things. Lead accumulates in the bones, where it replaces calcium. Normally about twice as much lead is derived from food as from either air or water, and about 75% is excreted in the feces, a little is passed in urine, and a little more than 10% is stored in the bones. Up to 50% of the lead inhaled reaches the blood, whereas less than 10% of ingested lead does.

In 1967–1971, the United States Public Health Service surveyed several hundred thousand children in 21 cities and found that 26% of them had dangerously elevated lead levels. Treating for chronic lead poisoning is difficult, since the process of chellation, by which a substance combines with the lead and causes it to be excreted from the body, can result in more lead being pulled out of the bone storage. This can precipitate an acute blood lead crisis that did not exist before treatment. Thus prevention has to be given priority over approaches that emphasize treatment.

Pica, the eating of nonfood substances, has long been a major source of lead poisoning. Although children in old houses in poor sections of a city may still be poisoned by eating paint flakes, lead has been banned from internal paints for decades. Automobile emissions are the main focus of concern. The vertical (from the ground) and horizontal (from the road) gradients of lead levels in the air, as well as isotopic matching of local aerosol lead and local gasoline lead, provide irrefutable evidence of the importance of automobile emissions. The transportation network constitutes the broad geographical pattern of lead pollution. Levels of lead in the dirt of agricultural fields and in the dust of urban apartments decay sharply away from traffic (distance decay is discussed in Vignette 6-1). Hunter (1977) has reported a strong summer seasonality to childhood lead-poisoning crises. He suggests that this is due not only to seasonal exposure to gutter dirt or automobile emissions, but to solar radiation and biosynthesis of vitamin D, which results in mobilization of the body's burden of lead. Another source of lead is the use of sewage sludge as fertilizer in irrigation systems. Domestic sewage has almost everywhere been contaminated by industrial effluents, with the result that heavy metals such as lead and cadmium are present at relatively high levels. With repeated application of sewage, lead can quickly build up in the soil and thus pass into the human food chain.

Table 6-1. Some Environmental Pollutants and Health Effects

Agent	Affected Organs	Health Effects	Economic Source
Sulfur oxides	Lining of respiratory tract	Chronic burning, phlegm, susceptibility to viruses	Burning of wood, coal, petroleum
Carbon monoxide	Hemoglobin	Damage to heart, brain, blood vessels from waterlogging and hemorrhage; death	Incomplete combustion of petroleum, coal, or wood; cigarettes
Lead	Nerve tissue; accumulates in bones, active in blood, liver, pancreas, kidney	Anemia fatigue, headaches, abdominal pain; kidney disease; clumsiness, convulsion, paralysis of extensor muscles; brain damage	Pica of crayons, paint; solder of cans, pipes; leaded gasoline aerosol; smelters, storage batteries, typesetting, moonshine equipment
Mercury	Brain, bowels, transplacental transmission	Minamata disease, liver and kidney disease, diarrhea; lack of coordination, numbness, convulsions, death	Chloralkali plants, pulp and paper processing, electrical industries; fungicides in water and food supply
Cadmium	Blood vessels, kidney	Hypertension; bone softening and fractures; kidney disease; cadmium emphysema	Mining many metals; electroplating; stabilizer for polyvinyl chloride; batteries, engine alloys, cigarettes, pigment
Chlorinated hydrocarbons	Fat tissue, liver	Hydrocarbon toxicity from photochemical reaction products	Vehicle emissions; processing, storage, transfer of petroleum products; organic solvents: rubber, plastics, paints, lacquer; dry cleaning
Organophosphates	Nerve–muscle synapsis	Chlorinated hydrocarbons: dizziness, headache, muscular weakness, incoordination, liver and cardiovascular diseases, convulsions, bone-marrow disease; blocks breakdown of acetycholine, which transmits nerve impulses; accumulation leading to twitch convulsions, blurred vision; diarrhea; stillbirth	Insecticide (DDT, lindane, dieldrin, aldrin, chlordane, toxaphene), hexachlorophene: shampoos, deodorants, insecticides, herbicides (malathion, diazinon, parathion)

Table 6-1. (*continued*)

Polychlorinated biphenyls (PCB)	Fat tissue, liver	Inhibit growth of cells and interfere with enzymes; enhance action of organophosphates; yusho disease, growth disturbance; fatigue, nausea, jaundice, diarrhea, cough, asthma; acne, loss of hair, numbness; nervous system disturbance, joint deformity at birth	Noninflammable, high plasticizing ability, high dielectric constant, insoluble in water; used as heat-transfer media, transformers and capacitors, solvents in adhesives, sealants, anticorrosion; paints and rubber, ink and brake linings, antioxidants
Dusts of quartz, silica, carbon, asbestos, cobalt, iron oxides	respiratory interstitial tissue	pneumonconiosis (scarring of lungs); silicosis, black lung	mining, sandblasting, quarrying, pottery and ceramics, stone masonary
Beryllium	lungs	sarcoidosis	metal alloys for heat stress, coal-burning
Hydrogen sulfide	respiratory center in brain	paralysis of respiration, consequent edema, hemorrhage, death	oil wells and refineries, sulfur and protein decay in sewers and mines
Fluoride	bones, teeth	binds to magnesium, manganese, and other metals to interfere with endocrine function and enzymes, damage to calcium metabolism and pituitary water balance; dental and skeletal fluorosis and osteomalacia if calcium intake inadequate	food and water; aluminum and other smelting
Asbestos	pleura and peritoneum	mesothelioma, rapidly fatal once symptomatic	mining asbestos; brake linings, fireproofing, and talcum powder; cement, ceiling tiles, clothing

Note. Derived from *Health Effects of Environmental Pollutants* by G. L. Waldbott, 1973, St. Louis: C. V. Mosby.

CHEMICAL POLLUTION

Air Pollution

It took a series of episodes in foggy Christmastime London (1952) and in the industrial valleys of Pennsylvania (1948) to raise public and official consciousness about air pollution. In these episodes, smog (a word coined from "smoke" and "fog") resulted in numerous deaths "in excess of normal." In discussing the policy and health implications of air pollution, one needs to distinguish between two quite different forms with different locational characteristics.

In Figure 6-1 the chains of *industrial*, or "gray sky," smog and *photochemical*, or "brown sky," smog can be distinguished. In many industrial processes, sulfur is released during the burning of large quantities of coal and other fossil fuels. Sulfur, combining with oxygen and eventually with water vapor, can become dilute sulfuric acid and damage buildings and cloth, as well as lungs. Small particles of metals may also be emitted by some industries, and smelting of ores may emit heavy metals such as lead. Iron, manganese, and titanium in the air have been most often associated with industrial sources. Particulates can include these metals, but more commonly are carbon, or soot, based. The several ways of modeling these point sources of pollutants usually focus upon mathematical models of plumes. Trigonometry can help predict

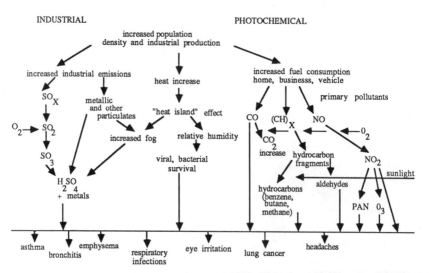

Figure 6-1. Air pollution chains. The two major types of air pollution are comprised of different chemical reactions and caused by different environmental and economic circumstances. The survival of infectious agents in the air may also be affected.

where the pollutants will settle by calculating from the height of the ejection point (smokestack), wind speed, thermal stability, topography, and so forth. As numerous point sources merge into the areal base of a large metropolitan area, the dynamics of atmospheric mixing and movement become more relevant. The rate of emission is a major concern, of course, but so are wind speed and mixing height.

Photochemical smog is often described as a soup. Some chemicals in the chain reactions, including some of the most dangerous, exist for only fractions of a second. Carbon monoxide is deadly because the blood's hemoglobin binds to it, instead of oxygen, so readily; but it eventually becomes carbon dioxide. Nitrogen dioxide causes the red color of smog and some of the choking and eye burning, but it is most dangerous when it forms secondary pollutants by combining with hydrocarbon fragments in sunlight. The emissions of automobiles are the most important and widespread source of these pollutants; sunlight is needed to catalyze their reactions. Photochemical smog is a pollution of sunlight and daytime; industrial smog is a pollution of fog, clouds, and stagnation at different times of day or night.

Both forms of smog can be common health hazards in places characterized by frequent atmospheric *inversions*. Normally the temperature decreases with altitude up to 40,000 feet (12 kilometers). The regular rate of decrease is 3.5° Fahrenheit per 1,000 feet (6.5° Celsius per kilometer) and is known as the adiabatic lapse rate. Since warm air rises, the polluted and heated air from the surface rises and is dispersed through atmospheric mixing. There are several ways, however, in which the air at lower altitude can become cooler than the air above it (see Figure 6-2). Then the normal temperature gradient is inverted, and warmer air above cooler air forms a "ceiling" that prevents dispersion of the surface pollution. On clear nights the earth's surface can radiate heat rapidly, causing air in contact with the surface to cool more rapidly than the air mass above it. Cool breezes can blow inland from adjacent water, causing the surface air to become cooler than the air above. Cold air can drain down hillsides into valleys, moving under the warmer valley air. Such conditions cause frequent nighttime inversions and morning fog, which disappear with the warmth of morning sunshine. They are shallow, surface inversions.

An inversion at higher elevation can occur when there is atmospheric subsidence and divergence. Such upper air inversions can occur under stagnating anticyclones, and they tend to occur in certain latitudinal zones where subsidence is common at least seasonally (see Vignette 3-1). Los Angeles became notorious for its smog not only because it has so many automobiles, but because its location is characterized by onshore breezes; downhill air drainage; radiational cooling on clear, cloudless nights; lots of sunshine; and summertime atmospheric subsidence. Of all these factors, only the automobile emissions are subject to control.

Frequencies of inversions and the associated health hazards of smog vary greatly not only from place to place, but seasonally. Industrial emissions and

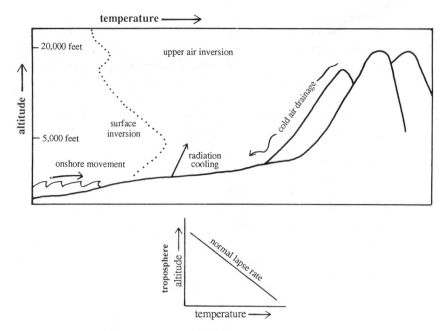

Figure 6-2. Temperature inversions. The normal lapse rate may be reversed either near the ground or at high elevations.

automobile emissions vary over the course of a week, as well as diurnally. Some pollutants persist for days, and others for seconds or less. The various cycling times of years, seasons, weeks, days, hours, and minutes form a sampling nightmare for the researcher. Some particulate matter, such as lead, has a steep distance decay that may move from dangerous to inconsequential within 200 yards of an intersection; some gases, such as sulfur dioxide, may persist for thousands of miles. Some pollution-related diseases may take years, or even decades, to develop. Sampling methods depend on the purpose of one's study. How fine a grid should be used? What periodicity of monitoring? It has proved difficult to connect specific types of air pollution with health hazards such as bronchitis or lung cancer in part because of the difficulties of valid sampling and appropriate scale. Researchers have often had to resort to classifications of urban or rural, for example, as surrogate measures for air pollution. These classifications, unfortunately, bring with them many confounding factors (see Vignette 6-2).

In 1985 a study by the United States Environmental Protection Agency added a new dimension to the air pollution picture (Shabecoff, 1985). The purpose of the study was to develop ways to measure individual exposure levels to toxic substances in the air (Mage & Wallace, 1979). This technology would certainly be useful for the kind of micromobility studies of exposure described in Chapter 4. The startling finding, however, was that exposures to toxic chemicals were sometimes 70 times higher indoors than outdoors. In addition, the high indoor exposures were as likely to occur in the study sites in rural North Dakota and Greensboro, North Carolina (places without major chemical industry), as in Elizabeth and Bayonne, New Jersey (cities in the heart of an enormous petrochemical and refinery zone). Only 11 chemicals were studied in 1985: chloroform, 1,1,1-trichloroethane, trichloroethylene, benzene, carbon tetrachloride, perchloroethylene, meta-para-dichlorobenzene, meta-para-xylene, styrene, ethylbenzen, and othoxylene. Known indoor toxins such as formaldehyde (from insulation and furniture) and radon (from building materials) were deliberately left out. Yet the study found it probable that such consumer products as paints, cleansers, propellants, plastics, and cosmetics as well as adhesives, fixers, resins, and building insulation materials were the sources of the pollution. This landmark study, which measured daily the chemicals in the participants' bodies, found significant correlation betwen blood levels of these chemicals and visits to gas stations and dry-cleaners, use of paint or solvents at home, and smoking. Important studies of air pollution in domestic space are sure to follow, but the habitat dimension of pollution hazards has been interlocked again with the behavioral dimension of house type and clothing, hobbies, and customs of personal and domestic hygiene (see Chapter 2).

Water Pollution

Several episodes have been responsible for attracting research attention and public interest to the hazards of water pollution. One of the first occurred in Minamata, Japan, during the 1950s when more than 200 people died or suffered severe brain and nerve damage. The cause was methyl mercury polluting the water and being passed through the food chain in fish to the population. In the 1970s Love Canal, near Niagara Falls, New York, focused national attention on the problems of hazardous waste disposal and the dangers of groundwater contamination. Since then, gasoline leaks from underground tanks, PCB dumps into rivers and roadsides, and leakage from municipal landfills have poisoned the water supplies of scores of communities and forced Congress to create a superfund for toxic-waste cleanup. As in the case of air pollution, episodes of extreme manifestations serve as warnings of chronic, low-level exposures of unknown consequence.

Although most people in the world get their domestic water supply from surface water or shallow wells, many municipal water supplies in the United States and large cities elsewhere tap artesian sources. Figure 6-3 illustrates the strata of water sources. Runoff on the surface fills streams, lakes, and reservoirs. Water that percolates into the soil eventually reaches the water table, or saturated zone of ground, above an impervious stratum of rock. This ground-

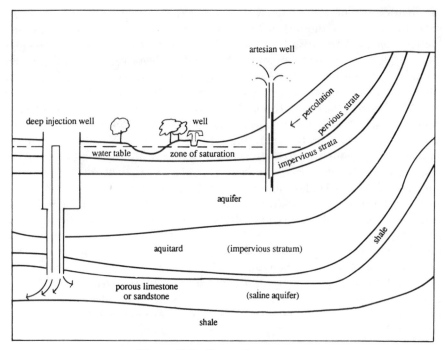

Figure 6-3. Groundwater strata.

water may also intersect depressions and form springs or help fill lakes. Traditionally the water table has been universally tapped by more or less shallow, hand-dug wells. Today bore holes are drilled deeply to form tube wells which, with their engine-powered pumps, are becoming important sources not only of irrigation water, but also of uncontaminated domestic water. In lands where schistosomiasis is prevalent, for example, tube wells are providing parasite-free village water sources for bathing and laundry.

Below the stratum of impervious rock there often is another layer of pervious rocks, such as sandstone, and a layer of impervious rock below that. Most rock strata are tilted. Where the pervious layer has contact with the surface, precipitation percolates into it to form an *aquifer* (see Figure 6-3). The source area of this precipitation may be many hundreds of miles from a desired well site, so aquifer water may be under pressure if the rock strata are even slightly tilted. An artesian well drilled through the rock cap into the aquifer may, if the location is right, flow under its own pressure. Although aquifers are filled by precipitation, in some places the water that originally filled the aquifer fell in other climatic periods and other continental locations. When an aquifer is tapped for water that cannot be replaced by precipitation, the water is being "mined."

There may be several layers of aquifers. Often the lowest one is saline. Injection wells are sometimes used to dispose of toxic wastes in saline aquifers. Under great pressure, and through wells constructed in layers and sealed to prevent leaching into higher strata, liquid wastes are injected into the saline aquifer. The Environmental Protection Agency (1977) estimates that about 260 billion gallons of toxic wastes are disposed of in this way each year.

The Council on Environmental Quality (1983) estimates that in 1981 the chemical and petroleum industries produced 71% of the hazardous waste in the United States, the metal-related industries 22%, and other industries the remainder. Hazardous by-products included heavy metals, solvents, and organic residues from the paint industry; organic chlorine compounds from plastics; organic solvents from medicine; and heavy metals, dyes, organochlorines, and solvents from the textile industry. Not all toxic chemicals entering groundwater are waste products. Figure 6-4 illustrates some of the sources of water

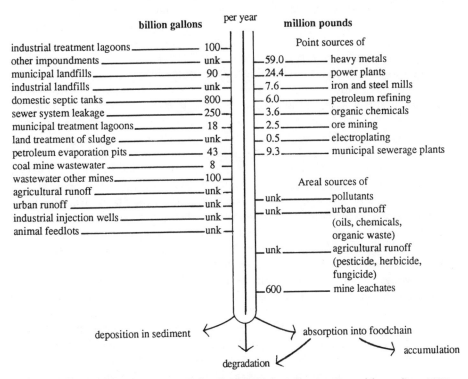

Figure 6-4. Sources of water pollution in the United States. Quantities reflect 1977 rates. Data from *The Report to Congress: Waste Disposal Practices and Their Effect on Ground Water* (p. 265-364) by the U. S. Environmental Protection Agency, 1977, Washington, DC: U. S. Government Printing Office and from *Environmental Trends* (p. 257) by the Council of Environmental Quality, 1981, Washington, D.C.: U.S. Government Printing Office.

pollution. Point sources include landfills, industrial treatment lagoons, dumping or leakage, and sewerage plants. These sources can be identified and more easily subjected to regulation and inspection than areal sources can be. Insecticides, herbicides, and fungicides leach and wash off fields over thousands of square miles. Rainfall in cities runs off roads and parking lots and eventually washes oil, chemicals, and animal feces into water systems. Abandoned mines provide an extensive source of leachates rich in heavy metals, sulfur, and sometimes radiation. Groundwater also dissolves common minerals from the rock through which it moves, and some sources have high natural levels of amorphous silica, fluorite, dolomite, calcite, gypsum, or carbonates. Some water in the southwestern United States is naturally high in lead and cadmium.

As with any pollution, the critical parameters are amount, mixing and dispersal, cleansing, and decay. The fate of chemicals in surface or groundwater is very different in these respects. Surface waters can be easily polluted, and the pollution can spread rapidly, be focused into high concentrations in time, and be flushed and removed. Large lakes and estuaries require a longer time to accumulate the pollutant, and perhaps decades to turn over and be cleansed. In contrast to these, groundwater quantities are enormous and spread over entire basins and even continents. The water is so slow moving that sometimes it is difficult to determine the direction of flow. It may take years for the pollutant to move more than a few feet from its source. It is difficult for pollutants to get access to groundwater because of filtration by the earth, but once contaminated groundwater is nearly impossible to cleanse and very long periods (up to hundreds of thousands of years) may be required for its renewal. The actual size of the groundwater supply in the United States is not known because extensive mapping of the nation's aquifers has not been completed. Concern about pollution of groundwater is growing. The California Department of Health Service reports that one fifth of the state's large drinking water wells are fed by groundwater that exceeds the state's pollution limits (Sun, 1986). Pesticides have been detected in half of Iowa's city wells. More than 1,000 wells in Florida have been shut down because of contamination with EDB, a chemical used to kill soil nematodes. The common beliefs that pesticides would decompose in the soil and that nature would cleanse itself through percolation can no longer be accepted.

Risk Assessment and Prevention

With more than 3 million chemicals registered, 70,000 chemicals in general use, and new ones appearing at a rate of more than 1,000 a year, it is not surprising that the environment in the industrialized countries has become contaminated. Analyses of the spatial pattern of chemical contamination in New Jersey showed a tendency for gross contamination of pesticides in agricultural areas and of light chlorinated hydrocarbons in industrial–commercial areas (Greenberg, 1983). Most of the organic chemicals and heavy metals occurred only at several orders of magnitude below the level that is considered hazardous for industrial workers. Usually they were found at the lowest detectable levels. Five of 11 organic substances were consistently found in urban sites: benzene, trichloroethylene, tetrachloroethylene, 1,2-dichloroethane, and 1,1,1-trichloroethane. Other groups of chemicals manufactured together were found in certain urban locations. Lead was found in relatively high concentrations near major highways, but otherwise there was no spatial pattern. As expected, surface waters were more commonly polluted than groundwaters, but the highest levels of concentration were found in groundwaters where dispersal is slow. If this study were duplicated, a similar spatial profile would probably be found in most places with advanced economies.

Exposure to pollutants can be either acute or chronic. A person may be exposed once for a few minutes at work to a leakage or spill or consume over a lifetime undetectable levels of contaminant in food or water. Toxic waste from a dump may contaminate a nearby residential water supply to very high levels; or low levels of benzene may be breathed throughout a lifetime in the ambient air of an urban area. Even for most of the chemicals thought to be hazardous, little is known of threshold levels or dosage effects. Threshold limit values (TLV) for major chemicals have been announced by government agencies responsible for protecting workers. These values establish time-weighted average concentrations that are acceptable for occupational exposure, in the same way that standards have been set for exposure to radiation, noise, and airborne dust. Few standards have been estimated for nonoccupational exposures. Research is complicated because reliable, subnational-scale data do not exist for etiological study. There are, for example, no microarea data for industrial exposures, cigarette or alcohol consumption, or dietary composition. One must work between small, local surveys of specific points and broad, socioeconomic classification of national associations and trends. It is not surprising that point and area data often are improperly mixed, that researchers generalize across scale, or that spatial autocorrelation is ignored (see Vignettes 7-2 and 7-3 for an explanation of these). The personal monitors, currently being tested, that identified indoor pollution may, in the future, provide reliable data for researchers interested in studying microscale exposure through daily and weekly activity patterns. For now, most studies rely on surrogate but measurable variables, such as substituting ethnicity for diet.

One way the Environmental Protection Agency assesses population at risk is by drawing concentric circles of various distances around hazardous sources such as dumps or nuclear power plants. Prevailing wind direction is only occasionally considered and then without accounting for diurnal or seasonal variation. Changes in humidity, particle nuclei for rain formation, local turbulence and mixing, air drainage, or frequency of inversion are not in the formulae. Ignoring the underground movement of water means that a population situated up-flow is counted at higher risk from the hazard location than a population situated down-flow a little further away. Differences in soil, rock type, and topography are not in the formulae. Neither are the transportation system and routing, labor circulation, or seasonal changes in population density or industrial production. Studies by political scientists, sociologists, and geographers have shown that often the most critical locational determinant for siting a hazardous facility is the political weakness of the local population.

Geographers can help in the detection of old waste disposal sites, the planning of new sites, and the etiologic analysis of diseases in developed economies. Processes of siting, transporting, storing, processing, concentrating, or dispersing occur in a differentiated environment of short-term micrometeorology, local geomorphology, transportation systems, settlement patterns, and economic and social activity. Risk assessment of population exposure to known and unknown hazards is a multiscale and very complicated business of intrinsic geographic interest, but there has been limited geographic contribution thus far.

RADIOACTIVE POLLUTION

The ultimate technological hazard is nuclear war. Besides the five major nuclear powers (the United States, the Soviet Union, the United Kingdom, France, and China), India, Israel, and South Africa have the capacity to build nuclear bombs, and Argentina, Brazil, Iraq, Libya, and Pakistan (among others) could have nuclear weapons within a decade. These countries include not only the most populous countries, but some of the most politically unstable and belligerent as well. The United States and the Soviet Union have more than 170,000 nuclear warheads between them. The Doomsday Clock of the *Bulletin of the Atomic Scientists* at the time of writing is set at 3 minutes to midnight, the closest it has been in 30 years. Scientists of the National Academy of Sciences and National Research Council agree that not only will target areas be devastated, but even a limited nuclear war will drastically affect the climate, ecosystem, and human genotype everywhere on earth (National Academy of Sciences, National Research Council, 1975). Here we look at the hazards of peaceful nuclear power.

Radiation Hazards

Naturally occurring radiation hazards are cosmic radiation and decay of radioactive elements. Cosmic sources provide a continuous background of eternal radiation with which all life has evolved. There remain after 4 billion years those original radioactive materials that have extremely long half-lives. Three of these are especially important: uranium 235 (U-235), uranium 238 (U-238), and thorium 232. Knowledge of some of their decay sequences and by-products helps in understanding the nature of certain environmental hazards.

The decay sequence of U-238 provides a significant portion of natural radiation. As U-238 decays it produces radon 222, an inert gas that diffuses into the atmosphere. Radon 222 is the source in turn of lead 210, a global constituent of atmospheric fallout. It decays to polonium 210. Lead 210 and polonium 210 are adsorbed by vegetation and consumed by animals. These products of the decay of U-238 are also produced by nuclear power plants that use the fission power of U-235. Since U-235 is not abundant on the earth's surface, nuclear power will eventually have to use the more abundant, but nonfissionable, U-238 and thorium 232 to produce fissionable plutonium 239 and uranium 233. Among the radioactive daughter nuclei that result, cesium 137, iodine 131, and strontium 90 are particularly important because they accumulate within organisms. Iodine 131 is an important fission product of U-235. Although it has a half-life of only 8.05 days, it is readily accumulated by vertebrate thyroid glands. (Therefore, after the Chernobyl nuclear accident in the Soviet Union, people in eastern Europe were advised to take iodine pills for a few weeks, so that their thyroid glands would not take up radioactive iodine.) Cesium 137 is chemically similar to potassium and especially affects the muscles. Strontium 90 is similar to calcium and becomes part of the bones, where it affects the production of red blood cells in the marrow.

Several radioactive gases are also produced by current nuclear technology and released to the atmosphere to become part of the background, or ambient, radioactivity. Principally these are the inert gas tritium and the isotopes krypton 85, xenon 133, and xenon 135. These do not enter into biological or chemical activity and so are not accumulated by organisms. The release of radioactive gases and other products is regulated to stay within limits of normal variation in background radiation. One often hears the reassurance after an unscheduled release of radiation that there is no more exposure for the population than getting a dental X ray. The exposure is, of course, in addition to the dental X ray.

Fusion power, it used to be believed, held the promise of providing a clean source of nuclear power. Then it was recognized that tritium would be produced in large quantities, and when it escaped to the atmosphere would increase the levels of ambient radioactivity. Next it was realized that the anticipated neutron flux in a fusion reactor would soon render all structural materials intensely radioactive. Although a hybrid fission–fusion reactor may be technologically viable in the future, it would almost certainly leak radioactivity.

Radiation can cause cancer and mutation. Damage to DNA may take generations to be expressed. As with any disease agent, some people will be affected at lower levels of radiation than others who remain apparently untouched. Little is known, however, of dose relationships either at the individual or the population levels. Is the relationship linear, the more radiation the more mutation? Is there a threshold level, below which the body usually can cope and above which damage begins? Are there different levels of tolerance, depending on whether the radiation is inhaled, ingested, or contacted? There seems to be considerable variation among species in resistance to radiation. Much of our knowledge of effects on humans comes from Japan, especially knowledge of the effects of lower level radiation after decades have passed. Accidental and occupational exposures provide some data, but generally about acute effects. The soldiers exposed to watching nuclear explosions in the early days of atmospheric testing are claiming compensation for their cancers and reproductive problems, but it has proven difficult for all the experts in court to separate these problems from those occurring in the general population (see Vignette 6-2).

Johnson (1981) investigated the case of the radiation hazard around Rocky Flats nuclear weapon plant near Denver, Colorado. He demarcated a region of hazard using the plutonium content of the soil for miles around the plant to measure cumulative exposure from 1953–1971. Figure 6-5 shows the results of calculating the relative risk of excess cancer deaths from 1969–1971. The cancers involved were those associated with Japanese survivors of the atomic bombs: leukemia; lymphoma; myeloma; and cancers of the lung, thyroid, breast, esophagus, stomach, and colon. The distance decay effect (see

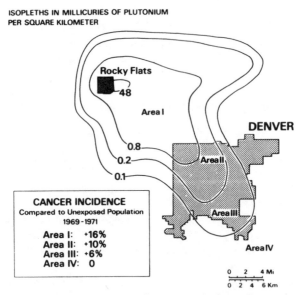

Figure 6-5. Cancer incidence around a nuclear weapons plant. An isoline map of plutonium that has accumulated in the soil provides a regionalization for exposure and correlation with cancer incidence. From *Technological Hazards*, (p. 48) by D. J. Zeigler, J. H. Johnson, and S. D. Brunn, 1983, Washington, DC: Association of American Geographers. Copyright 1983 by Association of American Geographers. Reprinted by permission.

Vignette 6-1) associated with plutonium released through the plant's smoke-stack (note the westerly wind direction configuration) clearly creates a hazardous region. The hazard zone was not perceived for two decades; otherwise the release of radionuclides might have been reduced or the population warned. Populations may be exposed to hazards without knowing it and may experience symptoms and not relate them to environmental conditions.

Other Sources of Radioactive Hazard

The hazards associated with nuclear power plants and nuclear weapons have been emphasized above, but there are other sources of hazard. One is the use of radiation in hospitals and research laboratories and the disposal of the low-level radioactive wastes they produce. Another is the mining of uranium and the manufacture of nuclear fuel. Yet another is the transport and storage of all the used material.

Uranium ore typically contains 2 to 5 pounds of uranium oxide per ton. After uranium is mined, therefore, large piles of waste and tailings remain. Efforts have been made to grade the slag heaps and stabilize them with vegetation, but in the Colorado River basin alone they cover several thousand acres and contain tens of millions of tons. From the heaps thorium 232, U-238, and radon can be leached by rainfall and eroded into water systems. The occupational exposure of the miners and subsequent processors has to be carefully controlled. The ore is crushed, ground, and leached with acid, and uranium is recovered from the leached liquid by such procedures as ion exchange and solvent extraction. Radioactive wastes are produced with each step.

The need to transport radioactive material creates linear hazard zones. Uranium from the mines must be transported to enrichment and fuel fabrication facilities. The fuel must be transported to nuclear reactor sites. Spent fuel must be transported to storage sites. There are presently four temporary spent-fuel storage sites in the United States for the millions of cubic feet of high-level wastes, and 22 sites for low-level wastes. Within a decade the Department of Energy will select one of six possible sites (in Washington, Nevada, Utah, Texas, Louisiana, or Alabama) for the first national, permanent, high-level–waste burial site. Most nuclear power plants are east of the 100th meridian, in the more densely populated parts of the country. The linear hazards of transport are not going to diminish as the amount of spent fuel from later generation plants increases and as the older plants themselves begin to be dismantled and transported for storage.

The health consequences of nuclear power are difficult to assess and easy to exaggerate. Clearly, however, the spread of nuclear power for the generation of electricity to low-income countries that, because of lack of money and skilled personnel, are less able to regulate it and maintain the highest standards of safety practice is a matter of concern. It necessarily involves greater international shipment of nuclear fuel and, because of efforts to control the general availability of plutonium, which can be used for nuclear weapons, the return shipment of spent nuclear fuel. We are at the beginning of the nuclear age.

GLOBALIZATION AND PERCEPTION OF THE HEALTH HAZARDS

Pollution flows without regard to international boundaries. Acid rain, the greenhouse effect of increasing levels of carbon dioxide, and the fallout of strontium 90 as a result of atmospheric nuclear testing are problems shared by humans in general, regardless of the sources. There has been a less generally recognized but concomitant diffusion of the useful technology that produces hazardous by-products. Among the most sought-after industrial capacities are petroleum refineries, the manufacture of plastics and synthetic materials, electronics, pharmaceuticals, and the production of fertilizers and insecticides. Mining and smelting are being extended into previously undeveloped areas. Agriculture is being intensified through application of more chemicals. Research is progressing into such useful areas as recombinant DNA and the capacity to produce strains of food crops resistant to fungi and insects—without consideration of what effects the "naturally occurring" plant chemicals that produce resistance will have on human health.

The experience the industrialized countries gained as they passed through their technological transition should be of great preventive value to Third World nations, but there is little evidence of its application. Modern health hazards are diffusing rapidly. Most Third World countries perceive the immediate benefits of technology more acutely than they see the distant, nebulous hazards. Their national budgets are stretched thinly to cover the needs of education, health, and infrastructural development. Civil servants with the education and skills required to inspect and regulate plants that use advanced chemical technology are in limited supply. The general populace is usually uninformed and naive about modern chemicals. Farmers, for example, commonly contract urticaria and more severe poisoning from handling insecticides and herbicides, and sometimes rinse their equipment in streams that provide water for drinking and washing. If they are illiterate, they cannot read labels. Farmers in Southeast Asia have been known to dump insecticide in streams to kill fish for harvest, as they used to do with native plant toxins. Sometimes, too, corporations export to the Third World poisons and other substances that have been banned for use in developed countries. According to expert testimony in Congressional hearings (United States Congress, 1978), a pesticide never registered in the United States for domestic use, leptophos, was sold to Egypt where it killed farmers and water buffalo. Herbicides similar to Agent

Orange, used in the Vietnam War, have been marketed to countries like Colombia, resulting in reports of miscarriages and birth defects. Drugs with serious side effects or carcinogenic hazard are marketed in the Third World after being banned in their country of manufacture. The attitude in developed countries is often that other countries must regulate themselves, but seldom are the expert knowledge and resources available in the Third World.

The Geometry of Hazard, Power, and Policy

There are many spatial dimensions to environmental health hazards. Atmospheric and water pollution constitute areal hazards; mines, manufacturing sites, refineries, and waste and storage dumps constitute point hazards; and the roads and sea-lanes of transportation constitute linear hazards. Each is of concern at a different scale, and involves different kinds of regulatory and preventive policies and different levels of government. Zeigler, Johnson, and Brunn (1983) also make a useful differentiation between the location of maximum benefits and maximum risks. Technology control and alternatives must be carefully considered in terms of social justice and compensation, private investment, and tax costs. Geographically, the costs and benefits may or may not coincide in one place. In the case of automobile accidents or pharmaceutical side-effects, for example, the zone of maximum benefits from the technology and the zone of maximum risk of its hazards coincide. Zones at risk from atmospheric or water pollution usually contain, but extend well beyond, the place that economically benefits from production. In contrast, the hazards of agricultural technology are usually felt most by farm workers, but the benefits extend to a much wider area. Toxic wastes, radioactive wastes of all kinds, and their transportation form a class by themselves, because the maximum risk is usually borne by a place far removed from the zone of maximum benefit. For example, the Three Mile Island nuclear power plant whose malfunctions endangered the local population in Pennsylvania was making electricity for New Jersey. Most of the productive chemical industry that fills South Carolina's major toxic waste dump benefits other states.

Whether the noxious facility at issue is the permanent, national, high-level radioactive waste dump or a mental hospital, they must be located somewhere. At what scale should democratic control of technology be exercised? Noxious facilities are usually located where people have the least political power. For urban facilities, this usually means areas where ethnic minorities or the poor live, rather than the upper-class suburbs. It makes sense to locate hazardous facilities in areas of sparse population, if the geology is compatible. The sparseness of population, however, also means that there are fewer people to oppose the will of the numerous people elsewhere who have reaped the benefits. Political geographers are beginning to address this issue.

CONCLUSION

The hazards faced by urban and rural areas, or by old industrial and newly developing places, are converging as technologies and pollution spread. As many industries decentralize and spread to other countries, so do certain hazardous occupational exposures. As trucks and automobiles, insecticides and herbicides, radios and air conditioners become universal, so do the products of their manufacture. Less than a decade after the 1978 declaration in the United States of the first national emergency due to technological, and not natural, causes (at Love Canal), came the disaster of thousands of deaths from the release of poisonous gas by a pesticide plant in Bhopal, India, in 1985. Agricultural intensification, proliferation of industrial occupations, production of toxic and radioactive wastes, and consumption of petrochemicals, pharmaceuticals, synthetic textiles, electronics, glues, and paints are now almost universal. The scale of environmental change does not match the scale of contemporary regulation and control.

It once seemed that modern, urban people were removed from the environment. Virtually all that was of consequence for health was subsumed under human behavior: smoking, driving, diet, mental stress, and so on. Urban people were safe from malaria produced by forest clearance or drainage modification and cared little about the nutritional consequences of agricultural cycles. In fact, modern people have so altered the environment that they dwell within their own creation. As we have seen, however, the system of population–behavior–habitat interactions has not changed, only the hazards to health.

REFERENCES

Bryson, R. A., & Kutzbach, J. E. (1968). *Air pollution.* Washington, DC: Association of American Geographers.

Council on Environmental Quality. (1980). *The 11th annual report.* Washington, DC: U.S. Government Printing Office.

Council on Environmental Quality. (1981). *Environmental trends.* Washington, DC: U.S. Government Printing Office.

Council on Environmental Quality. (1983). *Environmental quality 1983: 14th annual report.* Washington, DC: U.S. Government Printing Office.

Dubos, R. (1965). *Man adapting.* New Haven, CT: Yale University Press.

Girt, J. L. (1972). Simple chronic bronchitis and urban ecological structure. In N. D. McGlashan (Ed.), *Medical geography: Techniques and field studies* (pp. 211–231). London: Methuen.

Greenberg, M. R. (1983). Environmental toxicology in the United States. In N. D. McGlashan & J. R. Blunden (Eds.), *Geographical aspects of health* (pp. 157–174). London: Academic Press.

Hunter, J. M. (1976). Aerosol and roadside lead as environmental hazard. *Economic Geography, 52,* 147–160.

Hunter, J. M. (1977). The summer disease: An integrative model of the seasonality aspects of childhood lead poisoning. *Social Science and Medicine, 11,* 691–703.

Johnson, C. J. (1981). Cancer incidence in an area contaminated with radionuclides near a nuclear installation. *Ambio, 10*, 176–182.

Kleinbaum, D. G., Kupper, L. L., Morgenstern, H. (1982). *Epidemiologic research*. Belmont, CA: Wadsworth.

MacMahon, B., Pugh, T. F., Ipson, J. (1960). *Epidemiologic methods*. Boston: Little, Brown.

Mage, D., & Wallace, L. (1979). *Proceedings of symposium on the development and usage of personal monitors for exposure of health effects studies*. Chapel Hill, NC: U.S. Environmental Protection Agency.

Mill, J. S. (1856). *A System of Logic*. London: Parker, Son, & Bowin.

Miller, D. (1977). *Waste disposal effects on ground water*. Washington, DC: U.S. Government Printing Office.

National Academy of Sciences & National Research Council. (1975). *Long-term worldwide effects of multiple nuclear-weapons detonations*. Washington, DC: National Academy of Sciences.

Risebrough, R. W. & Jacobs, S. A. (1980). Anticipated effects upon wildlife of an increase in environmental levels of radioactivity. In Committee on Nuclear and Alternative Energy Systems, National Research Council, *Energy and the fate of ecosystems* (pp. 358–392). Washington, DC: National Academy Press.

Shabecoff, P. (1985). U.S. calls toxic air pollution bigger threat indoors than out. *New York Times, CXXXIV* (46, 437), 1.

Shapiro, F. C. (1981). *Radwaste: A reporter's investigation of a growing problem*. New York: Random House.

Stern, A. C., Lowry, W. P., Wohlers, H. C., & Boubel, R. W. (1973). *Fundamentals of air pollution*. New York: Academic Press.

Stoker, H. S. & Seager, S. L. (1972). *Environmental chemistry: Air and water pollution*. Glenview, IL: Scott, Foresman.

Sun, M. (1986). Ground water ills: Many diagnoses, few remedies. *Science, 232*, 1490–1493.

U.S. Congress, House of Representatives. (1978). *Hearing on U.S. export of banned products*. 94th Cong., 2nd sess. Washington, DC: U.S. Government Printing Office.

U.S. Environmental Protection Agency. (1973). *Proceedings of the International Congress on Noise as a Public Health Problem*. Washington, DC: U.S. Government Printing Office.

U.S. Environmental Protection Agency. (1976). *Some considerations in choosing an occupational noise regulation*. Washington, DC: U.S. Government Printing Office.

U.S. Environmental Protection Agency. (1977). *The report to Congress: Waste disposal practices and their effects on ground water. Executive summary*. Washington, DC: U.S. Government Printing Office.

U.S. Environmental Protection Agency. (1984). *National air pollution emission estimates, 1940–1982*. Washington, DC: U.S. Government Printing Office.

Waldbott, G. L. (1973). *Health effects of environmental pollutants*. St. Louis: C.V. Mosby.

Wilson, J. (1982). *Ground water: A non-technical guide*. Philadelphia: Academy of Natural Sciences.

Zeigler, D. J., Johnson, J. H., & Brunn, S. D. (1983). *Technological hazards*. Washington, DC: Association of American Geographers.

Further Reading

Bowie, S. H. U., & Thornton, I. (1985). *Environmental geochemistry and health*. Report to the Royal Society's British National Committee for Problems of the Environment. Dordrecht, Holland: D. Reidel.

Goldman, A. (1980). The export of hazardous industries to developing countries. *Antipode, 12*, 40–46.

Lenihan, J. & Fletcher, W. W. (1976). *Health and the environment*. New York: Academic Press.

Miller, G. T., Jr. (1975). *Living in the environment: Concepts, problems, and alternatives*. Belmont, CA: Wadsworth.

Muschett, F. (1981). Spatial distribution of urban atmospheric particulate concentrations. *Annals of the Association of American Geographers, 71*, 552–565.

Neiburger, M., Edinger, J. G., & Bonner, W. D. (1982). *Understanding our atmospheric environment*. San Francisco: W.H. Freeman.

U.S. National Institute of Environmental Health Sciences. (1977). *Human health and the environment—Some research needs*. Report of the Second Task Force for Research Planning in Environmental Health Science. Washington, DC: U.S. Government Printing Office.

Whyte, A., & Burton, I. (Eds.). (1980). *Environmental risk assessment*. New York: Wiley.

Vignette 6-1

A DISTANCE DECAY CURVE

Distance decay means that the intensity of a phenomenon is inversely proportional to its distance from a certain point, line, or area. Lead levels decline with increasing distance from a road, air pollutants decrease with distance from an industrial smokestack, and the incidence of disease may decline with distance from a focus. Here distance decay is applied to patients' use of a hospital.

We map a series of concentric rings around a hospital at one-mile intervals and estimate the number of people who reside in each ring. Using patient addresses from hospital records, we count the number of visits that people in each ring made to the hospital in one year and calculate patient visits per 10,000 people in each ring (Vignette Table 6-1).

Frequency of visits is plotted against distance. Each concentric ring is represented by its midpoint; for consistency, and because the numbers are small, the last ring is represented by 5.5 miles. The result of joining the plotted frequency–distance points would be a typical distance decay curve.

We can try to fit a curve to the data points. The data could be expressed by a formula, $f = k/d^b$, with f representing frequency, k a constant, d distance, and b the friction of distance. This is the equation of a curved line. To use linear regression techniques to fit the curve, the equation can be transformed to a linear form by taking natural logarithms of both sides. This produces the equation $\log(f) = \log(k) - b(\log(d))$. The natural logarithms of f and d for the six concentric rings are shown in Vignette Table 6-1.

A common statistical technique, linear least squares, is used to find values of $\log(k)$ and b that will produce a line that is the best fit to the hospital use

Vignette Table 6-1. Data for Constructing Distance Decay Curves

Distance Ring (miles)	Midpoint	Number of visits per 10,000	log(d)	log(f)
0.00–0.99	0.5	2,553	−0.69	7.85
1.00–1.99	1.5	247	0.41	5.51
2.00–2.99	2.5	101	0.92	4.62
3.00–3.99	3.5	42	1.25	3.74
4.00–4.99	4.5	34	1.40	3.53
5.00 or more	5.5	18	1.70	2.89

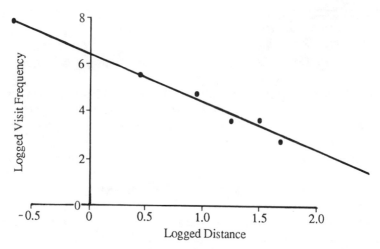

Vignette Figure 6-1. Regression line for logged frequency–distance data. A distance decay curve that plots declining visits per population base against decreasing distance from a hospital.

data. This can be done by using a computer program or by calculating b and $\log (k)$ from the formulas

$$b = \frac{n\Sigma(\log(d))(\log(f)) - (\Sigma\log(d))(\Sigma\log(f))}{n\Sigma(\log(d))^2 - (\Sigma\log(d))^2}$$

and

$$\log k = \frac{\Sigma\log(f) - b(\Sigma\log(d))}{n},$$

where n is the number of data points (six in this case).

We find that $\log (k) = 6.42$ and $b = 2.04$. The logged frequency and distance data points and the least squares regression line are plotted in Vignette Figure 6-1. Apparently, the fit is quite good. How good the fit is can be determined by using a statistical test of significance.

Vignette 6-2

EPIDEMIOLOGIC DESIGN

Determining the cause of a disease with a long latency period is very difficult, but the same logic is followed as for infectious or nutritional diseases. Four of the logical canons developed by John Stuart Mill (1856) underlie causal reasoning in health studies.

1. *Difference*. When all conditions among the study populations are alike except for one, it is implicated as either causal or preventive of the disease. This is the classic logic of laboratory experimental studies. Rats with the same inheritance are kept under identical conditions, except for exposure to the chemical that is being studied. The best application of such experimental design to study of human populations can be found in the clinical trial, in which one group of people is given a new drug and another group is given a placebo, such as a valueless sugar pill. In this way psychological attitude as well as the healing or exposure risks of time are the same for both groups except for effects of the drug being studied.

2. *Agreement*. When all circumstances are different except for the variable being studied, it is implicated as causal.

3. *Concomitant Variation*. When a factor varies systematically with the frequency of the disease, it is implicated as being causal. When more or less of the variable is associated with more or less of the disease, it is varying systematically.

4. *Residue*. When the effect of the known causal factor is removed in order to isolate and measure the variation remaining, successful explanation of the remainder supports factor causality. This is the method that geographers use when they map the residuals of a regression or the factor scores of a factor analysis in order to see if the pattern of unexplained variation remaining elicits any further hypotheses.

When repeated studies support these logical canons, causation is gradually established. Studies of cigarette smoking, for example, have found that when groups are matched for age, sex, ethnicity, income, education, occupation, personality, and activity and differ only in smoking cigarettes or not, there is a great difference in the incidence of cancer between them (difference). If people are studied who live in totally different cultures and environments—if they are Moslems and Buddhists and Christians, poor and rich, literate and illiterate, if they do and do not eat meat, if they live in cold, dry places and hot, wet places or in urban places and rural places—always the smokers have a higher incidence of lung cancer than the nonsmokers (agreement). When populations of smokers are subdivided according to how many cigarettes they smoke, at every increment there is associated greater risk of getting cancer (concomitant variation). When the effect of cigarette smoking is removed, it is possible to identify patterns of exposure to air pollution. Thus, although statistics cannot *prove* anything, the accumulated and consistent evidence along different logical paths makes doubt that cigarette smoking causes lung cancer unreasonable.

It is very difficult to be sure, however, that *all* other factors are the same. Many early animal studies of the dietary effects of water hardness, for example, had to be redone when scientists learned how to measure trace elements. The water had not only differed in calcium carbonate, as intended, but also in the trace amounts of molybdenum, cadmium, selenium, and other elements.

The amounts of trace elements had not been held constant and plausibly were causally related to the outcome.

A *confounding variable* is one that varies in a systematic way with the hypothesized causal relationship being studied. Although A seems to cause B, in fact another variable, C, is affecting both A and B. The relationship between A and B is therefore *spurious*. Sometimes one is aware of a confusing interaction. For example, soft water has been associated with higher risk of stroke. In the United States, soft water occurs in the South on the coastal plan. When one studies stroke in soft water regions and hard water regions, one also finds strong associations with altitude and with the range of temperature changes. Which is truly causal? Or do they affect each other and merely happen to vary in the same way as stroke, perhaps because of still another variable that has yet to be identified? One of the most difficult tasks of social science research on health is identifying which variables need to be controlled so that they will not confound the relationship being studied. This, of course, is the role of theory; but at early levels of understanding of disease etiology, theory may be inadequate.

Confounding can be controlled by both analysis and by research design. The most confounding variable of all is age. Whether one studies the prevalence of antibodies to a virus, life stresses, activity pattern, cholesterol deposition, or public health knowledge, one needs to know the age of those involved. We have seen how analytical technique can control for age (Vignette 2-1). The devilish thing about confounding factors, however, is that the researcher may not know they exist. One of the major purposes of epidemiological research strategy is to control for confounding factors.

The ideal epidemiological evidence is to find that the different disease frequencies in two populations are dependent on a difference in a certain factor; and that furthermore, within each of the populations, that factor is more common among those with the disease than those without it. There are two broad strategies used to address these questions, cohort studies and case-control studies, diagrammed in Vignette Figure 6-2. In a cohort study, a population is studied. It is divided into those with and without a particular exposure, and the frequency of the disease outcome is noted. When this study is started before the exposure and people are followed forward in time, it is called a *prospective* study. When the disease outcome has already happened and the history is reconstructed through interviews and records, it is a *retrospective* study. The case study starts with people who have the disease or who died of it, and their behavior and exposures are compared with the rest of the population. Although all the nondiseased population may be used, usually it is sampled. The whole population may be sampled randomly, or cases may be paired in a systematic way with people from the general population. One may interview the neighbor next door to the right, for example, or the next patient admitted to the hospital for a nonrelated reason. Alternatively, one

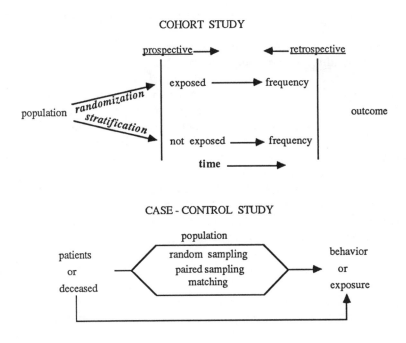

Vignette Figure 6-2. Epidemiologic design.

may pair the sample deliberately by matching the patient on a series of criteria, such as age, sex, ethnicity, income, education, and so on.

Each of these research strategies has advantages and disadvantages. A prospective cohort study has the great value of allowing direct estimation of the risk associated with causal factors. As one follows the population through time, for example, one can measure how many times they ate a certain food or drank a certain source of water, and how long they spent in a certain occupation. There are no spurious relationships added by the means of collecting data, such as a bias introduced when people who are suffering are motivated to remember events more carefully. The great disadvantage of such a study, however, is the cost in time, skilled personnel, and money. Following a population is laborious, time-consuming, and expensive, especially if the disease is not too common and large numbers of people must be involved in order to get enough data for a definitive analysis.

A case-control study is relatively quick and inexpensive, and these characteristics make it easily repeatable and able to include large numbers of people. There can be certain biases introduced to data collection, however, and the researcher is less likely to find things he or she did not set out to find.

Questions are formed by the researcher, and people recall their specific experience in reaction to the question.

In both strategies of study, randomization and stratification are used to convert confounding variables into control variables. When a population is randomly assigned to two groups or individuals are randomly chosen for comparison, the effects of the confounding variable are converted into residual variation equally distributed between groups, so there is no need for analytic control of it. For example, consider a test of a new vaccine. Some children are naturally resistant to the disease, and some may have had it already. Some children are more in contact with other children, some have better diets, some have more psychological stress. All of these things might affect whether a child gets the disease or not. One wants to compare the proportion of children given the vaccine who contract the disease with the proportion of unvaccinated children (given a placebo) who contract the disease, in order to assess the vaccine's effectiveness. By assigning children randomly, the difference in contracting the disease that is due to relative natural resistance or to differences in exposure is converted into unexplained variation common to both groups equally, and so is removed as a confounding factor. Matching cases or stratifying the population with regard to certain characteristics has the same effect of making sure those characteristics do not introduce confounding variance. Stratifying or matching by ethnicity or education, for example, eliminates not only variance due to those characteristics, but to every variable associated with them, such as diet, home environment, and neighborhood influences.

Geographers have implicitly used such causal logic in several of their methodologies. *Analog area analysis*, for example, matches characteristics of regions so that they are as similar as possible, except for the variable being studied. The method of agreement is frequently used. Japanese and Icelanders are different in almost every respect, except that both populations have a high incidence of stomach cancer and both populations consume large quantities of smoked fish. Therefore, other populations that consume large quantities of smoked fish should have a high incidence of stomach cancer. Similarly, Girt (1972) has explicitly based a sampling frame upon theories of urban structure. He used knowledge of how the location of neighborhoods is related to the age of structures, economic activity, ethnic composition, sex ratio and age structure, income and commuting, and land value to stratify his sample population. Instead of stratifying by socioeconomic groupings as other social scientists would do, he stratified by structural location within the city and sampled within areas. The use of theoretical understanding of the environmental, economic, and demographic development of cities over time should be especially important for studying degenerative and chronic diseases, which take decades to develop.

7

Geographies of Disease in Economically Developed Areas

Economically developed places have the trained human resources and the financial capacity to investigate the distribution, causation, treatment, and prevention of the techno-pathogenic complexes described in Chapter 6. At the end of the mortality transition (Chapter 4), death is caused mainly by degenerative and chronic diseases such as heart disease, stroke, and cancer. Because most infectious diseases are under control, there is little death in childhood and there is a long life expectancy; therefore, the population has an old age structure. Conditions in the equivalent stage of the mobility transition include very high population circulation rates and active urban–urban migration. Much of the research literature in medical geography, and some of its most sophisticated methodology, is concerned with the patterns of health conditions and disease etiology in North America, Europe, and Japan. These patterns are increasingly shared by recently industrialized countries such as Korea or Yugoslavia.

This chapter describes some of the major disease processes of the economically developed world, and reviews some of the most basic and innovative geographic research.

Population–behavior–habitat interactions provide a framework that can integrate and support many research directions. Genetic susceptibility and resistance affect the results of many exposures. Diet can be both cause and treatment. The sophistication and technology of health care systems, part of the habitat, are critical to resulting patterns of mortality. For the health of people living in industrialized societies, the built environment (and the pollution syndrome it includes) is the main habitat of exposure.

The older age structure and long-standing mobility of the population mean that exposures to hazards are difficult to trace or determine. Often exposure occurred in a place different than where the disease becomes manifest. Most of the diseases are manifest a long time after the susceptible person is exposed to hazard. Trying to understand the spatial pattern of chronic and degenerative diseases is thus very difficult. Some of the studies described in this chapter use statistics with which the reader may not be familiar. We have tried

195

to present them so that the substance is understandable, even if the precise methodology or significance may be obscure.

Categories of chronic and degenerative diseases are often hard to define precisely. Sophisticated radiology and laboratory procedures may not be available for diagnosis. Terminology fads mean that definitions change. The latency of these diseases may be decades. Cause is often difficult to determine because we do not know if there is a dose or a threshold relationship between disease expression and original hazards. There are many epistemological and methodological problems with scale of analysis, units used as observations, spatial autocorrelation, confounding, and other issues. These are discussed in Vignettes 1-3, 6-2, 7-1, 7-2, and 7-3.

Degenerative and chronic diseases have multiple causes. Often a single cause has multiple effects. Careful research design, based on solid knowledge of the biological disease process and of the spatial relationships of the units of observation, is absolutely essential in geographic research on these complex disease patterns.

THE POVERTY SYNDROME

Many diseases seem to occur consistently more often among the poor than among the affluent. There seems to be a socio-pathologic complex made up in large part of stress, life-style, diet, housing, polluted air, old paint, and old pipes. Physical, social, and mental diseases have similar patterns within urban areas. Eyles and Woods (1983) point out that diseases traditionally associated with low socioeconomic status (SES) include respiratory tuberculosis, rheumatic heart disease, bronchitis, and stomach cancer; those normally associated with high SES include breast cancer, leukemia, and cirrhosis of the liver. Their analysis and review suggests that diabetes, coronary heart disease, lung cancer, and duodenal ulcer, which used to be associated with high SES, have now joined the diseases of the poor.

Housing is probably the easiest part of the socio-pathologic complex to understand. The poor are crowded. Older residences are often subdivided into tenements for numerous families and individuals. Fertility rates are high, so that small children add to the high room-density measure frequently associated with influenza, bronchitis, and tuberculosis. Old pipes may have lead joints and occasionally are composed of lead; they tend to be made of metals whose cadmium and other trace elements are easily eroded. Although lead is no longer allowed in house paint in the United States, older buildings still have lead paint chips. Exposed asbestos is a more recently recognized hazard. Central heating systems frequently are obsolete or in poor repair. In Britain the poor frequently live with damp and drafts. Girt (1972) and others have suggested that damp might be associated with fungi whose spores generate an immunologic reaction of catarrh in the lungs. Inner apartments with few or no windows, outside windows opening onto brick walls, and higher levels of soot produce dark interiors suitable to tubercle bacilli, which are highly concentrated because of crowding and poor ventilation.

A diet high in starchy foods, sugar, and fat contributes to a greater tendency toward diabetes and coronary heart disease, but there is remarkably little small area–based dietary information with which to confirm or study relationships. Obesity is much more frequent among black females than other segments of the American population, and more frequent among poor whites than among rich ones. The role of diet is closely tied to life-style factors. When people in the United States started jogging and aerobic dancing for health reasons, these cultural practices, like most others, diffused down the socioeconomic scale from people with higher levels of access to information to those more tied to traditional ways.

Usually the innovation, or fad, diffuses from metropolitan areas in states such as California and New York to other urban centers in the country and out to small towns and rural areas. Diffusion also occurs among neighborhoods of a city, and blocks within neighborhoods. Knowledge about the benefits of steaming vegetables instead of over-boiling them, about the hazards of bacon fat, whole milk, salt, eggs, or beef, or about the uses of exotic substitutes like tofu, spreads slowly. A new definition of healthy diet seemed to appear among the educated just when dietary behavioral patterns once associated with the well-off became widely practiced among the poorly educated.

There are also socioeconomic impediments to change of life-style. Health spas, gyms, and safe jogging paths are not readily available to the poor. Night shifts, double jobs, and lack of child care impede many changes. Crime and fear can lead to deprivation of social life, especially for the elderly. Some social scientists attribute mental illness to the poor social climate and housing conditions of inner-city areas; others believe that certain environmental circumstances attract people with certain personality structures or problems and note that, when poor people are moved out to peripheral housing developments, the physical, social, and mental health problems usually move, too (Giggs, 1983).

Many aspects of the housing, neighborhood, and economic *milieu* cause stress: they include anxiety over jobs, fear of inscrutable institutional policies, insecurity over social security or aid to dependent children payments and medical care coverage, and hassles with landlords. In-migrants to a city have additional stresses of adjustment.

There are strong suggestions that all the above stresses are tied to the economic system. As unemployment goes up, infant mortality, suicide, and homicide often increase. One way of viewing the whole socio-pathologic complex is in terms of the accumulation of insults to health (Chapter 2). Some people with high levels of health can cope with the insults and emerge stronger and more creative for the struggle. Others, their health depressed by psychological and social insults, are unable to rally from infectious, chemical, or further mental insults. They become part of that small percentage of the population that accounts for a large percentage of illness and repeated health care needs.

CANCER

Cancer is a family of diseases. There is no one cause or one cure for cancer; there are only multiple effects of multiple causes and multiple treatments and outcomes. Our knowledge and classification of cancer is analogous to our knowledge of infectious disease in the early 19th century. Then, cholera was not differentiated from yellow fever or typhoid from diphtheria. There was only fever, "flux," vomiting, rash, flatulence, cramps, and so forth. The best doctors of the day could write of malaria "becoming typhous in its course." It was only after germ theory gave us a microbe for each disease that the separate entities were classified and causation and treatment defined appropriately. Classification of cancer by the site of occurrence is similar to classification of "fever" in the 19th century. Stomach cancer and cancer of the pancreas, both caused by a chemical, are classified as different diseases: pancreatic cancers that are caused (for all we know) by minerals and viruses are lumped together in spite of their different agents. As we learn more of the different causes and of appropriate prevention and treatment, our classification scheme is bound to change.

The long latency of the disease has bedeviled etiologic research because of the many changes over time and space that occur. A new synthesis in cancer research presents a model of cancer causation. Figure 7-1 illustrates why it takes cancer so long to develop. When a carcinogen enters the body, it is usually detoxified by enzymes, broken down, and excreted. Occasionally, these enzymes and related processes can activate the carcinogen and enable it to enter a cell. If it binds to anything in the cell except the DNA, it can only affect that individual cell. If it binds to the DNA, many repair mechanisms attack it. The bonds between the carcinogen and the DNA are broken, and the carcinogen is broken up, transported, and excreted. Only if the DNA is replicated before the carcinogen is removed can the carcinogen affect the new cell information, creating a cancer gene. The evidence clearly indicates that one altered gene is usually not enough to cause cancer. The process must be repeated, seemingly against all odds, by another carcinogen in a cell that already has the altered genetic material. The odds are changed if substances known as "promoters" (sometimes metal elements, fractions of hydrocarbon chains, fatty acids, or substances such as saccharin) encourage the altered cells to proliferate faster than normal cells. This process creates an increasing number of targets for another carcinogen to enter. The rare chance that such a contact will occur, among all the normal cells of the body, helps to explain why

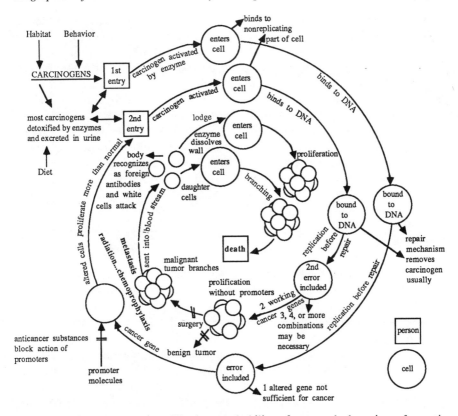

Figure 7-1. Cancer causation. The low probability of repeated alteration of genetic material, usually required to cause cancer, is affected by the presence of "promoter" substances and the dosage and duration of carcinogens, as well as the presence of protective nutrients for repair and buffering mechanisms.

decades may elapse before expression of the disease. It also explains why cancer is most common in tissues with normally high rates of proliferation, such as skin or the lining of the intestines and uterus, and rare in nonproliferating tissues such as nerves. Proliferation increases the number of cells containing the altered genetic information. Every time a carcinogen enters the body, the whole chain of chance events must repeat. Perhaps three or five alterations to the same genetic material are necessary before the potent combination occurs and the cell begins to proliferate without control. It may form a benign, noninvasive tumor that can be removed by surgery. It may, however, send out branches and envelop other tissue. The malignant tumor can send "daughter" cells into the bloodstream, an event known as metastasis. The daughter cells drift with the bloodstream, in danger of being recognized as foreign and attacked by the body's antibodies and white cells, until they lodge. They produce an enzyme that dissolves the blood vessel wall, which allows them to enter normal tissue and begin proliferation and invasive branching in the new site. Eventually vital tissues and processes are involved, and death results.

Cancer's need for repeated defiance of the odds in an orderly but improbable sequence of events explains the long latency period. The need for separate carcinogens, promoters, and overcoming of at least four defense mechanisms of the body explains the often baffling etiology and outcome. The adequacy of vitamins in the diet is important for the body's ability to detoxify the carcinogen, especially after it has bonded within a cell. The status of the immune system is critical because it must identify cells with altered genetic information as foreign and destroy them. All the interactions and balances of this complex process change over time.

The Genetic Base

Some kinds of cancer have long been suspected of being genetically linked. Some are more common in a family than in the general population. Others are common in some population groups (for example, nasopharyngeal cancer among the southern Chinese) and rare in others. There is evidence that some cancers affect men more than women when there has been equal exposure.

Several mechanisms of genetic involvement are implied in Figure 7-1. The importance of enzymes for detoxifying, or sometimes activating, the carcinogens makes it plausible that certain genetically given enzyme combinations may make people more susceptible. The importance of enzymes in DNA repair suggests that genetic variation can result in different levels of repair efficiency or even the inability to repair damaged DNA. People suffering from a rare disease called xeroderma pigmentosum, for example, have been found to have a mutation that results in a lack of the enzymes necessary to break the abnormal bonds formed among subunits of a gene under ultraviolet light. They have very high rates of skin cancer. There is growing evidence that men may be more susceptible to many mutagenic agents because they bind more carcinogen to their chromosomes. This has been most studied with regard to lung cancer and carcinogens and promoters in cigarette smoke. Men have been found to have greater stimulation of DNA repair synthesis, resulting from the increasing amount of damage. Damage and repair activities increase with age for both sexes but are higher at all ages for men. On the other hand, Calabrese (1985) points out that women may be more susceptible to benzene and other fat-soluble organic chemicals implicated as carcinogens.

One of the most active areas of cancer research concerns oncogenes and the breaking of chromosomes. Overwhelming circumstantial evidence is accumulating that oncogenes, genes contained within a human's chromosomes, are capable of transforming normal cells into cancerous ones. Molecular biologists have identified about 20 of these mysterious genes, which normally are inactive. They are found in such diverse organisms as flies, fish, mammals, and yeast. Thus, although their functions are unknown, they are presumed to be vital because they have been so carefully conserved throughout evolution. Oncogenes seem to be found consistently in tumorous cells, often in multiple copies. The cytogeneticists have found that the chromosomes of cancer victims are frequently broken and scrambled. There is growing evidence that chromosomes can be genetically predisposed to break, under radiation or chemical assault, in certain places. If an inactive oncogene is detached on a fragment from one chromosome and reattached to a fragment of another chromosome next to certain other genes, it can be activated to transform the cell into cancer. Many oncogene changes may be involved before that effect occurs. It is thought that certain genetically controlled processes normally limited to particular growth stages, such as the rapid proliferation of cells in a young fetus, may be turned on again by the new placement of the oncogene. Whatever the role of the oncogene and chromosome-breaking may be, it is clear that chromosomes have a propensity to break in certain places. Genetic control of these places may be involved in susceptibility to cancer.

Behavioral Base

Cultural risk factors for cancer involve a variety of customs and practices as well as economic goods and occupations. Given the same solar radiation by place and the same genetic susceptibility by population group, the propensity of modern youth to sunbathe aggravates risk of skin cancer, compared to earlier practices of wearing long sleeves and broad hats and modestly covering the body. Tobacco smoking is the most obvious endangering behavior, but mouth cancer has been linked to chewing betel, and lip cancer has been linked to smoking pipes. Diet may be endangering; a diet poor in fiber means that undigested food passes too slowly through the intestines, allowing toxic bacterial products to accumulate and become concentrated. Fatty acids may be a risk factor for breast cancer, and the hydrocarbons and nitrosamines of smoked fish (and presumably other smoked products) and charred meat seem to be risk factors for stomach cancer. Diet may also be preventive; carotenoids from green and yellow vegetables serve as antioxidants to protect cell walls. The human diet contains a variety of natural carcinogens, promoters, antimutagens and anticarcinogens. Science has barely addressed the relative risk or protection of the enormous variety of dietary habits.

The behavioral base in economic structure, occupation, industrial processes, and regulatory control has dominated risk assessment. Workers exposed to asbestos, to the manufacture or use of industrial chemicals, to agricultural sprays, or to the other multiform dusts and radiations connected with earning a livelihood in an industrialized society have long served as the guinea pigs for most of our threshold and effect data. The larger scale behavior of locating certain enterprises, such as oil refineries or nuclear power plants, in certain populous areas and not others, has rarely been evaluated from such an experimental perspective.

Patterns and Change

There is extraordinary geographical variation in the occurrence of cancer. For decades it has been recognized that differences of incidence rates, not only between men and women, but also between urban and rural populations and between countries, hold etiological clues. Table 7-1 illustrates some of this variation. For males the death rate from all kinds of cancer in Uruguay is more than seven times that in Nicaragua (special statistics area only). The highest female cancer rate by country is far less than the top 10 male rates. Japan has the highest stomach cancer rates but is not ranked in the top 10 for lung, breast, or rectal cancer. Given the country's biometeorology, the population's genetics, and such behavioral patterns as ranching and surfing, one can suggest reasons why Australia is the highest ranked in skin cancer; but why should countries as diverse as Uruguay, France, and Singapore be ranked so highly for esophageal cancer? Furthermore, within each country, urban and rural rates can differ as greatly as international rates.

Because most forms of cancer are relatively rare, most data have been mapped at the state or international scale. (Vignette 7-1 discusses the importance of selecting a research scale.) Small numbers of deaths tend to fluctuate randomly and give misleading associations. The National Cancer Institute (Mason, 1975) has published an atlas of cancer mortality at the county level by aggregating cases over 20 years. Figure 7-2 presents a few selections from a recent effort to map some cancer patterns in the United States. Cancer rates for blacks are mapped at the scale of state economic areas because small numbers of site-specific black deaths occur in many counties.

An important research question regarding disease distribution concerns the degree of spatial clustering. Does clustering occur at the county scale or at the household scale? If three cases of leukemia occur on a city block, is that more than could occur by chance? If there are inimical effects from a chemical factory or a power plant, at what scale would the health consequences cluster around it? Map patterns alone cannot indicate whether chance dictated the amount of clustering. Observed patterns can be compared to patterns based on randomized time and space, using stochastic approaches and chi-square tests.

The pattern of cancer incidence has changed over time. Lilienfeld and Lilienfeld (1980) have outlined several possible reasons. The changes may be *artifactual*; that is, they may result from errors due to changes in the recognition, classification, or reporting of the disease or from errors in enumerating the population. *Real* changes may result because the age structure of the population has changed, because people survive diseases that were once incur-

Table 7-1. International Cancer Mortality for Selected Sites: Age-Adjusted 1977 Rates per 100,000 World Standard Populations

Rank	Country	Rate	Country	Rate
	All Sites, Male		All Sites, Female	
1	Uruguay	202.58	Denmark	133.82
2	Scotland	201.81	Scotland	132.04
3	Hungary	201.13	Hungary	129.38
4	France	201.02	No. Ireland	127.52
5	Netherlands	195.47	New Zealand	125.78
6	F. R. Germany	189.79	Uruguay	125.15
7	England and Wales	189.03	England and Wales	124.59
8	Hong Kong	187.26	Ireland	124.13
9	Austria	185.97	F. R. Germany	123.71
10	Singapore	182.74	Austria	122.89
43	Nicaragua	27.34	Egypt	19.67

Rank	Country	Rate	Country	Rate	Country	Rate
	Stomach, Male		Rectum and Recto-Sigmoid Junction, Male		Esophagus, Male	
1	Japan	52.33	Denmark	10.09	Uruguay	14.80
2	Costa Rica	48.68	Austria	9.97	France	14.40
3	Chile	46.44	F. R. Germany	9.48	Singapore	12.21
4	Hungary	34.47	Hungary	9.34	Argentina	10.92
5	Poland	33.94	New Zealand	9.08	Hong Kong	10.85
6	Singapore	32.53	England and Wales	8.95	Chile	9.53
7	Romania	27.93	No. Ireland	8.06	Japan	7.09
8	Bulgaria	27.80	Scotland	7.59	Paraguay	7.02
9	Iceland	27.59	Norway	7.41	Scotland	6.88
10	Austria	27.43	Ireland	7.34	Switzerland	6.83
43	Egypt	1.24	Thailand	0.11	Nicaragua	—
	Trachea, Bronchus and Lung, Male		Skin, Male		Breast, Female	
1	Scotland	80.62	Australia	5.95	No. Ireland	30.31
2	Netherlands	74.63	New Zealand	5.58	England and Wales	27.91
3	England and Wales	73.70	Norway	4.12	Malta	27.31
4	Finland	66.57	Switzerland	3.18	Denmark	26.04
5	No. Ireland	55.66	United States	2.85	Scotland	25.94
6	United States	54.38	Finland	2.65	Netherlands	25.74
7	Hong Kong	53.46	Sweden	2.65	New Zealand	25.73
8	Austria	51.49	Hungary	2.60	Uruguay	24.84
9	Hungary	51.42	Denmark	2.59	Ireland	24.54
10	New Zealand	50.70	Ireland	2.12	Switzerland	24.16
43	Nicaragua	1.62	Nicaragua	—	Nicaragua	0.53

Note. Data from *Age-Adjusted Death Rates for Cancer for Selected Sites (A-Classification) in 43 Countries in 1977* Segi Institute of Cancer Epidemiology, 1982, Nagoya, Japan: Author.

Cancer of the Stomach
White Males: 1950 - 1959

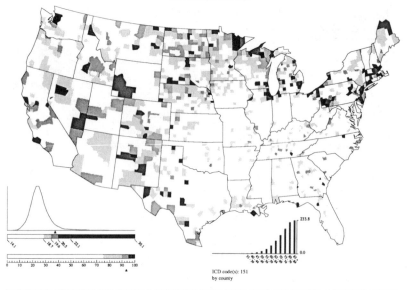

ICD code(s): 151
by county

Cancer of the Trachea, Bronchus and Lung including Pleura and Other Sites
White Males: 1970 - 1979

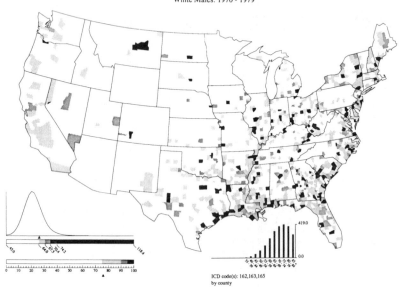

ICD code(s): 162,163,165
by county

Figure 7-2. Cancer patterns in the United States. The density trace of rate distribution identifies the number of data units (counties) in each shading category. The bar graph depicts the age-specific death rates for each cancer. From *The United States Cancer*

204

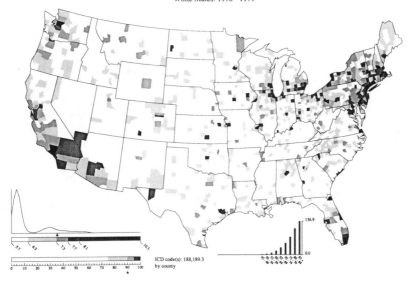

Cancer of the Bladder and other Urinary Organs
White Males: 1970 - 1979

ICD code(s): 188,189.3
by county

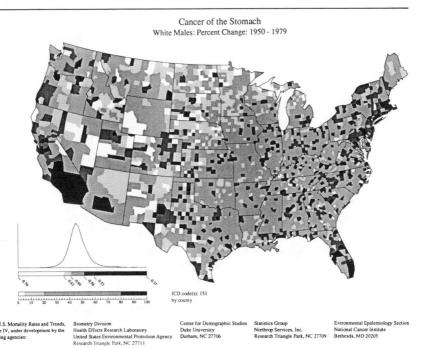

Cancer of the Stomach
White Males: Percent Change: 1950 - 1979

ICD code(s): 151
by county

from U.S. Mortality Rates and Trends,   Biometry Division   Center for Demographic Studies   Statistics Group   Evironmental Epidemiology Section
volume IV, under development by the   Health Effects Research Laboratory   Duke University   Northrop Services, Inc.   National Cancer Institute
following agencies: United States Environmental Protection Agency   Durham, NC 27706   Research Triangle Park, NC 27709   Bethesda, MD 20205
                                          Research Triangle Park, NC 27711

Mortality Rates and Trends 1950–1979. Vol. 4. Atlas by the U. S. Environmental Protection Agency, 1987, p. 62, 65, 156, 248, Washington, DC: U. S. Government Printing Office.

able and thereby live long enough for cancer to become manifest, or because genetic or environmental factors have changed. (As epidemiologists, the Lilienfelds include behavioral changes under environmental factors.)

Greenberg (1983b) has found a spatial convergence of cancer mortality in the United States between 1950 and 1975. He divided counties into those that were strongly urban (10%), moderately urban, and rural (69%). The strongly urban–rural ratio for cancers of the bladder, kidney, larynx, esophagus, lung, tongue, rectum, and large intestine was 1.9 overall and locally as high as 2.4 in the 1950s. By the early 1970s the ratio was only 1.2 overall and lower for white females.

There were other important changes as well. Rates for some sites of cancer increased, especially lymphoma, multiple myeloma, and melanoma, and connective tissue, lung, brain, pancreas, and central nervous system cancers. Other types decreased greatly, including five that had been among the most important: stomach, rectum, liver, cervix uteri, and corpus uteri. Females had the least increase and greatest decrease, so that the gap between males and females widened. Nonwhite males had a trend opposite to white females. In 1950–1955, white males had a cancer rate 7% higher than nonwhite males and 18% higher than white females. By 1970–1975, nonwhite male cancer rates were 22% higher than white male rates, 65% higher than nonwhite female, and 86% higher than white female cancer rates. There was a strong parallel trend for most types of cancer. Cancer rates diverged by sex and race subgroups of the population but converged geographically. Greenberg's monograph concludes with the strong statement that the white population is "moving toward a homogeneous pattern of cancer mortality for almost every type of cancer" (Greenberg, 1983b, p. 102). Elsewhere, Greenberg (1980) presents a method, for allocating components of the incidence changes to county, regional, and national scales, that builds upon scale differences in variance, discussed in Vignette 7-2.

Greenberg contends that this spatial convergence of cancer mortality in the United States and other industrialized countries is caused by change in the geography of risk factors associated with the diffusion of urban culture. These risk factors include air and water pollution, cigarette smoking, alcohol consumption, diet, occupation, socioeconomic status, stress, and medical practices. One might add to this list the last decade's metropolitan-to-rural net migration and regional exchanges of people.

Some Studies of Cancer

Geographers have studied spatial patterns to identify risk factors. Often the research questions and methodology involve determining regions and investigating common or different factors (such causal reasoning and analogue area analysis is described in Vignette 6-2). For example, several regions can be identified in which stomach cancer has a high incidence. Iceland, Japan, and northern Minnesota and Wisconsin are three such regions. Copper seems to be unusually concentrated in the soils of one of these—is it also high in the others? People in Iceland eat smoked and salted fish and are of Scandinavian descent. Do people in Japan eat salted and smoked fish, and do people of Scandinavian descent in Minnesota and Wisconsin continue ethnic dietary practices? Some people in eastern Africa have a high incidence of esophageal cancer in a culture area in which home-distilled alcohol is made. People outside of that culture area have lower cancer rates: perhaps the alcohol is involved. The region around the Caspian Sea in Iran, however, has some of the world's highest rates of esophageal cancer, and people there do not consume alcohol. Research suggests that the Iranian risk area is characterized by nutrient-deficient, saline soil, halophytic vegetation, and poor crops. Alcohol could easily absorb elements from the pottery and cans in which it is distilled. Could trace element deficiency or toxicity be common to both regions?

Geographers in Japan, Europe, and North America have analyzed the pattern of cancer occurrence in relation to environmental, socioeconomic, and demographic factors, by means of multivariate statistical procedures. The Japanese have found associations of lung cancer with air pollution, urbanization, and living in fishing ports and near refineries and mines. A cartographic analysis of cancer mortality in the British Isles also identified the urban affinity of lung cancer (Howe, 1981). However, it did not support the epidemiological hypotheses that mortality from cancer of the large intestine and rectum is related to a carbohydrate-rich diet, or that risk of gastric cancer is associated with blood group (genetic heritage). Others, struggling to relate aggregate data to exposure patterns and microenvironments, have inconclusively studied the relationship of various classifications of cancer to hardness and trace elements of water in Quebec (Thouez, Beauchamp, & Simard, 1981). Many geographers have analyzed the way various sites of cancer (cancer of the stomach, pancreas, esophagus, lungs, blood, and so forth) form groups when their patterns of spatial distribution are compared. Types of cancer having the same pattern of spatial distribution may share a common etiological factor.

Studies that are based on general populations, do not select disease categories for research carefully and deliberately, and target fortuitous places, run into many data and methodological problems. They so often suffer from the ecological fallacy (Vignette 7-2) that ecological studies as a whole have earned a reputation for scientific weakness. Such multifactor, population-based studies are, however, a powerful means for choosing places for case-control and other detailed studies.

A different approach has been taken by Glick. He opened up a new methodology by regarding spatial autocorrelation as an etiological tool instead of a statistical nuisance (Vignette 7-3). *Spatial autocorrelation* is the association between a variable's values taken from two adjacent places. What Glick has done is use the scale and intensity of that association to test etiological hypotheses.

Glick (1979a, 1979b, 1980, 1982) has used transects across the counties of the United States and distance/location relationships among states to study the spatial organization of cancer rates. Since different cancer-inducing processes act most strongly at different scales, knowledge of the relationship of scale and cancer variation can help identify the process. He starts with theoretical models of carcinogenesis and infers the distance relationships.

There are two major classes of skin cancer. Basal or squamous cell carcinomas are fairly common and relatively innocuous because they can be successfully treated without hospitalization. Malignant melanoma, a rare form of skin cancer, is often fatal. Glick (1979a) had data from four special surveys of skin cancer. He used regression techniques to estimate the effects of age and of ultraviolent radiation levels and found that nonmelanoma skin cancer has a strong age effect, as one would espect if the carcinogenesis process required repeated exposures and alterations of the cell (as illustrated in Figure 7-1). Glick reasoned that because nonmelanoma skin cancer involves more stages of exposure, it should have a steeper spatial gradient of mortality. Indeed, he found that the gradient in mortality rate for a transect (line) through Minnesota-Iowa-Missouri-Arkansas-Louisiana was very steep and constant for nonmelanoma skin cancer, going from about 1.0 to 2.6 per 100,000, whereas mortality from melanoma ranged from 1.1 in Minnesota to 1.8 in Louisiana, with a less constant gradient.

Figure 7-3 presents a spatial correlogram for two types of cancer that illustrates the spatial pattern of point and area exposure to risk. Across the

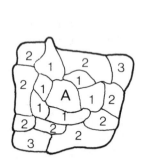

Lags from District A
Each District in Turn

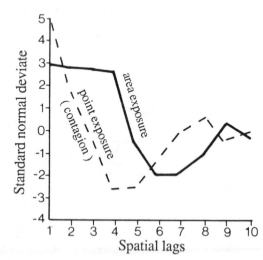

Figure 7-3. Spatial correlograms. Point and areal (regional) sources of exposure are reflected in different levels of correlation between spatial units when the units are lagged incrementally.

x-axis are the spatial lags in association. For example, at lag 1 the average autocorrelation of each county in the state with adjacent counties is expressed, whereas at lag 5 the autocorrelation is with five counties away. The autocorrelation is expressed as a normal (z) score on the y-axis. The farther away from zero, either positively or negatively, the stronger the association among counties. The line that shows point exposure has a steep distance decay in autocorrelation from high positive (similar to adjacent county) to high negative (different from nearby county). It represents a spatial pattern of isolated foci of incidence, such as might result from contagious disease occurring in large cities, but not yet diffused across the county, or from behavior limited to ethnic enclaves. Stomach cancer in Pennsylvania shows such a pattern, whereas bladder cancer is characterized by a nearly horizontal correlogram for lags 1–3 (Glick, 1980). Such a pattern, as illustrated, portrays a large region of similar rates and suggests an environmental carcinogen found over large areas, such as water catchment basins, rather than a diffusion process.

In a different geographic approach, the study of microscale mobility and exposure to different environments can be used to identify possible etiological factors and form new hypotheses. Armstrong (1976, 1978) studied nasopharyngeal cancer, a disease of unknown etiology, among the Chinese in Malaysia. Using a case-control method (Vignette 6-2), he compared the exposure patterns of people with the disease and of controls, people without the disease, to what he called self-specific environments. He measured how much time was spent daily in agricultural areas, in squatter housing or middle-class housing, in shop workplaces, factories, shopping on the street, in various means of transportation, in public places, and so forth, and also how much time was spent in smoky places, crowded places, around chemicals and under conditions of bad air pollution. He examined such indices of traditional or modern/assimilated cultural patterns as having an altar in the house, eating meals in a formal manner, and details of what people ate and how they prepared it. He found that among genetically susceptible Chinese, important stimuli are associated with industrial or trade occupations, smoky workplaces, poorer housing, traditional lifestyle, a diet with less variety of foods, and childhood consumption of salted fish. This methodology can be extended to many searches for unknown causes of disease.

CARDIOVASCULAR DISEASE

The group of diseases known as cardiovascular disease constitutes the major cause of death in economically developed countries. It includes ischemic heart disease, cerebrovascular disease, atherosclerosis, hypertension, and rheumatic heart disease. Ischemic heart disease occurs when all or part of the heart has reduced blood supply. There can be many causes for this, but in more than 90% of cases the blood vessels to the heart are blocked with the fatty deposits of atherosclerosis. When the blood supply is drastically affected, a myocardial infarction (heart attack) occurs. If the blood supply is only partially restricted, a chronic coronary thrombosis usually results. If the blood supply to the brain is blocked, cerebrovascular disease can culminate in a stroke (apoplexy).

The resistance of the blood vessels to the flow of blood through them is blood pressure. This is affected by the volume and composition of the blood, by the dilation and constriction of blood vessels, and by narrowing of the arteries caused by the build-up of fatty deposits. Systolic blood pressure is the pressure on the arteries when the heart contracts and pumps a surge of blood out; diastolic blood pressure is the pressure on the arteries between contractions, the minimum pressure that the arteries feel continuously. High blood pressure, or hypertension, is so often followed by a heart attack, stroke, or kidney failure that it is regarded as a serious disease by itself. High blood pressure for which the cause is unknown is referred to as *essential* hypertension.

These diseases all have a wide variation geographically. Age-specific rates of ischemic heart disease, for example, vary more than 500-fold among nations. Cardiovascular diseases are degenerative diseases, important because of the aging of the population structure in developed countries. The etiology of cardiovascular disease rests on the familiar triangle of population–behavior–habitat interactions.

The Population Base

A wide range of genetic factors is involved in cardiovascular disease. Several kinds of congenital defects may be involved, such as improperly closed heart valves. Genetic controls on metabolism are also important: people process, store, and remove fats in different ways. There may be genetic predisposition to hypertension, acting through the renal enzymes that control fluid volume and excretion. Individuals react differently to chemicals (in drugs and foods) that promote vasodilation. The facility with which blood clots form has a strong genetic component. One of the most marked genetic differences is expressed in the lower incidence of ischemic heart disease among premenopausal women as compared with men. Sex hormones influence how the body metabolizes and stores cholesterol and other fats. The level and composition of hormones differ not only between males and females but also from individual to individual and family to family.

Personality is also partly based on genetic combinations, although it has a behavioral expression and environmental impress. It forms a bridge from genetic to behavioral considerations. There have been many intervention studies designed upon the distinction of type A and Type B personalities (Rosenman *et al.*, 1970). Type A personality involves aggressiveness, excessive drive, compulsive behavior, impatience, and usually sensitivity and insecurity. It is considered a coronary risk factor, and intervention programs try to get people to slow down and relax. More recently, studies have suggested that it is the dimension of hostility and anger that is most destructive in the Type A personality syndrome (Chesney & Rosenman, 1985).

The Behavioral Base

The behavioral base for cardiovascular disease involves cigarette smoking, drinking alcohol, diet, exercise, occupation, psychosocial insults, stress, and stress management. Cigarette smoking is clearly a risk factor. The cadmium in cigarettes raises blood pressure, and chemicals they introduce to the blood circulation affect vasodilation. A little alcohol consumption is protective; apparently it favorably affects cholesterol balance. Heavy alcohol consumption, however, interacts with nutrition, weight, and exercise to constitute a risk factor. Many dietary factors in cardiovascular disease are still unknown. The high potassium levels found in some vegetables may be protective, for example, and calcium may be more important than previously recognized. Two dietary hazards are clear: high salt consumption and cholesterol.

The role of cholesterol and other blood lipids (fats) was long obscured by the fact that the liver makes some cholesterol itself, regardless of diet, and by the need to differentiate many subcategories of lipids. High-density lipid cholesterol (HDL) acts to transport fats and clean out the circulatory system. Low-density lipid cholesterol (LDL) builds up in deposits. The metabolic changes associated with exercise, with a little consumption of alcohol, and with some drugs increase the ratio of the protective HDL to the endangering LDL and so lower the risk of atherosclerosis. Exercise, then, has the dual impact of increasing the capacity of the circulatory system to pump blood so that it can better cope with sudden demands upon it, and inducing a more favorable lipid balance through metabolic changes.

Stress is a rubric that covers many evils. Psychosocial insults come from situations as diverse as occupation and working conditions, neighborhood blight, domestic problems with spouse or children, and prolonged fear and anxiety. Many studies have found sudden myocardial infarction to be the illness most related to life-event stress. Loss of job, upward mobility, marriage, divorce, death of a spouse, change of residence, and trouble with children characteristically are common for heart attack patients. Similarly, the important occupational characteristic seems not to be specific job categories so much as lack of control, repetitiveness, inability to use one's skills, and similar frustrations in one's work.

All the stresses related to the poverty syndrome are especially relevant to cardiovascular disease. It has, however, a rather complicated relationship with socioeconomic change in the Third World. Cross-cultural studies have found that increased hypertension is often related to greater participation in the money economy. This generally entails major changes in diet, occupation, mobility, and other conditions. Economic change and loss of traditions add to stress, and economic growth and upward mobility are associated with increased risk of cardiovascular disease. Nevertheless, numerous studies within industrialized countries have shown a positive health effect from a prosperous and growing economy. Times of recession and unemployment add tremendously to the life stresses and to mortality.

The effects of stress on health have illuminated the importance of stress management. Intervention programs are being implemented to teach people how to relax, to use exercise as a stress release, and to soothe themselves by caressing pets and smelling roses. The role of social support and close relationships in coping with stress has come to the fore of recent research.

The Habitat Base

There are three important components of the habitat vertex of cardiovascular diseases. Available health services are especially important. The detection, treatment, and control of hypertension are among medical science's greatest interventions. The development of emergency medical systems and of cardiopulmonary resuscitation techniques has been important for heart attacks. The other two critical aspects of the habitat are weather and geochemistry.

Several physical mechanisms of acclimatization and adaptation to changes in pressure and temperature were discussed in Chapter 5. Cardiovascular disease has a peak occurrence during periods of cold weather. Incidence seems to be responsive to the passage of fronts and other meteorological events. Because of physiological conditions of vasodilation and vasoconstriction, peripheral capillary networks for heat dissipation or conservation, and changes in the composition and thickness of blood, it is also likely that different climatic circumstances have something to do with the wide range of disease incidence.

An active area of research concerns the etiological significance of trace elements. These elements occur in such small amounts that only recently have scientists been able to measure them accurately, in parts per billion. The study of trace elements is further complicated by the constant interaction among them. Zinc, for example, is important to tissue healing. After a heart attack, most of the body's zinc is concentrated in the heart, and it seems reasonable that deficiency in zinc may play a role in fatal outcome. Zinc, however, is blocked by copper, so that the adequacy of zinc in the diet cannot be studied separately. Cadmium, a heavy metal and systemic poison like lead, is known to raise blood pressure in laboratory animals when they are constantly fed minute amounts of the metal. It might well be important in essential hypertension. Selenium, however, seems to block the action of cadmium and is a constituent of blood enzymes important for blood clotting and cholesterol metabolism.

Selenium could be measured only recently, by expensive techniques of nuclear absorption. Similarly, lead, nickel, chromium, manganese, lithium, potassium, sulfur, cobalt, and iron are variously involved in nerve transmission to the heart, muscle healing, cholesterol metabolism, blood clotting, and hypertension. They all combine, enhance, block, and precipitate each other in various combinations. In addition, abundant elements such as calcium and magnesium are important electrolytes and have been implicated in arrhythmias and sudden death.

A geographer has to be concerned about the pathways that these elements can take to the individuals at risk. The *water factor* has been implicated in many studies. More than a score of international studies have found water hardness to be inversely related to cardiovascular disease mortality. Hardness is usually based on the amount of calcium carbonate, but sometimes other minerals are included. The suggestion is that soft water has a different composition of trace elements than does hard water. Only some elements, however, can be consumed in water in sufficient quantity to make any difference. Thus, water may be a significant source of copper, but thousands of gallons would have to be consumed daily to obtain significant nutritional input of iron or potassium. Food is a much more concentrated source of most elements. Further complications can result from using analysis of surface water for people who drink from artesian sources, or from the alterations to water supply brought about by water treatment plants and municipal pipes.

Figure 7-4 illustrates some of the geochemical pathways that trace elements taken from the environment to humans. The original rock is very important. Limestones are high in calcium and magnesium and are accompanied usually by strontium, phosphorus, and zinc, and sometimes by cadmium and lead. Shales are high in aluminum silicates and, locally, in molybdenum, vanadium, and uranium. Sandstone is rich in silica, but it is usually leached of trace elements. Serpentine rock is high in iron and magnesium, and often in chromium, nickel, and manganese. How these elements become bound and available depends on temperature, other minerals available, and groundwater and surface water supplies. The uptake of these elements by vegetation may be controlled by the plant's metabolism, regardless of soil concentration, and the elements may be stored in plant parts that humans do not consume. Heating often causes the elements to combine and precipitate, while the passage of time during food processing and water piping permits them to absorb and bind to other elements. Herbicides, insecticides, fungicides, and preservatives, which usually derive much of their potency from metals such as mercury or selenium, add to the concentrations of elements associated with the original soil or river water. After all these sources have contributed, the elements ingested by humans are still measured in parts per billion. It is no wonder that the effects of water hardness and various trace elements remain a controversial subject.

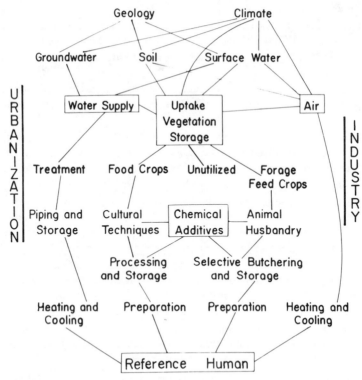

Figure 7-4. Geochemical pathways. Human consumption of trace elements is strongly influenced by the complex steps of processing and distributing of soil or water.

Patterns and Analysis

The high prevalence of cardiovascular disease in industrialized countries is a basic component of the mortality transition (Chapter 4). This generality, however, obscures important variation among places. In European countries the major cause of death is heart disease: stroke is the major cause in Japan. The incidence of stroke is especially high in the northeastern part of the main island of Honshu, the region of Tohoku, but lower around the Inland Sea. Takahashi (1981) found that in Tohoku salt consumption is higher and that the water hardness ratio is low in Tohoku and high in the Inland Sea area. In several European studies, stroke rates have been positively correlated with the age of rocks in the study area. Leaching of minerals is assumed to be involved.

The element most related to cardiovascular disease in British water studies is calcium. In Britain as well as Ontario, Canada, calcium has been associated with sudden death due to arrhythmia, which results from imbalance in the electrical activity of the heart. There is a strong north–south gradation of cardiovascular disease in the United Kingdom. The death rates of Scotland have been long recognized as mysteriously high. There has been great interest, therefore in the inverse association with water hardness. The lower southeastern incidence coincides with the hard waters of Tertiary and Cretaceous formations, the higher northwestern incidence with the soft waters of Precambrian rock formations. West and Lowe (1976), however, have also found the same northwest–southeast gradient for temperature and rainfall. Others have suggested the importance of genetic distribution (marked by blood types) in Britain, due to its invasion and settlement history (Howe, 1972). More plausibly, it has been noted that the industrial revolution began in the regions with soft water partly because of their water quality. Early urbanization and industrialization might involve confounding factors. The pattern persists, however, even without the industrial cities.

Cardiovascular disease in the United States also has a strong regional pattern. Schroeder (1966) was the first to study cardiovascular disease and the water factor there, using data on finished water supplies in 1,300 municipalities across the country. At the national scale, significant inverse relationship was found for male and female deaths and water hardness. Cerebrovascular disease is strongly clustered in the southeastern counties along the coastal plain of Georgia and South Carolina. Again, the water factor has been implicated, as well as the deficiency of certain trace elements, such as selenium, in the soil and water. Temperature, population settlement and inheritance, agricultural patterns, poverty, late urbanization, and dietary factors all seem to be related as well.

The spatial correlation of a geochemical factor and cardiovascular disease would be most convincing if it held consistently at several scales and for all subgroups of the population. It would be further strengthened if geochemicals as etiological agents could be plausibly connected with a biophysical mechanism. Researchers have noted that the more rigorous the study and the more focused the study area, the weaker and more contradictory the geological connection. The relationship may be weak, or it may turn upon some factor that has not been identified (for example, the water factor). As Comstock (1979) noted, however, associations may be spurious, coincidental, or causal, and the association of a geochemical factor with cardiovascular disease is *not* spurious. Many independent studies in many countries have found statistically significant correlations.

Most geographers studying cardiovascular disease have been concerned with mapping and modeling its spatial variation. The identification of patterns of clustering and the changing associations at different scales has attracted considerable study. Ziegenfus and Gesler (1984) identified incidence of major cardiovascular, acute ischemic, chronic ischemic, and cerebrovascular diseases in four distinct clusters of counties in the urban corridor from New York through New Jersey to Philadelphia. Pyle (1971) mapped patterns of heart disease and stroke in Chicago, using trend surface analysis to form his regions (see Vignette 7-4). Maps of cardiovascular disease in the United Kingdom present detailed information on the relationship of rates to urban and rural areas and population size of units, on the statistical significance of rates, and on absolute standard mortality rates (Howe, 1970). Learmonth and Grau (1969) mapped heart disease in Australia, using isolines of standard mortality ratios. Recognizing the rare event that ischemic heart disease mortality is for small units and populations, McGlashan and Chick (1974) pioneered the use of maps based on Poisson probability for analysis of mortality rates.

Multivariate statistics analyze associations with measures of housing, income, ethnicity, and variables that make up the poverty syndrome. At a macroscale, many of the risk factors have become well established. In an impressive demonstration, Brenner and Mooney (1982) developed a model, including variables of long-term economic growth, deleterious behavioral risk factors, economic instability related to unemployment and income loss, health care, and very cold temperatures, that closely predicts the trend and fluctuations in male and female cardiovascular disease in England, Wales, and Scotland from 1955–1976.

Most of the extreme regionalization of cardiovascular disease within countries remains unexplained. A microscale approach through study of population mobility may allow the identification of new environmental risk factors. The high death rates from cardiovascular disease on the southeastern coastal plain of the United States, mentioned above, cluster in a way that cannot be explained by the conventional risk factors. Analyzing the associated distribution and covariation of a series of overlays of geochemistry, agriculture, economics, ethnicity, industry, age of building construction, and various sources of water might suggest some new hypotheses. Meade and Gesler have selected census tracts within the city of Savannah, Georgia, on the basis of socioeconomic variables that predict stroke and blood pressure. Tracts that are well predicted serve as controls for those where stroke is significantly higher than predicted. Interviews ascertained the population's exposure to habitats within the city and defined the water and food pathways of trace elements from the local environment. Differences in environmental exposures and behavioral patterns may suggest new etiological clues.

MENTAL ILLNESS

Mental illness is certainly not limited to industrialized societies. Various forms of it are common in isolated, pastoral, and rural areas. Because of rapid social change and urbanization, mental illness is common in Third World countries. Its appropriate and effective treatment by traditional medical systems (Chapter 9) argues that all societies are familiar with mental illness. In industrialized countries, however, there are demographic and social processes that have hampered the ability of society to absorb and cope with affected individuals.

The need to cope with technological society may, through stress, isolation, and lack of emotional support, induce certain forms of mental illness.

Many forms of mental illness are organically based, perhaps resulting from genetics, alcoholism, syphilis, or senility. Some regard that alcoholism, suicide, and such social pathology as homicide *are* mental illnesses. Some mental illness has long been regarded as *functional*, meaning that the familial and social environment have had a role in its development; the illness may develop only in susceptible individuals. These forms of mental illness illustrate levels of health struggling under the barrage of new insults (Chapter 2). For example, McNeil says,

> The form and frequency of mental derangement has a host of correlations with other life circumstances and we cannot attribute causation exclusively to any one of these. . . . In this welter of discrete items related to psychosis must be hidden a single overriding factor that can connect all the disparate parts into a meaningful whole. Being white, having a stable marriage, being gainfully employed, being educated, and being intelligent may all summate to a condition called security and freedom from anxiety. Having all these 'advantages' may be the critical factor in determining how easily and successfully one copes with the tasks of living. The absence of any one of these conditions may diminish the human capacity to adjust. The absence of a great number of them may spell psychosis (1970, in Giggs, 1973, p. 71).

Studies of Mental Health

Ecological (aggregate, multivariate) studies of mental disease have not kept up with advances in social science methodology. Various forms of factor analysis, for example, are superior to the social area analysis units used by the National Institute of Mental Health. Geographers have made limited contributions thus far, however, to etiological studies of mental health. One area that has been studied is the incidence of schizophrenia.

Schizophrenia is a common (in some estimates, it affects 1% of the population), chronic, serious psychosis that can produce lifelong invalidism (Giggs, 1973). It usually strikes young people: 75% of schizophrenics have a first attack before age 25. There are problems with subdiagnosis, but it is one of the most replicable classifications of mental illness. There is strong evidence of a biochemical base and some evidence of genetic causation, but mysteries remain. There is, for example, strong seasonality of birth for schizophrenics (Chapter 5).

One of the classic ecological studies of schizophrenia related its incidence in Chicago to the newly developed Burgess model of urban structure, which identified concentric rings of land use and development history (types of industry, residential density, business, decay, and transition, etc.). Faris and

Dunham in 1939 regionalized 120 subcommunities of Chicago into 11 types of milieu. They related paranoid schizophrenia, for example, to patient origins in rooming house districts of the city. These stretched along the main transportation arterials and housed single, white-collar workers who commuted to the central business district. They noted the transient nature of the neighborhoods, and suggested that the isolation and lack of communication might precipitate the psychosis (Faris & Dunham, 1965).

Giggs (1983) addressed the same questions in Nottingham, England, using contemporary methodology. He used factor analysis to determine 10 components of the city's ecologic structure. One dimension, for example, he interpreted as "social and material resources," another as "urbanism/familism," and a third as "family life-cycle axis." Using Poisson probability, he tested the incidence of schizophrenia in 15 areas based on the ecological structure of the city. He demonstrated that the distribution of patient origins was very localized and that there were strong and statistically significant links between his characteristics of schizophrenic patients, such as age, sex, marital status, and birthplace, and the five leading components of the city's structure.

Smith (1977) addressed the neighborhood environment context of mental health. He and other geographers have been concerned with the deinstitutionalization of mental patients, which started in the 1970s and spread quickly in the United States. They are particularly concerned about the location of community mental health facilities. Smith identified several relevant dimensions of neighborhood environments by studying the incidence of recidivism (return to the mental hospital) among deinstitutionalized patients.

Smith found a spatial dimension and landscape expression to community mental health. Landmarks and the layout or spatial design of a community provide orientation and facilitate a sense of identity. Facilities such as community centers, places for parties, churches, and libraries promote rootedness and provide means to transcend the daily conditions of life. People can be separated from noise, vehicular traffic, and pollution.

By considering such environmental characteristics, Smith thought he could identify how neighborhoods function as therapeutic communities. Living in commercial and industrial areas, characterized by high traffic and industrial buildings, was a good marker for recidivism. Being old and living alone, however, was not as reliably associated with recidivism. He suggested that neighborhoods that offer many patients a familiar setting and a place to be alone are therapeutic. Overall, it was easier to predict nonreturn on the basis of positive neighborhood characteristics than return on the basis of negative ones. People seemed able to tune out what they did not want to see.

Neighborhood studies have moved into another dimension. Dear (Dear, Taylor, & Hall, 1980; Isaak, Taylor, & Dear, 1980) has investigated community attitudes toward mental health facilities and patients and the fit between community characteristics and type of facilities. He has developed methodology for investigating mental health service delivery and has connected medical geography to geographic studies on location of noxious facilities and on cognition and perception of environments. Dear's work has especially served to connect medical geography with the movement in social science away from positivist epistemology and toward greater social phenomenology.

UNKNOWN ETIOLOGY AND OTHER QUESTIONS

A geographic perspective on a disease can result in discovering patterns of association. Developing etiological hypotheses for diseases of unknown cause is potentially a fruitful, and perilous, endeavor. Mayer (1981), for example, has focused attention on the dramatic geographic pattern of multiple sclerosis (MS). The disease has a latitudinal gradient from high northern rates to low southern ones in Europe and North America. Only Japan is anomalous, having very low rates even at high latitudes. The low incidence in Japan and among people of Asian descent in the United States suggests a lack of genetic susceptibility.

The most obvious latitudinal environmental factor is solar radiation, but it may be spurious. Less obvious variations north to south in both Europe and North America include agricultural crops, development of industry, level of urbanization, and economic standards. Migration streams to Israel from MS high-rate areas of Europe and low-rate areas of Africa have been studied for MS incidence later in life. Childhood exposure to some unknown factor is important: adults who migrate have the risk of their place of origin, whereas small children take on the risk of their destination. The new generation of native Israelis, surprisingly, has high MS rates despite low latitude, implying that some environmental condition associated with industrialization and developed economies is important (Lowis, 1986). In the United States, however, California, though more industrial, has lower MS rates than Washington state. Migrants from the low-rate South and high-rate North to California and Washington have been studied. Children assume the high or low rates of their destination, as in the Israeli study. Adult migrants from the South to Washington have higher MS rates than in their birthplace but lower than for native Washingtonians. This implies that some protective factor continues to shield the migrants in the presence of greater environmental hazard. In the absence of a United States registration system, however, it is very difficult and expensive to get a large enough stream of migrants to be able to analyze rare diseases.

Rheumatic diseases, responsible for widespread suffering and loss of work time, have received little attention from geographers. Rheumatic diseases include degenerative conditions such as osteoarthritis, rheumatoid arthritis, spinal disk degeneration, gout, and ankylosing spondylitis and diseases of the connective tissue, such as lupus. Etiological hypotheses include the causal impact of wear and tear, calcium deficiency, metabolic deficiency in regulating blood uric acid levels, infectious agents, "bad" water, diet, occupation, and emotional stress. Genetic heritage has often been implicated, but the best-designed studies have found the least evidence for familial clustering. The rheumatic diseases are notoriously difficult to confirm or detect in any repeatable manner, especially at subclinical levels. Even the blood serum test for rheumatoid factor (anti–gamma globulin) is known to flip capriciously across the arbitrary border that separates positive and negative results. In a review of knowledge about these diseases, Cobb (1971) noted that "for any disease, classification will remain inadequate until etiology is known" (1971, p. 18).

Another neglected disease is Parkinson's disease (paralysis agitans). It is due to degeneration and death of cells in one region of the brain (substantia nigra). The dysfunction of these cells results in a loss of the dopamine that they produce, resulting in blocked nerve impulses in the basal ganglia of the brain. Semiautomatic movements (such as movement of the tongue while speaking or swinging of the arms while walking) become difficult, while involuntary movements (tremors) become common. Balance is disturbed, stiffness occurs, paralysis develops, and the disease progresses slowly but inexorably as regional brain cells continue to die. The reason those particular brain cells die is unknown. A recent discovery shows that a toxin (MPTP), a by-product of the manufacture of illicit drugs, produces a parkinsonism-like disease, even in young people (Lewin, 1984). Furthermore, MPTP can induce the symptoms in monkeys and so has created the first animal model for studying the disease and its treatment. This discovery strongly implies that exposure to one or more unknown environmental toxins may cause the disease. Incidence data for parkinsonism is hard to get because, at the early stages, it is not diagnosed properly, but it is a reported cause of death reflected in mortality statistics. Unfortunately, there has been little, if any, geographic analysis of the spatial distribution and environmental circumstances of this serious disease.

CONCLUSION

This chapter has reviewed some of the geographic research on diseases of primary importance in the industrialized countries. As the mortality, fertility, and mobility transitions continue and the world's population urbanizes and ages, the etiology, prevention, and treatment of degenerative diseases will come to dominate world health concerns.

The spatial patterns, causal relationships, and time and scale parameters of these diseases are complex matters. If hypotheses are to have value, the geographic researcher needs to understand the disease processes as founded in the population's biology and its interaction with habitat and behavior. The complexities of spatial pattern and disease etiology demand competence in statistics. Upon this foundation rests the ability to project population health needs, as distinct from economic demand, and to plan for the delivery of appropriate health services.

REFERENCES

Ames, B. N. (1983). Dietary carcinogens and anticarcinogens. *Science, 221*, 1256–1264.

Armstrong, R. W. (1971). Medical geography and its geologic substrate. *Geological Society of America Memoir, 123*, 211–219.

Armstrong, R. W. (1976). The geography of specific environments of patients and non-patients in cancer studies, with a Malaysian example. *Economic Geography, 52*, 161–170.

Armstrong, R. W. (1978). Self-specific environments associated with naso-pharyngeal carcinoma in Selangor, Malaysia. *Social Science and Medicine, 12D*, 149–156.

Brenner, M. H. & Mooney, A. (1982). Economic change and sex-specific cardiovascular mortality in Britain 1955–76. *Social Science and Medicine, 16*, 431–442.

Calabrese, E. J. (1984). *The environmental gender gap: Differences between males and females in response to pollutants.* Paper presented to the Institute for Environmental Studies at the University of North Carolina, Chapel Hill.

Calabrese, E. J. (1985). *Sex differences in response to toxic substances.* New York: Wiley.

Chesney, M. A. & Rosenman, R. H. (Eds.). (1985). *Anger and hostility in cardiovascular and behavioral disorders.* Washington: Hemisphere Publications.

Cleek, R. K. (1979). Cancer and the environment: The effect of scale. *Social Science and Medicine, 13D*, 241–247.

Cobb, S. (1971). *The frequency of the rheumatic diseases.* Cambridge: Harvard University Press.

Comstock, G. W. (1979). The association of water hardness and cardiovascular disease: An epidemiological review and critique. In U.S. National Committee for Geochemistry, *Geochemistry of water in relation to cardiovascular disease* (pp. 46–68). Washington, DC: National Academy of Sciences.

Dear, M., Taylor, S. M. & Hall, G. G. (1980). Attitudes toward the mentally ill and reactions to mental health facilities. *Social Science and Medicine, 14D*, 281–290.

Dear, M. & Willis, T. (1980c). The geography of community mental health care. In M. S. Meade (Ed.), *Conceptual and methodological issues in medical geography* (pp. 263–281). Chapel Hill, NC: University of North Carolina, Department of Geography.

Eyles, J. & Woods, K. J. (1983). Man, disease, and environmental associations: From medical geography to health inequalities. In J. Eyles, & K. J. Woods (Eds.), *The social geography of medicine and health* (pp. 66–114). New York: St. Martin's Press.

Faris, R. E. L., & Dunham, H. W. (1965). *Mental disorders in urban areas: An ecological study of schizophrenia and other psychoses.* Chicago: University of Chicago Press.

Florin, J. W. (1971). *Death in New England: Regional variations in mortality.* Studies in Geography No. 3. Chapel Hill, NC: University of North Carolina, Department of Geography.

Gardner, M. (1976). Soft water and heart disease. In J. Leniham & W. W. Fletcher (Eds.), *Health and the environment* (pp. 116–135). New York: Academic Press.

Gardner, M. J., Winter, P. D., & Acheson, E. D. (1982). Variations in cancer mortality areas in England and Wales: Relation with environmental factors and search for cause. *British Medical Journal, 284*, 284–287.

Giggs, J. A. (1973). The distribution of schizophrenics in Nottingham. *Transactions of the Institute of British Geographers, 59*, 55–76.

Giggs, J. A. (1983). Schizophrenia and ecological structure in Nottingham. In N. D. McGlashan & J. R. Blunden (Eds.), *Geographical aspects of health* (pp. 197–222). London: Academic Press.

Girt, J. L. (1972). Simple chronic bronchitis and urban ecological structure. In N. D. McGlashan (Ed.), *Medical geography: Techniques and field studies* (pp. 211–231). London: Methuen.

Glick, B. J. (1979a). Distance relationships in theoretical models of carcinogenesis. *Social Science and Medicine, 13D*, 253–256.

Glick, B. J. (1979b). The spatial autocorrelation of cancer mortality. *Social Science and Medicine, 13D*, 123–130.

Glick, B. (1980). The geographic analysis of cancer occurrence: Past progress and future directions. In M. S. Meade (Ed.), *Conceptual and methodological issues in medical geography* (pp. 170–193). Chapel Hill, NC: University of North Carolina, Department of Geography.

Glick, B. J. (1982). The spatial organization of cancer mortality. *Annals of the Association of American Geographers, 72*, 471–481.

Greenberg, M. R. (1980). A method to separate the geographical components of temporal change in cancer mortality rates *Carcinogenesis, 1*, 553–557.

Greenberg, M. R. (1983a). Environmental toxicology in the United States. In N. D. McGlashan & J. R. Blunden (Eds.), *Geographical Aspects of Health* (pp. 157–174). London: Academic Press.

Greenberg, M. R. (1983b) *Urbanization and cancer mortality: The United States experience, 1950–1975*. New York: Oxford University Press.

Haggett, P. (1976). Hybridizing alternative models of an epidemic diffusion process. *Economic Geography, 52*, 136–146.

Howe, G. M. (1970). *National atlas of disease mortality in the United Kingdom* (2nd ed.). London: Nelson.

Howe, G. M. (1972). *Man, environment, and disease in Britain*. New York: Barnes and Noble.

Howe, G. M. (1981). Mortality from selected malignant neoplasms in the British Isles: The spatial perspective. *Social Science and Medicine, 15D*, 199–211.

Isaak, S., Taylor, M., & Dear, M. (1980). Community mental health facilities in residential neighbourhoods. In F. A. Barrett (Ed.), *Canadian studies in medical geography* (pp. 231–256). Downsview, Ontario, Canada: York University, Department of Geography.

King, P. E. (1979). Problems of spatial analysis in geographical epidemiology. *Social Science and Medicine, 13D*, 249–252.

Learmonth, A. T. A. & Grau, R. (1969). *Maps of some standardised mortality ratios for Australia 1965–66 compared with 1959–63*. Occasional Papers No. 8. Canberra, New South Wales, Australia: Australian National University.

Lewin, R. (1984), Trail of ironies to Parkinson's disease, *Science, 224*: 1083–1085.

Lilienfeld, A. M., & Lilienfeld, D. E. (1980). Foundations of epidemiology (2nd Ed.). New York: Oxford University Press.

Lowis, G. W. (1986). Sociocultural and demographic factors in the epidemiology of multiple sclerosis: An annotated selected bibliography. *International Journal of Environmental Studies, 26*, 295–320.

Marx, J. L. (1984). What do oncogenes do? *Science, 223*, 673–676.

Mason, T. J., McKay, F. W., Hoover, R., Blot, W. J., & Fraumeni, J. F., Jr. (1975). *Atlas of cancer mortality for U.S. counties: 1950–1969*. Washington, DC: U.S. Department of Health, Education, and Welfare, Epidemiology Branch, National Cancer Institute.

Mason, T. J., McKay, F. W., Hoover, R., Blot, W. J., & Fraumeni, J. F., Jr. (1976). *Atlas of cancer mortality among U.S. nonwhites: 1950–69*. Washington, DC: U.S. Department of Health, Education, and Welfare, Epidemiology Branch, National Cancer Institute.

Mayer, J. D. (1981). Problems of spatial analysis in geographical epidemiology. *Social Science and Medicine, 13D*, 249–252.

McGlashan, N. D. (1972). Food contaminants and oesophageal cancer. In N. D. McGlashan (Ed.), *Medical geography: Techniques and field studies* (pp. 247–257). London: Methuen.

McGlashan, N. D. & Chick, N. K. (1974). Assessing spatial variation in mortality: Ischaemic heart disease in Tasmania. *Australian Geographical Studies, 12*, 190–206.

Meade, M. S. (1980). An interactive framework for geochemistry and cardiovascular disease. In M. S. Meade (Ed.), *Conceptual and methodological issues in medical geography* (pp. 194–221). Chapel Hill, NC: University of North Carolina, Department of Geography.

Meade, M. S. (1983). Cardiovascular disease in Savannah, Georgia. In N. D. McGlashan & J. R. Blunden (Eds.), *Geographical aspects of health* (pp. 175–196). London: Academic Press.

Minowa, M., Shigematsu, I., Nagai, M. & Fukutomi, K. (1981). Geographical distribution of lung cancer mortality and environmental factors in Japan. *Social Science and Medicine, 15D*, 225–231.

Ohno, Y., & Aoki, K. (1981). Cancer death by city and county in Japan, 1969–1971: A test of significance for geographical clusters of disease. *Social Science and Medicine, 15D*, 251–258.

Pyle, G. F. (1971). *Heart disease, cancer, and stroke in Chicago* (Research Paper No. 134). Chicago: University of Chicago, Department of Geography.

Rosenman, R. H. (1974). The role, behavior patterns and neurogenic factors in the pathogenesis of coronary heart disease. In R. S. Eliot (Ed.), *Contemporary problems in cardiology* (Vol. 1, pp. 123–141).

Rosenman, R. H., Friedman, M., Straus, R., Jenkins, C. D., Zyzanski, S. J. & Wurm, M. (1970). Coronary heart disease in the western collaboration group study: A follow-up experience of 4H years. *Journal of Chronic Disease, 23*, 173–184.

Schroeder, H. A. (1966). Municipal drinking water and cardiovascular death rates. *Journal of the American Medical Association, 195*, 81–85.

Segi Institute of Cancer Epidemiology. (1982). *Age adjusted death rates for cancer for selected sites (A-classification) in 43 Countries in 1977*. Nagoya, Japan: Author.

Smith, C. J. (1977). *Geography and mental health*. Washington, DC: Association of American Geographers.

Takahashi, E. (1981). Geographic distribution of cerebrovascular disease and environmental factors in Japan. *Social Science and Medicine, 15D*, 153–172.

Thouez, J. P., Beauchamp, Y. & Simard, A. (1981). Cancer and the physicochemical quality of drinking water in Quebec. *Social Science and Medicine, 15D*, 213–223.

U.S. Environmental Protection Agency. (1986). *The United States cancer mortality rates and trends 1950–1979. Vol. 4. Atlas*. Washington, DC: U.S. Government Printing Office.

Waldron, I. Nowotarski, M., Freimer, M., Henry, J. P., Post, N., & Witten, C. (1982). Cross-cultural variation in blood pressure: A quantitative analysis of the relationship of blood pressure to cultural characteristics, salt consumption and body weight. *Social Science and Medicine, 16*, 419–430.

West, R. R. (1977). Geographic variation in mortality from ischaemic heart disease in England and Wales. *British Journal of Preventive and Social Medicine, 31*, 245–250.

West, R. R., & Lowe, C. R. (1976). Mortality from ischaemic heart disease—Inter-town variation and its association with climate in England and Wales. *International Journal of Epidemiology, 5*, 195–201.

Yerushalmy, J. H., & Palmer, C. E. (1959). On the methodology of investigation of etiological factors and chronic disease. *Journal of Chronic Diseases, 10*, 27–40.

Ziegenfus, R. C., Gesler, W. M. (1984). Geographical patterns of heart disease in the northeastern United States. *Social Science and Medicine, 18*, 63–72.

Further Reading

Anderson, R. (1984). Temporal trends of cancer mortality in eastern New England compared to the nation, 1950–1975. *Social Science and Medicine, 19*, 749–757.

Bennett, R. J. (1979). *Spatial time series*. London: Pion.

Burbank, F. (1972). A sequential space–time cluster analysis of cancer mortality in the United States: Etiological implications. *American Journal of Epidemiology, 95*, 393–417.

Cliff, A. D., & Ord, J. K. (1973). *Spatial autocorrelation*. London: Pion.

Friedman, M., & Rosenman, R. H. (1974). *Type A behavior and your heart*. New York: Knopf.

Howe, G. M., Burgess, L., & Gatenby, P. (1977). Cardiovascular disease. In G. M. Howe (Ed.), *A world geography of human diseases* (pp. 431–476). London: Academic Press.

Kmet, J., & Mahboubi, E. (1972). Esophageal cancer in the Caspian littoral of Iran: Initial studies. *Science, 175,* 846–853.

Pyle, G. F. (1979). *Applied medical geography.* New York: Wiley.

Robinson, V. B. (1978). Modeling spatial variations in heart disease mortality: Implications of the variable subset selection process. *Social Science and Medicine, 12D,* 165–172.

Shigematsu, I. (1981). *National atlas of major disease mortality in Japan.* Tokyo: Japan Health Promotion Foundation.

Vignette 7-1

THE QUESTION OF SCALE

Geographic scale, the size of the area under consideration, influences the questions that may be asked, the form of the data needed to answer them, and the interpretation of results. Is it an entire country, a state, a county, a city, or a neighborhood? The scale at which we work helps determine the base maps we may use and the specific cartographic approach we follow.

The terminology used to identify different scale levels in cartography can be confusing. A *large-scale* map is actually one that covers a small area, while a *small-scale* map covers a larger area. The terms large and small refer to the scale fraction that identifies the relationship between one unit of distance on the map and the actual distance on the surface of the earth.

Often limitations in the available data determine the appropriate scale. Use of data collected by others, often government agencies, restricts the scale available. Most census and United States vital statistics data, for example, are available only at the scale of the county, city, or state. This is sometimes because of the expense involved in making larger scale (smaller area) data available and sometimes because the government does not wish to release confidential data about individuals. The United States National Cancer Atlas, for another example, presents mortality for the more common cancer sites by county, whereas the less common forms are presented by economic areas of a state. These are a census grouping of several counties that create an acceptable large data set for statistical reliability.

A related problem is the so-called ecological fallacy (see Vignette 7-2). The questions asked at one scale create answers that are appropriate only at that scale. We cannot use state-wide data to suggest what might happen in a specific neighborhood, nor will that neighborhood data enable us to say what is happening in the surrounding county.

The geography of stroke mortality in the United States offers an example of the implications of scale. At the national level, the state-by-state differences indicate high mortality in major urban-industrial states and suggest the importance of occupation and stress. County-level investigations, however, identify a pattern of much higher than normal mortality for the band of counties along the Atlantic coast from North Carolina through Georgia (note the importance of age standardization for identifying these patterns). Many possible explanations have been suggested, ranging from high levels of smoking, to the consumption of salt and fatty foods, to variations of trace elements such as cadmium or selenium. No satisfactory explanation of this "enigma area" was available at this level of aggregation, however. The next step has been to move to a smaller area, the city of Savannah. At this scale a variety of individual data can be collected that might help substantiate earlier hypotheses or suggest other associations not apparent at the larger scale.

This pattern of using large-scale studies to suggest associations, followed by smaller scale studies that allow for a more specific investigation of the suggested associations, represents a common approach to the understanding of disease causation.

Vignette 7-2

THE ECOLOGICAL FALLACY

The infamous ecological fallacy has bedeviled geographic comparisons for a long time. It crops up when statistics are compared across scale. You cannot use a state-level risk factor to predict what will happen to specific individuals, and you cannot interview friends and neighbors to predict opinion in the county population. Even given a bona fide cause and effect, measurable association varies differently at different scales.

Variance changes with scale. If your information is based on individual questionnaires, you are aware of the diversity of response. When these answers are aggregated to county level, central tendency leveling takes place. There will be less variation among 100 counties in a state than there was among the millions of individuals who contributed the information. There is less variance among state economic areas than among 3,000 counties, and less among states than among state economic areas. These changes in variance have little to do with association or cause and effect and a lot to do with the way we can measure and generalize our findings.

Cleek (1979) and others have pointed out that correlations are especially affected by changes in aggregation. Data are often aggregated by an independent variable. This inflates correlation coefficients but usually does not affect regression coefficients. For example, as the individuals are aggregated into the county in which they live, socioeconomic measures are generalized. Several adjacent counties are likely to be more similar with regard to such characteristics as median income than are the thousands of individuals who compose them. For example the coefficient of variation (standard deviation/mean) for mortality rates from leukemia was 7.0% comparing states at the national level and 20.9% comparing counties within the state of Wisconsin. Similarly, the coefficient of variation of mortality rates for cancer of the nasopharynx is 188.9% for Wisconsin at county level and 24.4% for the United States at state level. Colon cancer, however, has higher state-level variation, 25.8%, than county-level, 17.3%. The message, Cleek says, is clear: report regression coefficients, and treat correlation coefficients with care.

What this all means is that patterns of association are different at different scales of analysis, and it is an error to take an association that is true at one scale and infer that it will be true at any other scale. The change in pattern of association can, however, be used as an etiological clue as different causal factors find expression. In one case there may be great between-state variation but little within-state, between-county variation; in another there may be great local variation but little state-level variation. These patterns can be used to address different possible causes that relate to individual behavior, broader patterns of water sources or occupations, or still broader regional patterns of atmospheric pollution or population migration.

Vignette 7-3

SPATIAL AUTOCORRELATION

Spatial autocorrelation is often discussed among geographers because it can be a serious hindrance to their work, but it can also be used as an analytical tool. Spatial autocorrelation means that observations from places next to each other are influenced by each other, in the same way that the real estate value of one piece of property affects that of the property around it. If one city block is poor and black, the adjacent one is probably similar; if one county is affluent and has low unemployment, the adjacent county is probably better off than average as well. One important problem with spatial autocorrelation is that the assumption of independent observations, required for certain statistical procedures such as linear regression and correlation, may be wrong.

It is perhaps easiest to explain spatial autocorrelation by mapping a dichotomous (nominal data) variable, one that is either present or absent in each spatial unit of a study area. Consider Vignette Figure 7-3a, which diagrams three situations in which a certain disease is either present (black) or absent (white) in each of 16 square units. In the first situation, there is positive spatial autocorrelation: black and white units are grouped together. Perhaps

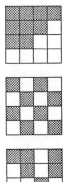

Vignette Figure 7-3a. Spatial autocorrelation. Three types of spatial autocorrelation with a dichotomous variable are illustrated.

the illness is quite contagious in the black part of the study area, but has come up against some type of physical or human barrier in the white area. The second situation illustrates negative autocorrelation; adjacent units are dissimilar. In the third diagram, a random pattern of black and white units indicates no autocorrelation, either positive or negative.

There are two important things to consider in measuring autocorrelation: whether units are adjoining ("have a join") and what the value of a variable or phenomenon is in each unit. One can say that there is a join if units have a common nonzero boundary (rook's case, from chess), a common vertex or point (bishop's case), or either of these (queen's case). In the figure the unit values were simple presence or absence of a phenomenon.

How can one tell, statistically, if there is autocorrelation in a particular situation? Basically, one counts the number of black–white (BW), black–black (BB), and white–white (WW) joins and compares these with the number of joins that would be expected if the black and white units were distributed randomly. In the third diagram of Vignette Figure 7-3a, there are 6 BB joins, 4 WW joins, and 14 BW joins (rook's case). The appropriate formulas can be found in texts that deal with autocorrelation. If there are significantly more BB or WW joins than expected, then there is positive autocorrelation, and if significantly more BW joins than expected there is negative autocorrelation.

Unit values need not only represent absence or presence of a disease; they could also represent high and low disease rates. In addition, more than two nominal data categories can be considered and the analysis taken from the "two-color" to the "k-color" case. Furthermore, definitions of a join can be altered in innovative ways. For example, Haggett (1976) in a study of measles diffusion in England defined joins in seven ways, based on different types of paths along which the disease might be diffusing (for example, along journey-to-work routes). If a certain path type indicated positive spatial autocorrelation, then that particular path type could have been important in measles spread.

Spatial autocorrelation techniques have been used in a constructive manner to determine links between spatial patterns and causal processes (Glick, 1979b). In particular, the techniques can help identify connections between disease rates and environmental and socioeconomic factors. Disease rates, which are interval data, have been most commonly used.

Thirteen countries from central and southern New Jersey were selected to illustrate how to use spatial autocorrelations. The first step is to construct a 13-by-13 join matrix (rook's case) that consists of 0s if two counties do not have a nonzero boundary and 1s if they do (Vignette Figure 7-3b). Note that a county does not join itself.

To examine the heart disease rates for spatial autocorrelation, Moran's I statistic, which is used for interval data, can be calculated. If units with similar rates (high or low) tend to be next to each other, the I statistic will be relatively large (positive autocorrelation); if the opposite is the case, then the I statistic will be relatively small (negative autocorrelation). The I statistic can be tested

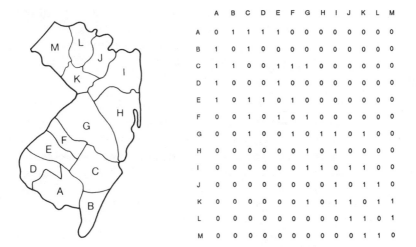

	A	B	C	D	E	F	G	H	I	J	K	L	M
A	0	1	1	1	1	0	0	0	0	0	0	0	0
B	1	0	1	0	0	0	0	0	0	0	0	0	0
C	1	1	0	0	1	1	1	0	0	0	0	0	0
D	1	0	0	0	1	0	0	0	0	0	0	0	0
E	1	0	1	1	0	1	0	0	0	0	0	0	0
F	0	0	1	0	1	0	1	0	0	0	0	0	0
G	0	0	1	0	0	1	0	1	1	0	1	0	0
H	0	0	0	0	0	0	1	0	1	0	0	0	0
I	0	0	0	0	0	0	1	1	0	1	1	0	0
J	0	0	0	0	0	0	0	0	1	0	1	1	0
K	0	0	0	0	0	0	1	0	1	1	0	1	1
L	0	0	0	0	0	0	0	0	0	1	1	0	1
M	0	0	0	0	0	0	0	0	0	0	1	1	0

Vignette Figure 7-3b. The join matrix for lag 1, neighbors, for a 13-county area in southern New Jersey.

for significance as a standard normal deviate (z-score) after determining the mean and variance of its distribution. (Appropriate formulas can be found in books dealing with this subject.) If the calculated z-score for a particular value of Moran's I is significantly positive, then one can say that a particular heart disease showed rate clustering at the level of adjoining counties.

The original join matrix can be modified or refined to reflect two factors simultaneously, the proportion of the boundary of one county that is common to another county, and the distance between county centers if there is a nonzero boundary. These modifications can be thought of as adding weights to the simple binary scheme of the original join matrix.

The original join matrix could be modified to reflect a particular potential risk factor. For example, one could assign a join matrix value of 1 if two counties were the same degree urban and within a certain maximum distance of each other. If the value of Moran's I were to increase following this modification, then the risk factor might be of importance.

The next stage in the analysis is to produce a *spatial correlogram* that is a rough indication of the areal extent of disease clustering. The original join matrix is modified to include 1's only if two counties are neighbors of neighbors; for example, county A's neighbors of neighbors are county F and county G. The new matrix can be tested for spatial autocorrelation. If it is significantly positive, then one can say that clustering is manifest at a larger scale than the original adjoining county scale. This process can be continued to neighbors of neighbors of neighbors, and so on (the general rule is to carry out approximately $n/4$ steps where n is the number of units of observation). The result is a series of z-scores for I statistics, which can be plotted as a spatial correlogram.

The *z*-score from the original binary matrix is called the first lag and succeeding *z*-scores are second, third, and fourth lags, and so on.

Vignette Figure 7-3c shows spatial correlograms for white male and female acute ischemic heart disease for two time periods for a 49-county area surrounding New York City and Philadelphia. For all four sets of rates, there is significant positive spatial autocorrelation at the 99% level for the first lag and significant positive spatial autocorrelation at the 95% level for the second lag; on succeeding lags significant positive autocorrelation is no longer evident. At some lags there is significant negative autocorrelation, indicating that, at these scales, adjoining groups of counties have dissimilar rates. It seems that, for whites, acute ischemic rates cluster up to the neighbor of neighbor scale, a fairly large area within the 49-county area. Still, there are groups of counties within the area that have dissimilar rates, as maps of acute ischemic heart disease rates attest (Ziegenfus & Gesler, 1984).

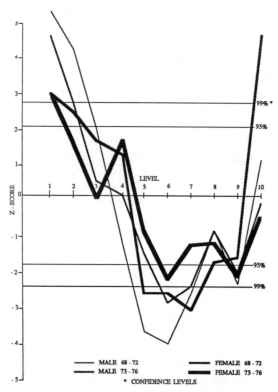

Vignette Figure 7-3c. Spatial correlograms for whites with acute ischemic heart disease in the New York/Philadelphia metropolitan area, 1968–1972 and 1973–1976.

Vignette 7-4

TREND SURFACE MAPPING

The geographic distribution of health-related phenomena may be examined through trend surface analysis. Let us look, for example, at the distribution of average ages for people who died in a sampling of Massachusetts towns for the period 1788–1792, part of a study of the diffusion of economic and mortality change in New England (Florin, 1971). The distribution in Vignette Figure 7-4a seems to suggest a concentration of highest ages in the northeastern part of the commonwealth, with declining averages to the south and west. Is that observation valid?

Trend surface analysis, some form of which is available in most general computational software packages, essentially fits a least squares regression surface to a three-dimensional (latitude, longitude, data value) map surface. This technique generalizes patterns by filtering local anomalies from the map. This tends to give a clearer picture of the broader regional trend. The simplest trend surface is a plane (Vignette Figure 7-4a).

It is possible to add warps to the surface through the addition of new terms in the regression equation. The more complex second order, or quadratic, surface retains the basic east-west pattern and adds a curve from a high in the northeast to a low in the northwest (see Vignette Figure 7-4b).

It is also possible to obtain residuals, or indications of the extent to which the actual values at individual data points differ from the value predicted by the

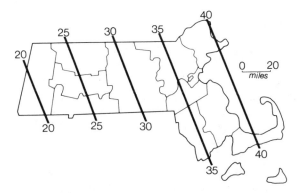

Vignette Figure 7-4a. Average age of death for selected Massachusetts towns, 1788–1792: trend surfaces. Map of the plane surfaces. From *Death in New England* (p. 58) by J. W. Florin, 1971, Chapel Hill, NC: University of North Carolina. Copyright 1971 by J. W. Florin. Reprinted by permission.

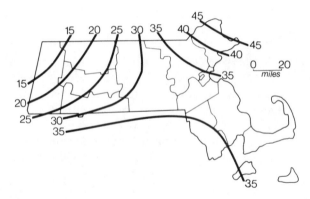

Vignette Figure 7-4b. Average age of death for selected Massachusetts towns, 1788–1792: trend surfaces. Map of the quadratic surfaces. From *Death in New England* (p. 59) by J. W. Florin, 1971, Chapel Hill, NC: University of North Carolina. Copyright 1971 by J. W. Florin. Reprinted by permission.

fitted trend surface. The examination of such residuals can suggest variables that might influence the data distribution but are not adequately incorporated into the existing surface. Negative residuals from the quadratic surface—that is, places with average ages substantially lower than predicted by the fitted surface—were concentrated in some of the older and larger cities near Massachusetts Bay (see Vignette Figure 7-4c). Positive residuals were mostly in towns somewhat farther from Boston. This suggests that something associated with

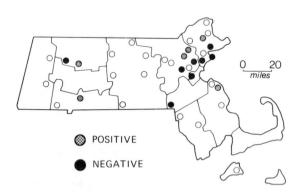

Vignette Figure 7-4c. Average age of death for selected Massachusetts towns, 1788–1792: trend surfaces. Map of residuals from the quadratic surfaces. From *Death in New England* (p. 60) by J. W. Florin, 1971, Chapel Hill, NC: University of North Carolina. Copyright 1971 by J. W. Florin. Reprinted by permission.

these older cities, perhaps the quality of life but perhaps just the existence of a younger population, might be an important additional variable.

Trend surface analysis should be used with care. Surfaces fitted to a very uneven geographic distribution of data points may not be reliable. The surface may statistically explain only a very small part of the total distribution of the data, although this can be checked through a coefficient of determination that states the percentage of the total variation in the data accounted for by the surface.

8

The Geography of Disease Diffusion

There are only two ways in which anything, whether it be an idea, a type of tree, cholera, or a sophisticated new plan for the financing of hospital care, can have found its way to a particular location. Either it developed there independently, or it somehow moved there from another place. Movement is by far the most usual explanation of the existence of phenomena at a particular location. An understanding of the mechanisms that influence the spread of any phenomenon and its spatial pattern is at the core of the geographic study of diffusion.

The term *diffusion* implies a spread or movement outwards from a point or beginning place. Diffusion research by medical geographers can be divided into two basic categories. Some studies focus upon the spread of medical innovations within a health care delivery system. Examples of such research are investigations of the diffusion of tomography scanners in the United States (Baker, 1979) or abortion facilities in the northeastern United States (Henry, 1978), wherein the central concern has been the development and spread of new ideas and goods. More common, however, are studies of the diffusion of infectious disease, especially nonvectored infectious diseases. In these studies many of the conceptual and methodological concerns of epidemiology and geography have been combined.

Western civilization has long had a general recognition of the existence and importance of disease diffusion. Disease quarantine, which could involve an individual, a household, or even an entire community or country, at least tacitly implies the existence of disease diffusion. In the United States, Noah Webster's *A Brief History of Epidemic and Pestilential Diseases* traces the spread of several late 1700s epidemics across New England (Webster, 1799). At about the same time William Currie was also describing the geography of a number of epidemics in the United States (Currie, 1792, 1811). The contagionists investigating the distribution of yellow fever in the early 1800s recognized that the disease was carried from place to place—diffused—by the infected individual. While we might find their medical explanations of causation amusing, the quality of their geographic descriptions were, given the severe data limitations, excellent.

Most studies of geographic diffusion have emphasized an explanation of events in a space–time context. The object has been to describe the history of a phenomenon, to identify general mechanisms operating to influence the diffusion, or to study some aspect of the *network* through which the phenomenon diffused.

The extent and form of human intercourse is critical for both innovation and diffusion. Variables such as distance, intervening opportunities, and the distribution of attractions and facilities all influence the level of interaction. Diffusion is patterned by the configuration of the networks that encourage movement and *barriers* that discourage it. A major prerequisite of disease diffusion, for example, is the existence of a sufficiently large susceptible population. Immunity might be gained through exposure to infection and recovery, through genetic immunity, or through vaccination or chemical prevention (Stock, 1976, p. 4). A sufficiently large immune population, or more precisely, a sufficiently small susceptible population, can serve as an effective barrier to disease transmission.

THE TERMINOLOGY OF GEOGRAPHIC DIFFUSION

Types of Diffusion

Two diffusion mechanisms can be illustrated by following the hypothetical spread of the acceptance of a new birth control device. These first individuals who learn of the device might tell their immediate neighbors and friends, who in turn tell their acquaintances, and so on. The spread of an idea or innovation through a population is *expansion diffusion*. Alternatively, people with knowledge about the device might migrate and introduce it at their new location in a *relocation diffusion*. Relocation diffusion often involves great leaps in the movement and may pass over intervening populations.

Contagious diffusion involves the close contact of individuals and the resultant passage of some variable, such as a disease agent, from one to the other. Contagious and expansion diffusion usually both have distance as the strongest variable and result in a clear distance decay of the element being diffused (see Vignette 6-1). Indeed, their patterns are so similar that many geographers choose to lump them together and call the process *contact diffusion*. Some innovations may jump over intervening people without relocation. This usually characterizes *hierarchical diffusion*, in which large (central) places first gain information, then transmit it to smaller places (Gould, 1969). Innovations usually move very quickly to the top of the urban hierarchy, and then *cascade* down.

Diffusion processes are often a combination of these types (Figure 8-1). A disease epidemic may follow a contagious pattern yet be affected by both migration (relocation diffusion) and the hierarchical structuring of movement. Indeed, most diffusion processes do not fall helpfully into distinct types, and in a single diffusion example one type of transmission may be important first, then another, and perhaps yet another at a different stage.

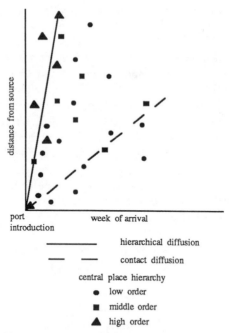

distance from source

port
introduction

week of arrival

——————— hierarchical diffusion

—   —   contact diffusion

central place hierarchy
● low order
■ middle order
▲ high order

Figure 8-1. An infection spreads by hierarchical diffusion and contact diffusion, simultaneously. The two vectors represent the average of distance traveled through time for the two simultaneous diffusions.

Networks

The *diffusion network* is the web of places, people, and the physical and communication links through which information, people, and goods flow (Alves & Morrill, 1975, p. 290). The diffusion of a given phenomenon might be studied at several times to gain knowledge about changes in network structure. As an alternative approach, research on the simultaneous diffusion of the same or similar phenomena in different networks could illuminate the impact of different network structures. The study by Stock examined later in this chapter is an excellent example of such an approach (Stock, 1976).

Barriers

In opposition to networks, which pattern and encourage diffusion, barriers slow and shape the process. Barriers have three basic effects. *Absorbing* barriers simply stop an innovation. A population thoroughly vaccinated against smallpox is an example of an absorbing barrier; isolated cases had no chance to diffuse, and smallpox disappeared. *Reflecting* barriers channel and intensify the local impact of a diffusion process while blocking its spread to another locale. For example, an insurance agent, selling a new program of individual hospital insurance, lives on the margins of a very large wilderness. The agent has a quota of individuals to contact daily. The wilderness serves a a reflecting barrier, so that the intensity of sales activity and thus the diffusion of the new plan is greater in the other, unblocked direction. Finally, the barrier may be *permeable*, allowing some diffusion but slowing the process.

Barriers are often *physical*—oceans, deserts, mountains, rivers, and so forth. Broad barriers such as an ocean or desert were very effective barriers to many kinds of contagious diseases during the period of slow transportation. The disease might run its life course during the journey and never reach the destination. Aboard a ship, the susceptible population might be too small to maintain the contagion through the course of a trip. Rapid transport has minimized the effectiveness of such barriers.

Cultural barriers, which may be permeable, reflecting, or absorbing, are especially important in the diffusion of styles of health care. The Chinese carried their medical system with them as they migrated to Southeast Asia (relocation diffusion). Comparatively few natives have chosen to adopt the Chinese system, however, an indication of the cultural separation of the two groups. The Soviet medical system has not penetrated the cultural–political barrier that splits east from west in Europe. As another example, religion has been important in restricting the diffusion of family planning programs.

DISEASE DIFFUSION

The characterization of infectious disease as either endemic or epidemic is often inadquate. While it is true that sometimes a disease heretofore nonexistent in an area will suddenly and often *virulently* arrive, more commonly a disease that is already endemic in a population develops into a health problem of epidemic proportions. Thus, most cases of disease diffusion represent a problem already well known to a society. This may involve a strikingly repetitive pattern in outbreaks of the problem. Influenza is one example of a contagious disease that can diffuse through a population with nearly predictable frequency.

The epidemiological character of infectious disease is dependent on several sets of factors. First are the qualities of the microorganism that influence its relationship with a human host. These include the number of organisms needed to initiate symptoms or to produce clinically visible infections rather than subclinical, but immunizing, ones. Second are the number of organisms shed by a carrier of the infection, the period of shedding, the method of shedding into the environment, and the survival time of infectious organisms in the particular environment. Third is the probability of contact with a susceptible host during the period of infectiousness. A fourth is the population's immunity as a result of genetic predisposition, prior infection, or immunization (Burnet & White, 1974, p. 118–119).

The cyclic pattern of epidemics for many diseases, such as influenza, often results from the gradual build-up of a sufficiently large susceptible population. Many of the common epidemics, such as measles and mumps, infected children almost exclusively because they were born after the previous epidemic and were therefore the only large susceptible population. The periodicity of measles epidemics, especially in less populated areas, was regulated by the fairly consistent number of children born each year and the substantial minimum susceptible population needed to maintain the infection. Artificial immunization has greatly reduced the number of measles cases; however, relaxing the completeness of vaccination could reestablish a susceptible population, of all ages, large enough to support a measles epidemic. Such an event in the United States in the early 1980s caused considerable concern among public health officials. The epidemic was brought under control and resulted in stricter enforcement of vaccination laws for children in many states.

MODELING DISEASE DIFFUSION

Most of the epidemiologic research in disease diffusion has involved microscale field investigations. They usually rely on relatively small samples (Pyle, 1979, p. 132). The range of scales from urban to international, used by geographers in diffusion studies, offers a view often ignored by epidemiologists.

Diffusion research is often based on the geographic diffusion models developed and applied outside the area of health problems. The general models are extended and expanded to fit the special circumstances of disease diffusion. The studies by Brownlea (1972) on hepatitis, Haggett and others on measles (Haggett, 1976; Cliff & Haggett, 1982, 1983; Cliff, Haggett, Ord, & Versey, 1981), Stock and Kwofie on cholera (Stock, 1976; Kwofie, 1976), and Pyle on influenza (1969, 1984, 1986) are some examples. Their contributions will be discussed later in the chapter.

A long-range advantage to broad-scale models of disease diffusion is their possible importance for forecasting. If we can identify those variables (carriers, barriers, and centers of innovation) that consistently influence the diffusion of a particular disease, or if the geographic diffusion pattern of a place is repetitive and thus predictable, health officials could focus vaccination programs and health education on blocking or minimizing outbreaks. A disadvantage of studies at this scale is that they often leave the question of causality unanswered. At the level of the individual or household or in the research lab specific associations between individuals and their environment can be addressed.

One methodological problem associated with geographic diffusion studies is spatial autocorrelation (see Vignette 7-3), which is at the heart of contagion diffusion. Another problem is that the repeated wave nature of many epidemics does not fit models that assume a single wave of diffusion (see Vignette 8-1).

Influenza

Influenza epidemics and diffusion result from the periodic appearance of forms of the disease agent that are new, modified, or old but long absent from a population. Influenza is highly infectious because large numbers of microorganisms are transmitted in droplets by coughing, sneezing, or talking. Although the incubation period of the disease is very short, people are infective before as well as after symptoms appear. Influenza can spread through a community rapidly. The course of an epidemic is environmentally influenced, with epidemic onset usually in the fall, followed by a winter peak (see Chapter 5). The many strains of influenza are divided into two types, A and B. Type A strains are responsible for more frequent and severe epidemics.

The type A virus has two kinds of protein surface spikes, H and N. The H spike provides the mechanism for a virus to enter a cell, and the N attacks mucoprotein and is released in saliva, causing the spread of infection. These spikes are matched by antibodies in an immune individual. Each year, however, the spikes "bend" and change form in a process called *drift*, and the antibodies no longer fit perfectly. The drift adaptation means that the virus can survive in relatively small populations. Every several years the ratio of H to N spikes changes dramatically as a result of genetic shifting between strains. Such a *shift* can lead to more substantial viral outbreaks.

Influenza epidemics cycle over roughly 10 years. After the great pandemic of 1918–1919, believed to have been type A influenza (the influenza virus was not isolated until 1933, but retrospective archival studies and serological analysis of antibodies have been traced to earlier major outbreaks), other epidemics occurred in the winters of 1928–1929, 1936–1937, 1946–1947, 1957–1958, 1968–1969, and 1977–1978. The pattern results from the repeated introduction of type A variants after increased population immunity to previous strains has lowered infection levels. The 1918–1919 pandemic was extraordinarily virulent, with total global mortality estimated at between 20 and 50 million. The 1957–1958 "Asian" flu outbreak, the result of an A strain virus so different from its predecessors that it was labeled A2, was perhaps the greatest pandemic in world history, in terms of total infection, although it did not have a major impact on mortality.

Pyle (1986) debunks the general belief that most major influenza pandemics originated outside of the United States, especially in China or central Asia: the 1918 "Spanish" flu, the 1957 "Asian" flu, and the 1968 "Hong Kong" flu. He mapped the 1918–1919, 1943–1944, 1947–1948, 1957–1958, 1976–1977, and 1980–1981 influenza diffusions in the United States.

Mapping cumulative percentiles of the total distribution of reported cases in one community during one epidemic season, normally from fall to spring,

Pyle (1986) could discern substantial differences between the disease pattern early in the epidemic and the pattern later on. Identifying the beginning of an epidemic outbreak of a common endemic disease is difficult. What percentage of incidence represents the initiation of an epidemic and not just the continuing endemic condition? How many cases constitute an epidemic? The United States Center For Disease Control defines the epidemic threshold as the point where more than 4.5% of all deaths in a given week are attributed to pneumonia–influenza (Pyle, 1986, p. 169).

Pyle (1979) has not identified a common spatial pattern of influenza diffusion in the United States. Maps of the 1918–1919 influenza identified a dramatic east–west progress, with rapid penetration into the interior and a lagged arrival in the southwest and northwest. The 1946–1947 diffusions, however, began in a number of centers along the East Coast, the upper Middle West, and the Gulf Coast, then spread outward in a radial diffusion pattern. In 1957 outbreaks seemed to follow two pathways, one down the Mississippi River system and a second from west to east across the country. The 1976–1977 epidemic followed a hierarchical diffusion pattern. That outbreak was examined through a series of maps that charted the months that cities reported influenza mortality well above the 10-year average. A statistical technique called harmonic analysis was used to identify seasonal patterns, which followed an almost classic diffusion pattern. The epidemic started in Megalopolis along the northeastern seaboard, spread to many large cities across the country and to smaller centers in the northeast quadrant, and eventually reached a number of scattered, generally smaller, urban areas. Distance decay was part of the pattern (Vignette 6-1). A "surprise" 1980–1981 epidemic, by comparison, had a complex, mixed diffusion pattern, with multiple sources scattered across the country.

Pyle (1986) noted that the 1918–1919, 1947–1948, and 1976–1977 epidemics all started in the East and spread generally westward. The viral strains of the 1947 and 1976 epidemics, and presumably also of 1918, are very similar. The complexity of the 1980–1981 pattern, by comparison, more closely resembles the epidemics of 1928–1929 and 1968–1969.

If the influenza outbreaks began outside the United States and diffused into the country, it is surprising that they started at several locations in the interior of the United States. Pyle (1986) suggests that spring *seeding* of preepidemic reservoirs may be responsible. He examined two preepidemic seasons, 1946–1948 and 1975–1977, and found that places with a spring peak of cases were often next to areas of initiation during the following epidemic year. He hypothesized that viruses in these places maintained a low and unreported level during the summer, quickly reestablished themselves at the start of the next epidemic season, and continued the diffusion begun the year before. Epidemiologists sometimes call the precursors "herald waves."

An earlier study of the 1957 influenza epidemic in England and Wales, using reported pneumonia as a surrogate for influenza because influenza cases were not systematically reported, also suggested that the role of the urban hierarchy in influenza diffusion is unclear (Hunter & Young, 1971). Seeding began at a port in the north of England. Early foci were northern industrial centers like Liverpool and Hull. London, the country's largest group of susceptibles and a major point of entry for people arriving from infected areas abroad, was minimally affected at first. Analysis of the shifting weekly *centrogram* (defined as the point of minimun total straight line distance to all cases) of newly reported cases indicated a southeast bias in the direction of London, with the epidemic center climbing up and then along a ridge of high population potential (Figure 8-2; see Vignette 8-2). Influenza incidence had a significant positive correlation with an index of persons per room, although the index did not correlate to the dates of epidemic onset or peaks. Distance from the seaport entry points was positively associated with onset and peak but not with total incidence.

These studies suggest the problems of studying the diffusion of a respiratory problem like influenza. It is often multinodal in character; diffuses rapidly because of its short incubation period and ease of transmission; is beset by a myriad of influences, with contagious, relocation, and hierarchical diffusion components; and is sometimes inaccurately counted or not recorded at all. In addition, individual outbreaks may represent the interplay of different strains, obscuring the impact of immunity barriers.

Measles

Measles is highly infectious and endemic throughout most of the world. Nearly all susceptible individuals contract measles after close contact with an infected individual. Epidemics occur regularly at intervals of about 3 years. The disease flares when a sufficiently large population of susceptible children is created. Before artificial immunization, over three fourths of the adults in most populations had caught measles in childhood. The highest incidence is normally in late winter and spring.

Studies have identified a relationship between population size and measles periodicity. It has been suggested that a population of 250,000 is the minimum necessary to maintain regular periodicity, although it could be less for urban areas with regular in-migration of infected individuals. Other research has observed breaks in the pattern for isolated communities of less than 500,000. The threshold depends on the intensity of population circulation and contact. There is a clear positive relationship between community size and measles endemicity and epidemic regularity (Cliff & Haggett, 1983).

Haggett (1976) first studied measles diffusion in the relatively isolated county of Cornwall in southwestern England. He posed seven possible models to explain the diffusion process. A *regional* model assumed two separate regional subsystems of diffusion. A *rural–urban* model split rural from urban

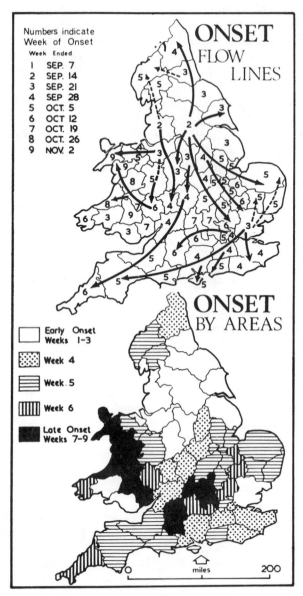

Figure 8-2. Population potential and the diffusion of influenza in England and Wales, 1957. From "Diffusion of Influenza in England and Wales." (p. 645) by J. M. Hunter and J. C. Young, 1971. *Annals of the Association of American Geographers, 6*(4, Dec.) p. 645. Copyright 1971 by Association of American Geographers. Reprinted by permission.

areas. The *wave-contagion* model assumed a shortest-linkage (pathway) diffusion from the endemic center at the county's largest city, and distance decay in intensity. The *local-contagion* model identified common geographic boundaries as the critical influence. A *journey-to-work* model suggested commuting as a surrogate for spatial interaction. A *population-size* model applied a size hierarchy to the county, while a *population-density* model applied community density. He applied these diffusion models to a 222-week series of measles occurrence data for the 28 political jurisdictions in the study area between late 1966 and the end of 1970, a period of two waves of measles epidemic. The seven models, using graph or network theory, suggested different linkages or pathways connecting the geographic units.

Most of the models identified the expected lower levels of contagion between epidemic periods. The exception, not surprisingly, was the population-size model, as the infection persisted in the larger population clusters. Thus, contagion levels between epidemic periods were higher. Haggett identified advance, peak, and retreat phases in the epidemic sequence. The advance phase is characterized by a rapid increase in intensity and spread. The wave effect is important, and population size is less important. Local contagion effects are important at the peak phase, resulting in strong regional contrasts between clusters of places. The retreat phase is associated with a decreased incidence throughout the area (not with a geographic contraction).

The study of measles diffusion has been extended into a more geographically isolated and less populated location, Iceland (Cliff *et al.*, 1981; Cliff & Haggett, 1982, 1983). The hypothesis was that in such an area an epidemic would move slowly and would take the form of a series of broken, spatially irregular outbreaks rather than a continuous wave. Measures of epidemic velocity took into account the discontinuous nature of Icelandic population distribution and focused on the time lag in the arrival of the epidemic episode at a community and the duration of its peak intensity.

The isolated nature and small population of Iceland means that measles is not endemic to the island. The researchers identified 13 free-standing measles epidemics that struck the entire country between 1896 and 1975. The average time lag for the epidemics was less for the population center of Reykjavik than for rural areas of the island. Rural epidemic periods covered a shorter time, and they were more sharply pointed. These findings are consistent with the hypothesis.

The difference between Reykjavik and the remainder of the country has declined, particularly since 1945. The velocity of diffusion has declined everywhere and especially in the capital. The peakedness of individual outbreaks has declined in Reykjavik but increased elsewhere. These changes may reflect reduction in both the internal and international isolation of Iceland. Air transportation has been improved and a road net established around the island, connecting once isolated communities and increasing the interaction between susceptible and infected individuals. Shorter intervals between epidemics since

1945 may be related to more frequent introduction of the infection from overseas. Also, education consolidation has brought students together at boarding schools, increasing the concentration of susceptible population.

Cholera

Cholera is an acute intestinal bacterial disease. Symptoms include severe diarrhea and vomiting. Mortality, resulting from severe dehydration, may be above 50% from some strains, if the symptoms are untreated. Most carriers of cholera are asymptomatic, greatly increasing the difficulty of controlling the disease. Humans are its only natural host. The *Vibrio cholerae* bacteria multiply in the gut of the carrier and are excreted. In areas of poor sanitation the feces may directly contaminate soil, water, or food. Transmission is completed when a susceptible person consumes contaminated food or water. Flies may carry cholera from contaminated nightsoil to food, or it may spread through close contact with an infected individual. The bacteria are sensitive to high temperatures, acidity, and dry conditions. The key to effective control is environmental sanitation. Cholera is largely a problem of the less developed world.

A cholera-like disease was described in India in the 5th century A.D. European travelers to India in the 1400s reported cholera. In about 1817, cholera diffused from South Asia to southeast and southwest Asia in the first recorded cholera pandemic. During the rest of the 19th century six pandemics, each originating in India, diffused across much of the world. Transportation pathways and trade centers played a key role in the spatial pattern of the pandemics. Mecca, focus of the annual Islamic pilgrimage called the *Hajj*, became a major center of infection. The pandemics spread across Egypt to Europe and North America. Cholera deaths totalled many millions.

Pyle (1969) studied the diffusion of three major cholera epidemics in the United States during the 19th century. The first, in 1832, was introduced at several points. From a Canadian origin cholera spread into a number of small cities in northern New York (Figure 8-3). The disease independently appeared in New York City and diffused westward along the Erie Canal, then down the Ohio and Mississippi rivers, reaching New Orleans late in the year at about the same time that cholera arrived in that city by sea. A second pattern seemed to extend from New York north and south to the other large cities along the eastern seaboard. This epidemic occurred when water transportation was of major importance in the country, and the influence of water routeways on its spread was apparent.

The second epidemic, in 1849, occurred when the hierarchical structure of the United States urban system was more clearly defined. Railroads were becoming important, although much of the interior was not yet connected to the East by rail. The epidemic was introduced in New York City and New

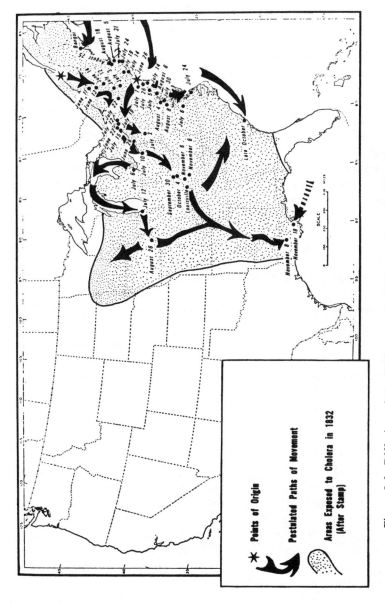

Figure 8-3. Diffusion of the 1832 cholera epidemic in the United States. From "The Diffusion of Cholera in the United States in the Nineteenth Century" by G. F. Pyle, 1969. *Geographical Analysis, I,* p. 66. Copyright 1969 by Ohio State University Press. Reprinted by permission.

Orleans. From these cities it apparently followed two routes: up the Mississippi and Ohio rivers, affecting the larger cities of the interior, and later, from New York to other large cities along the eastern seaboard.

By the time of the third epidemic in 1866 the rail system of the eastern half of the country was well established. Cholera was again introduced at New York City. The epidemic moved along the seaboard and the Ohio and Mississippi valleys, but diffusion was far more clearly structured by the country's urban hierarchy. A transportation system, offering a relatively rapid movement and focusing on larger cities, had clearly grown to dominate the diffusion pattern (see Figure 8-1).

Cholera ceased to be pandemic after World War I. It disappeared from the industrialized world and appeared only sporadically in Southeast and Southwest Asia and North Africa. It remained endemic in the Indian subcontinent but was largely restricted to the delta of the Ganges and became less virulent. After 1960, a new biotype of *V. cholerae*, El Tor cholera, emerged from Indonesia. It has a lower case fatality rate, produces many carriers, and has resulted in the abandonment of international vaccination and control regulations as ineffective. From 1960–1975 it spread over much of southern Asia and Africa, reaching West Africa for the first time in recorded history, and cases were reported in Europe, the southern Soviet Union, and North America.

The diffusion of El Tor in Africa in the early 1970s was investigated by two geographers. Kwofie (1976) focused his research on a trend surface analysis of cholera diffusion in West Africa. He hypothesized that cholera epidemics in the initial stages would spread outward in a contagious fashion. As the epidemic advanced, variations resulting from the interrelationship of transportation, sanitation, and treatment would develop. Kwofie divided the outbreaks into three periods corresponding to primary, saturation, and waning phases. Then he examined the trend surfaces of new occurrences during each period (see Vignette 7-4). Quadratic and cubic trend surfaces during the primary and waning periods indentified a clear west to east diffusion pattern, while the cubic surface for the saturation phase was far more complicated, having centers of diffusion in Mali and Chad along the Sahelian, sub-Saharan pathways. He identified two major routes of diffusion: comparatively regular contagion along the coast, and the more complex path in the Sahel, involving contagion but influenced by the region's network of movement.

One of the most comprehensive and conceptually thoughtful studies of diffusion is Stock's (1976) study of the diffusion of El Tor in Africa between 1970 and 1975. His goals were not only to map the diffusion of cholera but also to examine the applicability of existing diffusion models. He identified and developed four models to describe the overall diffusion pattern.

The first model was *coastal* diffusion, with fishermen carrying the disease along the coast of West Africa. This took a contact diffusion pattern. Once coastal villages were infected, the disease spread to nearby urban centers in a secondary phase and diffused hierarchically along inland roads in a tertiary phase. The *riverine* model described the diffusion process along the Niger River in interior West Africa. From its focus the epidemic spread rapidly along the middle course of the river. Distance from the focus was far more important than the urban hierarchy in determining the sequence of onset. The epidemic spread to the valley periphery in a secondary phase and into the hierarchical system only in the tertiary phase. These two models emphasize the importance of water routes in the less developed world.

The El Tor diffusion in Nigeria was a clear example of the *hierarchical* model. Cholera was introduced into the country at the capital city of Lagos, on the coastal diffusion route. A location's level in the urban hierarchy and distance from Lagos determined when the epidemic would appear. Overland transportation routes channeled diffusion. The least affected areas of the country had low population density, little urbanization, and undeveloped transportation.

The last model, *contagious diffusion*, follows a radial pattern and was identified around Lake Chad. Here the disease developed a particularly virulent intensity, driving people away from the lake. The lack of developed transportation routes, natural channels, urban centers, or barriers resulted in a clear pattern of contagious diffusion.

Stock's monograph emphasizes the complexity of the geographic structure of diffusion. All of the basic models had some applicability. Immune populations, deserts, and sparsely populated areas were permeable barriers, and rivers, the coast, and developed overland routes all helped channel movement. Stock suggested that the linear pattern along the Niger River loosely resembled the 1832 cholera diffusion along the internal waterways of the United States, (Pyle, 1969) while the hierarchical diffusion in Nigeria more closely resembled the 1866 United States epidemic's structure. The difference between Nigeria and Mali in economic development in the early 1970s thus seems similar to the economic progress of the United States between 1832 and 1866.

Infectious Hepatitis

There have been few attempts to apply to disease diffusion the structure of simulation formulated by Hagerstrand (see Vignette 8-3). This may be due to complexity of the influences on diffusion and to the multiple waves of many epidemics. The best example of the application of diffusion simulation of disease is a study of infectious hepatitis in Wollongong, Australia (Brownlea, 1972).

The infectious hepatitis virus is transmitted through fecal contamination and has much the same cultural ecology as ascariasis (see Figure 2-2). The virus is robust and tolerant of a wide range of environmental conditions. Water, fish, seafood such as oysters, and pet hair may be vehicles for disease transmission. Brownlea identified cyclic fluctuations in the number of cases reported. During epidemic years incidence rates of two per thousand or more affected mainly children and showed an equal sex ratio, strong spring seasonality, and no spatial concentration. During interepidemic years, rates were rather uniformly 0.3 per thousand, affected all ages, and showed little seasonality and no spatial concentration. Since hepatitis is often poorly reported, Brownlea aggregated cases over 15 years and used a Poisson probability test to determine where active spread was present. Brownlea identified this peak of aggregated reports as a wave that moved spatially through time and labeled it a *clinical front.*

Brownlea developed several models for hepatitis behavior at different scales within this industrial–suburban, rapidly growing region. The basic model was based on Hagerstrandian-type stochastic simulation of a random walk (see Vignette 8-3). Assuming a closed population, equal chance of diffusion in all directions, and essential community immunity as the epidemic passed, he simulated a random diffusion in which the clinical front would advance as a ring from Wollongong's initial node. Time periods for measurement were determined by incubation period. The bulges and bends in the actual advance of the clinical front he attributed to ecological parameters that operated as constraints (Figure 8-4). Modifying the model to fit the actual disease behavior showed that concentration of young families, light sandy soils, concentration of older people, and a polluted lake used for fishing and swimming were important parts of the physical and socio-demographic surfaces that provided friction or channels affecting epidemic diffusion. Recalibration of cell probabilities led to a close match between the simulated and real diffusion patterns.

These findings were applied to the movement of the clinical front among the settlement nodes in the study area. Brownlea found that the disease diffused from each population center, exhausting the susceptible population at the core and rippling to the surrounding nodes. If a peripheral node had developed a sufficient concentration of in-migrants and young children, it became a new center of infection and node of diffusion. The first exhausted node was reinfected from the surrounding, newly active nodes, which caused the in-filling epidemiologic pattern of the interepidemic years.

The model laid a basis for identifying the degree of underreporting within the study area and for explaining disease rhythmicity and spatial patterning.

McGlashan (1977) investigated viral hepatitis in the urban hierarchy of Tasmania, Australia. Unlike Brownlea, he identified a series of "discrete hepatitis regions" in Tasmania. The disease was influenced by distance decay from endemic centers but diffused in a combination of hierarchical and contagion patterns. Differences between the two sets of findings may reflect the impact of scale.

Other Examples

The diffusion patterns of other diseases, among them onchocerciasis, tuberculosis, and schistosomiasis, have also been investigated. The diffusion of oncho-

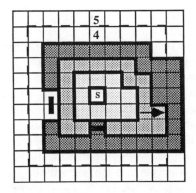

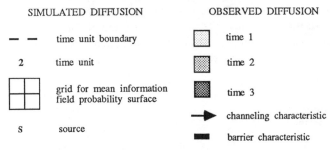

Figure 8-4. Simulation of the diffusion of infectious hepatitis.

cerciasis (river blindness) into the river valleys of northern Ghana represents the consequence of a cyclic pattern: occupation of the river valley, parasitic infection from a bite by the fly vector that breeds in the fast-flowing stream, gradual onset of blindness, abandonment of the unhealthy area, and resettlement of the valley as the memory of the health problem dulls and the need for good agricultural land grows (Hunter, 1966).

Recent research on the relationship among leprosy, tuberculosis, and urbanization in Africa suggests that some degree of antagonism exists between tuberculosis and leprosy, that the incidence of tuberculosis on the continent is promoted by growing urbanization, and that there is an associated decline in new cases of leprosy (Hunter & Thomas, 1984). This may be a partial explanation for the disappearance of leprosy from Europe in the 19th century.

The roles of the snail intermediate host and human host in schistosomiasis diffusion is clear (Chapter 3). Since 1920 the diffusion of the disease from its endemic core in northeast Brazil (where it arrived by relocation diffusion from Africa) has been associated with the distribution of the snail host, the migration of infected people from the impoverished northeast, the opening of new agricultural areas and irrigation programs, and the general expansion of the transport network (Kvale, 1981).

CONCLUSION

The diffusion process is defined by the specific environmental conditions that affect each disease (Brownlea's ecological parameters). However, certain environmental conditions, most obviously inadequate sanitation and crowding, affect many diseases and contribute to similarity of diffusion patterns.

Recent research suggests that the geographic models of diffusion have a substantial explanatory ability, whether used to study conditions at different locations or at different times. While it is obvious that the medical geographer needs knowledge of the epidemiology of disease diffusion, it seems equally clear that the geographic concepts and models offer opportunities not yet fully pursued. As researchers such as Pyle (1979, 1986) and Brownlea (1972) have demonstrated, an understanding of the geographic diffusion process of a disease can result in new insights into its environmental associations, means of survival between epidemics, and transmission. Such information is an important contribution toward more effective disease control.

REFERENCES

Alves, W. R., & Morrill, R. L. (1975). Diffusion theory and planning. *Economic Geography, 51,* 290–304.

Baker, S. R. (1979). The diffusion of high technology medical innovation: the computed tomography scanner example. *Social Science and Medicine, 13D,* 155–162.

Brownlea, A. A. (1972). Modelling the geographic epidemiology of infectious hepatitis. In N. D. McGlashan (Ed.), *Medical geography: Techniques and field study* (pp. 279–300). London: Methuen.

Burnet, M., & White, D. O. (1974). *Natural history of infectious diseases* (4th ed.). Cambridge: Cambridge University Press.

Cliff, A. D., & Haggett, P. (1982). Methods for the measurement of epidemic velocity. *International Journal of Epidemiology, 11,* 82–89.

Cliff, A. D., & Haggett, P. (1983). Changing urban–rural contrasts in the velocity of measles epidemics in a island community. In N. D. McGlashan & J. R. Blunden (Eds.), *Geological aspects of health* (pp. 335–348). London: Academic Press.

Cliff, A. D., Haggett, P., Ord, J. K., & Versey, C. R. (1981). *Spatial diffusion: An historical geography of epidemics in an island community.* Cambridge: Cambridge University Press.

Currie, W. (1792). *Historical account of the climates and diseases of the United States of America.* Philadelphia: Dobson.

Currie, W. (1811). *A view of the diseases most prevalent in the United States of America.* Philadelphia: J. & A. Y. Humphreys.

Florin, J. W. (1971). *Death in New England: Regional variations in mortality.* Studies in Geography No. 3. Chapel Hill, NC: University of North Carolina, Department of Geography.

Gould, P. R. (1969). *Spatial diffusion* (Resource Paper No. 4). Washington, DC: Association of American Geographers Commission on College Geography.

Hagerstrand, T. (1952). *The propagation of innovation waves.* Lund, Sweden: Gleerup.

Haggett, P. (1976). Hybridizing alternative models of an epidemic diffusion process. *Economic Geography, 52,* 136–146.

Haggett, P. (1979). *Geography: A modern synthesis.* New York: Harper & Row.

Henry, N. F. (1978). The diffusion of abortion facilities in the northeastern United States, 1970–1976. *Social Science and Medicine, 12D*, 7–15.

Hunter, J. M. (1966). River blindness in Nangodi, northern Ghana: A hypothesis of cyclical advance and retreat. *The Geographical Review, 56*, 398–416.

Hunter, J. M., & Thomas, M. O. (1984). Hypothesis of leprosy, tuberculosis, and urbanization in Africa. *Social Science and Medicine, 19*, 27–57.

Hunter, J. M., & Young, J. C. (1971). Diffusion of influenza in England and Wales. *Annals of the Association of American Geographers, 61*, 637–653.

Kvale, K. M. (1981). Schistosomiasis in Brazil: Preliminary results from a case study of a new focus. *Social Science and Medicine, 15D*, 489–500.

Kwofie, K. M. (1976). A spatio-temporal analysis of cholera diffusion in western Africa. *Economic Geography, 52*, 127–135.

May, J. M. (1958). *The ecology of human disease*. New York: MD Publications.

McGlashan, N. D. (1977). Viral hepatitis in Tasmania. *Social Science and Medicine, 11D*, 731–744.

Pollitzer, R. (1959). *Cholera*. Geneva: World Health Organization.

Pyle, G. F. (1969). The diffusion of cholera in the United States in the nineteenth century. *Geographical Analysis, 1*, 59–75.

Pyle, G. F. (1979). *Applied medical geography*. Washington, DC: V. H. Winston.

Pyle, G. F. (1984). Spatial perspectives on influenza innoculation acceptance and policy. *Economic Geography, 60*, 273–293.

Pyle, G. F. (1986). *The diffusion of influenza*. Totowa, NJ: Rowman & Littlefield.

Stock, R. F. (1976). *Cholera in Africa*. Plymouth, England: International Africa Institute.

Webster, N. (1799). *A brief history of epidemic and pestilential diseases*. Hartford, CT: Hudson & Goodwin.

Further Reading

Adesina, H. O. (1984). The diffusion of cholera outside Ibadan city, Nigeria, 1971. *Social Science and Medicine, 18*, 421–428.

Adesina, H. O. (1984). Identification of the cholera diffusion process in Ibadan, 1971. *Social Science and Medicine, 18*, 429–440.

Angulo, J. J., Haggett, P., Megale, P., & Pederneiras, C. A. (1977). Variola minor in Braganca Paulista County, 1956—a trend-surface analysis. *American Journal of Epidemiology, 105*, 272–280.

Cliff, A. D., Haggett, P., & Ord, J. K. (1986). *Spatial aspects of influenza epidemics*. London: Pion Limited.

Girt, J. L. (1978). A programming model of the spatial and temporal diffusion of contagious disease. *Social Science and Medicine, 12D*, 173–181.

Goldsmid, J. M. (1980). Imported disease: A continuing threat to Australia. *Social Science and Medicine, 14D*, 101–109.

Haggett, P. (1972). Contagious processes in a planar graph. In N. D. McGlashan (Ed.). (1977) *Medical geography: Techniques and field studies* (pp. 307–324). London: Methuen.

Vignette 8-1

DIFFUSION WAVES AND THE LOGISTIC CURVE

It is often convenient to think of diffusion as *waves* of innovation and acceptance spreading geographically (Hagerstrand, 1952). These innovation impulses tend to lose their energy with distance from the source of the innovation. If we plot the acceptance of a new idea, or the onset of a disease contagion, for a series of time periods against the distance from the source, we can see how the innovation gradually fades with increased distance (Vignette Figure 8-1a). During each successive time period the locus of greatest initial acceptance of contagion is further from the source. During the first several periods the total volume of acceptance increases; after that, the number of new acceptances decreases with each successive period. The summation of all of these time curves across space and through time results in a pair of *bell-shaped curves*. The *geographic* bell is centered over the original innovation point and identifies a declining share of the total population ever accepting the innovation with increasing distance. The second bell graphs the overall pattern of *volume of acceptance* (or contagion) from a small number of innovators through the great bulk of acceptors to a final few laggards.

    The course of a diffusion process may be described with a *logistic curve* (Vignette Figure 8-1b). The curve is described by the formula

$$P = U/1 + e^{(a - bT)},$$

where *P* is the proportion of adopters, *T* the time at some point in the diffusion

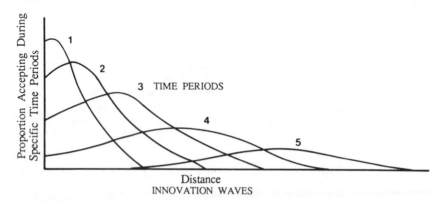

Vignette Figure 8-1a. Innovation waves. From *Spatial Diffusion* (p. 11) by P. R. Gould, 1969, Washington, DC: Association of American Geographers. Copyright 1969 by Association of American Geographers. Adapted by permission.

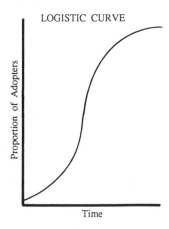

LOGISTIC CURVE

Proportion of Adopters

Time

Vignette Figure 8-1b. Logistic curve. Many diffusion studies have identified the characteristic s-shaped curve of volume of adoption through time. From *Spatial Diffusion* (p. 20) by P. R. Gould, 1969, Washington, DC: Association of American Geographers. Copyright 1969 by Association of American Geographers. Adapted by permission.

process, U the upper limit of the adoption (100 if everyone accepts the innovation), e the base of natural logs (2.7183), and a and b particular values that describe the location and shape of the curve (a identifies the height above the time axis where the S-shaped curve starts, and b how quickly it rises). Since a and b define a particular curve so that it is like no other, they are *parameters* (constants in a relational expression that determine how two or more variables change together). The logistic curve can be thought of as a cumulative frequency curve derived from the bell-shaped curve of innovation acceptors.

Vignette 8-2

POPULATION POTENTIAL

Population potential provides a useful measure of *interaction* possibilities or *accessibility*. As a measure of potential interaction among people it can help explain how a disease spreads. As a measure of accessibility, it can help determine the optimal location for a health care facility. Like density, population potential provides a measure of the distribution of population. Unlike density, it also takes into account the distribution of the population surrounding each unit of measurement and incorporates an aspect of directionality.

Population potential is based on the gravity model. The force of gravity is proportional to the product of the masses of two bodies and inverse to the square of the distance between them. It has been found that there is a remarkable regularity to human interaction in forms such as telephone calls, mail, and migration. The interaction between places tends to be proportional to the product of their populations and inverse to the distance between them. Put another way, the more people there are in two places, the more likely they are

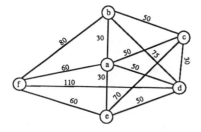

Vignette Figure 8-2a. Population clusters a–f
with distances between them.

to have some contact; the farther apart the two places are, the less likely they
are to have contact. The exponent of distance must be calibrated for each
research purpose. The friction of distance (how rapidly interaction decreases as
distance increases) is not the same for contact by electric wire, foot, and
airplane; nevertheless, the basic model serves well to create a surface of
population interaction.

In Vignette Figure 8-2a, *a* to *f* are six places. Their populations are listed
in Vignette Table, 8-2, and the distances between them are shown on the
connecting lines. The gravity model of population potential is

$$I_a = \frac{P_a \times P_a}{1} + \frac{P_a \times P_b}{D_b^2} + \dots \frac{P_a \times P_n}{D_n^2},$$

where I_a = the population potential of place *a*, P_n = the population of place *n*,
and D_n = the distance from place *a* to place *n*.

To calculate the population potential of place *a*, the population of *a* is
multiplied in turn by the population of all the other places and divided by the
square of the distance between them. Since the size of the population in *a* is
important to the overall population accessibility, it is multiplied by itself and
divided by one. The convention is to use a distance of one (of whatever units
are being used) as a divisor, which can be interpreted to mean that all the
people in a place are not really living at a point. It is usually most efficient to
construct matrices of the cross products when many calculations (as for all the

Vignette Table 8-2. Population Potential for Six Places

Location	Population	Population Potential
a	85	7,282
b	150	22,547
c	200	40,097
d	300	90,117
e	250	62,581
f	100	10,016

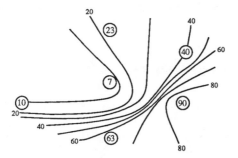

Vignette Figure 8-2b. Population potential surface. Once the population potential is calculated for points, isolines of potential can be interpolated.

counties in a state, or census tracts in a city) must be made. Vignette Table 8-2 shows the population potential in the six places.

One difficulty with using population potential comes from the usual necessity of "closing" the study area. Since there are almost inevitably places just over the border (unless there are bodies of water, and so forth), a bias may be introduced if adjacent places important for interaction are not included in the calculations, even though they lie outside the study area. In calculating the population potential surface of a county, for example, one may want to include major towns in surrounding counties in the calculation of population potentials for places within the county.

The population potential value may be the end in itself. It can serve as a variable in many analyses. To map a surface, the values are located in the appropriate places and isolines (connecting points of equal value) are interpolated between them. Thus, in Vignette Figure 8-2b, a population potential value of 62,581 at e has been generalized to 63. Since population potential is a continuous surface, a value of 80 must lie between e and d, which has a value of 90. The result is a surface that can be interpreted much like the contours of a topographic map. For example, influenza in Great Britain spread up a ridge of population potential and along it into London. There are several computer cartography programs that will interpolate isolines from point data and map contours (Vignette 9-1).

Vignette 8-3

DIFFUSION SIMULATION

The geography of disease diffusion can be very complex. Within a small group the probabilities of contact (and thus of infection) can be nearly random. However, the probability of contact between groups is often related to dis-

.0096	.0140	.0168	.0140	.0096
.0140	.0301	.0547	.0301	.0140
.0168	.0547	.4432	.0547	.0168
.0140	.0301	.0547	.0301	.0140
.0096	.0140	.0168	.0140	.0096

Vignette Figure 8-3. Mean information field.

tance. How can this idea of a contact field, with the likelihood of infection decreasing with increasing distance from the source of the infection, be turned into an operational model to predict future diffusion?

Hagerstrand (1952) used the bell-shaped curve of probabilities of contact (see Vignette 8-1) to determine a *mean information field* (MIF), or an area where contacts might occur (Vignette Figure 8-3). The greatest likelihood of contact is within the central cell fitted directly over the innovator. Distance decay in the chance of contact is equal in all directions. Hagerstrand used the MIF to initiate a stochastic *simulation* of the diffusion process. To prime the MIF, the range of numbers from 0 to 9,999 was assigned to each of its 25 cells on the basis of the cell's probability. Thus, each corner cell would receive 96 of the 10,000 digits, while the central grid gets 4,432.

The primed MIF can be used as the basis for the simulation process. The driving force that powers the model is a table of random numbers from 0 to 9,999. In a simple form of the model, one innovator exists. The MIF, which can be thought of as a grid floating over the underlying population, is centered over that innovator. He or she has the opportunity to pass the innovation to two other individuals. Those acceptors will be somewhere under the MIF grid. A random number is drawn to determine the grid location of the first acceptor. There is a nearly 45% chance that the acceptor will be in the same cell as the innovator and a less than 1% chance that the acceptor will be in one of the corner cells. Let us assume that the first acceptor is located under the cell immediately to the left of the innovator and the second acceptor is within the central cell. Thus, after the first *generation* (the first set of transfers), two individuals in two cells of the map now have the innovation. In each subsequent generation each individual with the innovation becomes an innovator and can pass it to two other persons. In generation 2, then, the floating MIF centers over each of the new acceptors, each time with a random number drawn to identify the locations of the next acceptors. In generation 3 these four will join the early acceptors, four individuals passing the innovation to eight others.

The ultimate pattern of diffusion may vary greatly from simulation to simulation. For example, if the first acceptor was to the left of the innovator, that might pull the pattern in that direction, which is known as direction bias. If the patterns resulting from many simulation exercises are combined, however, their overall pattern would approximate the bell-shaped curve with its high point centered over the innovator.

This *Monte Carlo simulation* model can be modified in many ways to better approximate reality. Is the innovator an especially powerful transmitter of the innovation? Then allow that individual several new contacts each generation while the other generations of innovators have only one. Is the disease very infectious? We can allow every innovator and acceptor to infect several new individuals each generation. Suppose the underlying population is unevenly distributed. This population can be used to create a set of *normalized probabilities*, with each cell's probability based not only on distance from the innovator but also on its share of the total population. Barriers can be incorporated into the model as well. A permeable barrier (such as immunity levels) might mean that only every fourth attempted passage of the innovation into a cell beyond the barrier is successful. Other types of barriers can be similarly incorporated.

The opportunities for modification of the basic model to more precisely simulate reality are nearly limitless. However, the goal should be to model the essence of the diffusion process and thus to reduce to a minimum the number of modifications to the MIF model. The critical essence of the model is that it enables identification of these primary influences and suggests some approximation of their relative importance. Even a simple model, however, is difficult to run by hand. High-speed computers make feasible the great volume of arithmetic manipulations in a simulation modeling.

9

Health Care Delivery Systems Worldwide

This chapter presents a perspective on health care issues that is broader than the usual geographic concerns. A process as complex as trying to prevent or cure illness cannot successfully rely on any one approach. Geographers will apply their own techniques to solving problems of disease and health, but they should be aware of other approaches.

This chapter defines a medical or health care delivery system and introduces medical pluralism and a multidisciplinary approach to studying health. The beginnings, development, and diffusion of the world's major medical systems are described. Health care delivery systems in several countries are outlined.

HEALTH CARE DELIVERY SYSTEMS

Combatting the insults that bring about illness requires appropriate physiological and immunological responses. All vertebrate life has evolved biological defense mechanisms: what makes humans unique is the addition of cultural responses to disease. Medical systems are part of the cultural response to disease.

What is a medical (or health care) delivery system? It consists of ill people and practitioners who diagnose and treat illness. A good medical system also tries to enhance health. Prevention, a healthy care environment, and good relationships between patients and medical personnel are important. Furthermore, patients and practitioners exist within a wide context of social institutions and beliefs involving health education, dietary taboos, government policies concerning the distribution of health resources, the social and financial status of population groups, and ideas people have about what causes disease. The study of medical systems should not be too narrowly focused: many studies have treated health problems, health personnel and facilities, the location of services, and the use of services as four separate entities, when they are interrelated in health care delivery systems.

A *system* consists of a variety of distinct elements and interactions among them. Disease systems involve elements such as disease agents, hosts, vectors,

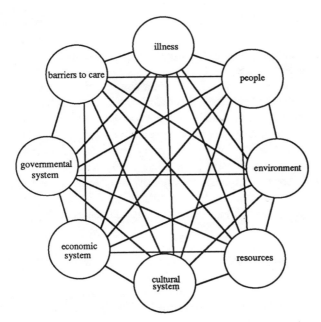

Figure 9-1. Interactions of the basic elements of a health care delivery system.

environmental conditions, and human behavior, plus the interplay among all of these. A medical system consists of these disease complexes as well as all of the factors and interactions shown in Figure 9-1. In any system, positive and negative feedback mechanisms are at work; for example, patients' experiences at a hospital could lead to either more or less use of the hospital.

THE MULTIDISCIPLINARY APPROACH

Because health care delivery involves or intersects with a large part of human culture and the environment, its study must involve contributions from many areas of inquiry. The scientific study of medicine is essential. The natural sciences are emphasized—basic fields like biology and chemistry and newly developed specialties like genetic engineering and molecular biology. The branches of public health—epidemiology, health administration and education, environmental health, and biostatistics—are involved in the study of health care. In the last few decades, social science has become very much involved.

Eyles and Woods (1983) state, "We thus see medicine and health as being truly embedded in the social system with the shape of that system significantly affecting the definition of health and the nature of health care provision" (pp. 11–12). There are social, economic, and political factors in society that help determine how disease is perceived and how treatment will be provided. Cultural norms are as important as the biological characteristics of a disease in defining health and ill-health.

Each of the social sciences has made a contribution to the study of health care delivery systems (Table 9-1). There are at least two reasons to be aware of these contributions. First, a knowledge of other approaches aids in understanding the spatial aspects of health care delivery. For example, the distance people are willing to travel to a hospital may be influenced by their social class. Second, the factors that each social science emphasizes in studying medical systems can be seen as constraints on how a system can actually function. Many political, historical, economic, cultural, and social factors place limits upon how treatment can and will be provided and accepted.

MEDICAL PLURALISM

Some medical systems tend to be spatially static, while others have diffused over wide areas. In either case, the result is that various systems overlap in the same space and that most people have a choice of health care systems.

Cultural pluralism exists within the boundaries of every country. The varieties of religions and languages in India, of tribal backgrounds in Nigeria, of social status in Brazil, and of ethnicity in the United States, are all examples of cultural pluralism. It should not be surprising that medical pluralism exists throughout the world as well. Still, we tend to be ethnocentric; that is, we feel that our culture traits are the only traits worth possessing. Certainly this is true of the attitude of most Westerners toward health care. The biomedical, or

Table 9-1. Social Science Contributions to Health Care Delivery Study

Discipline	Contribution
History	Evolution of the major medical systems
	Changes in illness prevalence and treatment modes
	Awareness of historical inertia
Political science	Impact of type of medical system
	Role of public and private power-wielding groups
Economics	Medical costs and cost–benefit analysis
	Private and public payment plans
	Health care and economic development
Anthropology/sociology	Beliefs about illness causes and effective treatments
	Characteristics of patients and practitioners
	Patient–practitioner relationships

modern or Western, system, with its firmly entrenched elements of physicians, nurses, technicians, hospitals, clinics, high-technology equipment, drug industry, and research laboratories, is *the* medical system for most people in the industrialized world and also for elite groups in the Third World. Alternative medical systems, however, do exist. In the United States these systems include holistic health care, chiropractic, lay midwives, *espiritismos* and *curanderos* among Hispanics, and root doctors among blacks.

Many of the first Western doctors to visit non-Western countries assumed that health care did not exist in these countries and that biomedicine would be filling a vacuum. This was the "fallacy of the empty vessels." Non-Western medicine serves most of the world's health needs. Despite attempts to extend biomedicine to all parts of the globe, it is too expensive for most people, and not culturally relevant for many. All medical systems "work" to one degree or another; if they did not heal they would not be tolerated by their societies. All systems have strengths and weaknesses and work in different ways, according to the beliefs and expectations of the people served.

HISTORY OF THE WORLD'S MAJOR MEDICAL SYSTEMS

Medicine in Prehistoric and Ancient Times

Evidence from paleopathology (the scientific study of disease in former times) shows that precursors of human disease are far older than humans. The bones of dinosaurs, for example, show evidence of damage from arthritis. When humans evolved over 2 million years ago, many potentially harmful pathogens already existed. We can surmise that before the neolithic revolution (around 8000 B.C.) humans suffered mostly from pathogens that could survive in small, nomadic populations or that were transmitted by animals. As humans settled into agricultural communities, they were exposed to a variety of new diseases to which they had immunological responses. These diseases, which included measles, smallpox, whooping cough, diphtheria, and tuberculosis, depended for steady transmission on the presence of relatively large groups of people living in close proximity.

It is assumed that humans have always made attempts to combat disease. Concrete records of specific treatments are scanty. There is evidence that trepanning (trephining) was practiced by neolithic humans. This technique, often the work of a skilled surgeon, was primarily used to bore holes into the skull. The main purpose for trepanning is not clear; it may have been performed to relieve headaches, to remove bone fragments following an injury, or to let evil spirits out of people afflicted with epilepsy.

The first solid evidence of medical practice has been found in the primary culture hearths that produced the earliest written records. The medicine of these ancient civilizations had roots in the supernatural, and practitioners were essentially priests. There is evidence, however, of empiricism, practical organization, and the beginnings of public health. Egypt was famous in ancient times for its medical techniques. Physicians and dentists practiced there as early as 2700 B.C. The earliest known legal code, the code of King Hammurabi of Babylon (2250? B.C.) contains laws on malpractice and setting medical fees. The Indus Valley and North China were the birthplaces of the *ayurvedic* and Chinese medical systems. The Yellow Emperor's medical treatise codified the thoughts of Chinese medical practitioners on disease causation and treatment. Written between 2600 and 1000 B.C., this is the world's oldest medical text. Medical practice in the New World, reported by the European explorers of the 15th and 16th centuries A.D., was on a par with the rest of the world. In particular, the Spaniards highly praised the sewerage systems of the Incas.

Origins of Professional Systems

As societies developed in complexity, a variety of systems of treating illness developed. Five systems, Chinese, ayurvedic, galenic, unani, and biomedicine, survive in some form. Biomedicine is sometimes also called Western medicine or modern medicine. All these systems are professional; they are highly organized and have an established array of techniques and codes of conduct. All have a highly developed pharmacopoeia, long and progressive histories, and serve very large populations today.

Chinese medicine developed during the Chou dynasty (1121–225 B.C.) and the Han dynasty (206 B.C.–A.D. 221). Disease was considered to be the result of disharmonies within the body, between humans and the environment, and throughout the universe. The treatment of disease arose from philosophical concepts such as *yin* and *yang*, which are the negative and positive principles of universal life. Yin stands for things such as earth, moon, darkness, and femaleness, and yang symbolizes heaven, sun, light, and maleness. The life energy *ch'i* flows through the "meridians" of the body. The meridians touch the skin at special points that can be stimulated by heat, pressure, or needles in the healing technique of acupuncture.

China remains the core area of the practice of Chinese medicine, but it is practiced wherever Chinese people have migrated. Its influence has been greatest in Korea and Japan, part of China's culture sphere.

The origins of *ayurvedic* medicine, which means "the science of living to a ripe age," can be traced to the migration of Aryans into the Indus Valley and the development of Harappa culture around 2000 B.C. By the 6th and 5th centuries B.C. the Indian medical system had approached its present form. In this system disease represents a disequilibrium of the humors (wind, bile, mucous, and blood), and cures attempt to reestablish a proper balance. Health is related to *karma,* or the effect of good and evil deeds, in both former lives and one's present life. Practitioners of ayurvedic medicine must have a comprehensive knowledge of pharmacopoea and understand well the influences on health of climate and morality.

Ayurvedic medicine is important for both rural and urban Indians. Its core area is North India; South India has many ayurvedic practitioners but has also established the *siddha* system, which employed substances that purportedly could transform base metals into gold as well as aid in rejuvenating the human body. Although siddha has its origins in the Dravidian culture of South India, in diagnosis and treatment it generally corresponds with ayurvedic medicine. Indian migrants have carried their ancient medical practices to other parts of the globe, particularly to the Arab world.

Galenic medicine, unani, and biomedicine all have their roots in Greek medicine. Hippocrates (460?–377? B.C.), from the island of Cos, represented the culmination of early Greek medicine. By his time, the idea of the four fundamental qualities in nature (hot, dry, wet, and cold) and the four bodily humors (blood, phlegm, yellow bile, and black bile) had been developed. Disease and treatment depended on proper or improper balances among the humors and qualities. Hippocrates' book, *On Airs, Waters, and Places* (see epigraph), was the first great classic of medical geography: it associated certain diseases with certain climates and recognized that cultural practices and social institutions can change or temper climatic conditions. Hippocrates denied that disease had supernatural causes and stressed careful observation of patients. One of the many statements ascribed to Hippocrates is "Persons who are naturally very fat are apt to die earlier than those who are slender" (Hippocrates, 1939, p. 305).

After the time of Hippocrates many humoral and nonhumoral schools of medicine arose in Greece. Only with the coming of Galen of Pergamum (A.D. 130–201) did humoral medicine gain the upper hand. Galen, the "father of experimental physiology," was a Greek residing in Rome. He was physician to the emperor Marcus Aurelius. His experimental insights became important texts, first for the Islamic Arab civilization and in the early Renaissance for Western Europe. He modified the Hippocratic idea, that an equilibrium of humors could be achieved, by teaching that physical, cultural, and demographic factors could cause one of the four humors to dominate and produce a unique "temperament"—sanguine, phlegmatic, choleric, or melancholic.

The *galenic* body of medical thought held sway in Europe for 1,500 years. Galenic practice began to lose its importance only over the last 300 years. When Europe began exploring and colonizing the world after A.D. 1500, galenic medicine diffused along with many other cultural traits.

The Arabs took over the system of Greek medicine in the 7th and 8th centuries A.D. and called it *unani* ("Greek" in Arabic). In common with other early medical systems unani had a strong ethical element and stressed the importance of the doctor–patient relationship in healing and the influence of beliefs on disease and health.

Western incursions into the Arab world in the 19th century led to the decline and eventual stagnation of unani. Biomedicine supplanted unani to a large extent. Unani is still important in South Asia, where it was introduced by Moslem conquerors. Unani exists alongside systems like ayurvedic, siddha, and biomedicine in India, Pakistan, Sri Lanka, and other South Asian countries. Some of these countries have established state and private pharmaceutical companies that produce unani drugs. Unani is also important in other areas of strong Moslem influence, Southeast Asia in particular.

Biomedicine arose out of the galenic medical tradition. Practitioners of galenic medicine in post-Renaissance times were sympathetic toward experimental physiology, which led to many important medical discoveries (such as the circulation of blood). Biomedicine advanced during the development of scientific method and inquiry. The 19th century discoveries in bacteriology gave scientific medicine its present high status. These discoveries, collectively called *germ theory*, focused on the idea that infectious diseases are caused by microorganisms. Two outstanding bacteriologists were Louis Pasteur (1822–1895) of France and Robert Koch (1842–1910) of Germany. Pasteur identified the organisms that produce anthrax and fowl cholera and worked on vaccines for these diseases; he also developed an antirabies vaccine, although viruses could not yet be identified. Koch identified the bacteria that caused wound infections, developed better techniques to identify bacteria, and discovered the germ that causes turberculosis. Several other scientists entered the new field of bacteriology, and between 1875 and 1905 about two dozen disease agents were found. During this period there were also discoveries in serology and immunology; disease vectors like the tsetse fly and *Anopheles* mosquito were identified; surgery and gynecology advanced; and anesthetics and antiseptic conditions became part of operating procedures.

Public health improvements were of great significance in the 19th century. In fact, modern sanitation movements preceded bacteriology. Increased life expectancy in the Western world between 1850 and 1950 was due more to preventive than curative medicine. Germ theory merely proved the scientific validity of the ancient idea that cleanliness was important to health.

In terms of biomedicine's organization, two movements are most important. Specialization was spurred by the rapid accumulation of vast amounts of new information in diverse areas of medicine. Professionalization tightened ethical and educational standards within established medicine, partly in response to proliferation of nonscientific, nonestablishment health practitioners. Organizations like the British Medical Association (1832) and American Medical Association (1847) were formed.

Of all the world's major professional medical systems, biomedicine has clearly been the most widely diffused. In Western countries the system dominates and the Third World has adopted it to varying degrees. Each European power took to its colonies its medical practices. Europeans (and some natives) in government and business were cared for first, then perhaps natives who worked for Europeans. When it was recognized that many diseases are contagious, programs were initiated to treat larger populations that were in contact with Europeans. Biomedicine had little real impact in the Third World until after World War II. Since then biomedicine has played a major role in the mortality transition, one of the primary reasons for today's population growth in developing areas (see Chapter 4). When the colonized countries gained their independence, a native elite had acculturated to Western medicine, and some had studied medicine in Europe or North America. This elite perpetuated the Western medical hegemony.

Traditional Medical Practice

The cultures of preindustrial, traditional societies usually are complex and deal with the environment in sophisticated ways. We shall call their medical systems either non-professional, indigenous, or traditional. Practitioners in these medical systems may be called traditional healers, native doctors, shamans, or medicine men.

Most traditional medical systems are confined to limited areas and specific populations, scattered throughout every continent. Various Native Americans of both North and South America, numerous African tribes, and a variety of groups in Asia practice traditional medicine. Although it is difficult to generalize about the medical practices of all these people, some ideas and techniques are widespread. For example, diagnosis and treatment can be carried out by immediate family, kin, and/or group leaders as well as healers. The various treatments held in common include the use of medicinal herbs, prayers, the sacrificing of animals, exorcisms, the wearing of sacred objects, and the transferral of disease from one person to another. Such treatments often depend on an intimate knowledge of intracommunity relations in order to be successful. They are especially efficacious in dealing with mental illness.

The traditional healers' roles, whether assumed by one person or by various specialists, illustrate how traditional medical practice is integrated into a cultural complex. The religious role, for example, has always been closely tied to healing in traditional societies. Supernatural beings control, among other things, illness and health. The healer mediates between the supernatural and this world. Eliade (1976) says this about a shaman: "Through his own ecstatic experience he knows the roads of the extraterrestrial regions. . . . The danger of losing his way in these forbidden regions is still great; but sanctified by his initiation and furnished with his guardian spirits, the shaman is the only human being able to challenge the danger and venture into a mystical geography" (p. 182).

A healer may be a judge; many diseases are seen as stemming from violations of the morals and mores of society. The sick person has broken relations with the supernatural or with other humans and his or her suffering is a social sanction. The healer's diagnosis is a kind of social justice, and treatment often involves a cathartic confession. Since village life is very close-knit, tensions must be resolved for the group's survival. Thus the healer also plays a role of creating "psychic unity." Treatment often includes having sick people, their relatives, and other people bring out their ill feelings toward each other. The healer's knowledge of community conflicts is important here. Another function of the healer is to entertain, to perform before an audience. This shows the group that the healer is up to the task and creates group solidarity. Music, drama, story-telling, mythmaking, dance, and fantastic costumes become part of the healing ritual. The healer may go into a trance. These efforts can inspire intense emotion in the audience.

Traditional medicine, although not as unified in its practice as biomedicine, is more ubiquitous and generally coexists with it. Highly educated and westernized people get help from traditional healers.

NATIONAL EXAMPLES

The following descriptions of health care systems provide examples from countries with a wide range of disease patterns, levels of industrialization, and government systems.

China

When China was opened up to trade with Europe in the middle of the 19th century, Chinese medicine had stagnated. China resisted westernization, but many intellectuals sought to replace Chinese practice with biomedicine. Following the Communist takeover in 1949, the Chinese tried to emulate the Soviet model. This meant introducing modern technology and capital-intensive medical care. Doctors were trained in Western medical specialties, and many large hospitals were built. However, given the immensity of disease problems in China (widespread malaria, schistosomiasis, typhoid, and tuberculosis), high infant mortality, and the scarcity of health service personnel (one doctor served over 7,000 people), there was a return to Chinese traditional medicine. The Cultural Revolution (1966–1969) witnessed a further shift in policy. Reacting to high technology, high cost, and unequally distributed health care, the Chinese began a campaign of delivering appropriate technology to rural areas. When China reestablished contact with the United States and other Western countries in the 1970s, public health measures and paramedics had raised life expectancy to 65 and lowered infant mortality to less than 4%. Most "experts" in biomedicine were in their 70s, however, and the quality of medical education and research had almost been destroyed. Thus China's first involvement with Western education in the 1970s was to send people abroad for training in fields of specialization such as biochemistry and microsurgery.

In recent decades, China has implemented several innovative health care ideas, many of which are the basis of primary health care (PHC). In 1978, the World Health Organization and UNICEF sponsored a conference on PHC at Alma-Ata, USSR. The conference set as its goal the achievement of health care for all by the year 2000. It advocated the determination of needs and making health decisions at the local level; a multifaceted approach to health that involved education, nutrition, water supplies, sanitation, immunization, assured drug supplies, and better housing; geographic, financial, cultural, and functional accessibility to health care; and a strong emphasis on paramedical personnel and community health workers. China is closer than any other country to achieving these goals. Therefore, many developing countries have watched with great interest China's progress.

Perhaps the most well-known aspect of the Chinese medical system today is that paramedics carry out most routine tasks. In rural areas they are known as "barefoot doctors." They are chosen by local communes and receive from 3 months to 2 years of training that is geared to their home villages' needs. There are also paramedical factory workers called "worker doctors." All paramedics work also at jobs in factories or fields. Their main task is prevention and health education, but they also provide curative care for minor problems and serve as the first link in a chain of referrals.

Another striking feature of China's efforts to improve health is the eradication campaigns. Instead of sending teams of experts to break the life cycles of disease or destroy pests, the entire population of an area is educated in basic ideas such as composting human waste to destroy pathogens or the destruction of snails that transmit schistosomiasis.

China has attempted to fit its health care into its social fabric. How successful has it been in providing health care for all its people? Rural areas are still disadvantaged in comparison to urban places, and the integration of traditional and biomedicine is not complete, but the Chinese have made great progress. In three decades they have passed through the epidemiological transition (Chapter 4). Transferring their type of system to other countries is problematical, however; the ability of the Chinese to adhere to a particular political and social ideology has not been matched.

India

When Europeans started going to the Indian subcontinent in the 15th to 16th centuries, three of the traditional professional systems discussed above already existed there. Ayurvedic medicine was found in most areas, siddha was concentrated in southern India, and unani, a Moslem import, was also practiced. The British and others introduced biomedicine, and immunization campaigns and treatment centers slowly began to reach the mass of people. As in many other parts of the Third World, Christian missions also supplied health care, often in remote areas.

Today, reflecting India's mixed economy, health care delivery is supported from both the public and private sectors. The federal government, which makes 5-year plans following the Soviet model, attempts to reach all the people. It has been far less successful than China in this regard. Both traditional professional and Western-trained doctors have private practices and charge fees for service on a profit-making basis.

Any discussion of cultural pluralism usually cites India as a premier example. It has many variations in language, religion, social status, and other culture traits. India's heterogeneity results in spatial and social imbalance in the quantity and quality of available health care. Regional differences in wealth, emphasis on health, and medical system mixes create spatial inequities. Social inequities are brought about by the caste system and other types of social stratification. Several medical systems may exist side by side, often in competition with one another. The traditional professional systems have government approval and thus official status. These systems parallel biomedicine in terms of professional organizations, practitioners' enjoyment of high social status, and training institutions. Integration of the various systems has been given little attention.

Zaire

The history of medical systems among the Kongo of Zaire in this century illustrates the interplay of political systems and medical practice. Before 1910 Kongo medical practice was entirely in the hands of traditional nonprofessional healers called *banganga*. Their methods included divination, ceremonial support of chiefs, bone setting, and the treatment of parasitic diseases. Formed into powerful ceremonial societies, the banganga acted as chiefs, healers, and judges; they controlled the social order and established social sanctions.

Although King Leopold of Belgium had acquired this (Belgian Congo) territory following the Berlin Conference of 1884–1885 and the Belgian government had taken over in 1908, Belgian colonial influence had little effect on Kongo culture until the 1910s. By 1920 the Belgians had destroyed the Kongo ceremonial societies by taxing their surplus funds and by replacing banganga control with Belgian armed forces.

In 1921, Simon Kimbangu, a Christian prophet, led an uprising that attempted to establish an indigenous church and restore native authority and control. One of the main thrusts of Kimbangu's religion was healing. Kimbangu rejected the magical medicine of the banganga and condoned biomedical practices. However, his messianic movement was soon crushed by the Belgians. In the ensuing decades, a few traditional practices like bone setting and herbal medicine were tolerated, but biomedicine became fully entrenched as the only true system of health care.

Amidst tremendous political unheaval that soon led to civil war, Zaire gained its independence in 1960. The new leaders administered biomedicine under Belgian laws. The banganga remained stigmatized and had no political power. The role of the prophets, descendants of Kimbangu, changed from an illegal status to one of acting as ritual counselors to the politicians and their parties.

When Colonel Mobutu came to power in 1965, the balance of power shifted again. As a result of his cultural authenticity campaign, prophet organizations lost power while the banganga gained new support and influence. The ministry of health began to think of these native healers as health resources and, through research on traditional medicine, legitimized the banganga again.

Healing among the Kongo is carried out by *therapy managing groups* who decide when a person is ill, how severe the problem is, whether outside help is necessary, and whether a healer's orders are appropriate. Illness beliefs include a major division between "natural" illness, amenable to biomedical treatment and some traditional treatments, and "human-caused" illness, treated by native medicines, rituals, and kinship analysis. The therapy managing group, taking into account the illness diagnosis and consultations with appropriate specialists, develops a package of solutions or cures for the patient.

Patterns of therapy choices can be discovered within this pluralistic health care system. Most cases begin with biomedical treatment by doctors, nurses, lab technicians, and specialists. A second set of therapies, the art of the Nganga, consists of healing by the banganga. Kinship therapy is carried out by matrilineal members of the patient's kinfolk or clan. This group convenes to reconcile kin conflicts and performs several socially integrating rituals. The fourth set of therapies is carried out by a purification and initiation group. Here nonkin meet to discuss a patient's problems and perform certain ceremonies of healing.

The Soviet Union

The health care delivery found in the Soviet Union is a product of the world's first state socialist system. Faced with the low level of health care provision in Russia before 1917, the leaders of the Communist revolution attempted to make health care an integral part of their new system. Health care was seen as important in establishing and helping to maintain a productive work force.

The overall aim of the Soviet system is centralized planning and decentralized control. Spatial organization follows rather closely the structure of central and local governments, which are based on population size. The spatial-functional organization has seven hierarchical levels. The lower three levels, where very basic care is provided, are comprised of clinics that serve rural farm areas (dispersed population, under 500 people), rural villages (500–2,500 people), and neighborhoods in towns and cities (3,000–65,000 people). In districts

(40,000–150,000 people), treatment is usually given in polyclinics, the first system contact for most people. Further up the hierarchy are regional health departments (which serve 1–5 million people), ministries of health in the 15 republics (5–50 million people), and at the top the central Ministry of Health (responsible for about 250 million people).

The Soviet health care system has strong and weak points. Serious attempts have been made to provide service to the entire population. Treatment is free for half the population, and the rest make nominal payments only for drugs and other healing aids. The country boasts relatively high facility-to-population and personnel-to-population ratios. Much of the service is provided in the workplace. There are inequalities and imbalances, however. Job status is important to quality of service received. The same level of service is not provided in all of the 15 republics, and as elsewhere in the world, rural areas are at a disadvantage. The Soviets have found it easier to provide facilities than people in remote places. Indeed, despite incentives to doctors and other medical personnel to go to these places, members of the medical profession can and do resist such assignments. In an effort to supply care to rural areas, paramedics called *feldshers* are used; they have been a successful addition to the system.

Health care in the USSR can be applauded for its emphasis on prevention, planning, evaluation, and encouragement of public participation, but almost all doctors are specialists, people cannot choose a doctor, and medical practice, psychiatry in particular, is sometimes used to control dissidents. Journalists and others have reported a lack of control over standards and incentives. Because most physicians are women, being a doctor is not considered a high-status occupation. Hospitals are often dirty, and postoperative infections are a problem. There is little use of high-technology machinery that could, for example, help to save premature babies. In recent years infant mortality has increased (not the case in other industrialized nations), and life expectancy has decreased by a few years. Although influenza and pneumonia are serious problems, much of the deterioration in health status may be due to a severe public health problem with alcoholism and associated conditions (such as fetal alcohol syndrome).

Sweden

Health care delivery in Sweden is often cited as the premier example of a welfare state system. Sweden has a compulsory health insurance scheme managed by the national government. Supporting funds come from individual graduated income taxes, fees paid by employees, and government contributions from other sources of revenue. Physicians in the system receive a salary. About 10% of the doctors are in private practice.

Sweden is relatively homogeneous culturally. The Lutheran Church provides a common religious background. The Swedes for many centuries have remained a close-knit group with focused national goals. Thus their health care system has had a long history of planning. The first fully organized health care delivery system was established in 1960. A geographer, Sven Godlund, was asked to conduct a study to determine which cities should become regional centers in the new system.

The Swedish system is notable for its attempt to create a functional-spatial organization. The functional aspect is based on the minimum or threshold populations required to support different levels of facilities, such as large and small hospitals, clinics, and health centers. At each level of service is a geographic space in which accessibility to facilities is optimal. At the highest, regional level are large hospitals that provide many specialist services. Smaller hospitals with fewer beds and specialists provide service at the county level. District hospitals comprise the third level; these are often too small or poorly located to be effective. The most basic care is provided at the commune or township health centers.

It is generally acknowledged that Sweden provides a very high standard of technically excellent health care to its citizens. Criticisms of the system include too much dependence on hospitals, which results in high costs, and the high taxes needed to support the system.

United Kingdom

Britain's National Health Service (NHS), which was established by law in 1946 and began to function in 1948, was seen as part of the new social order in Europe following World War II. The idea of equality of access that had been gathering force for some time in the United Kingdom was set out in four propositions. (1) The NHS should meet all acute and chronic medical needs. (2) Health care should be universal and free to all citizens and bona fide foreign visitors. (3) Funding would come from collective taxation rather than from payments by patients. (4) The medical groups within the system would have professional independence.

The NHS struggled with organizational problems from its inception, which is typical for such a large undertaking. The first system attempted proved to be inefficient and lacked proper communication channels. This was due mainly to the fact that the responsibility for health care was divided among hospital authorities, local governments, and 134 executive councils. Thus, not only was communication poor, but many services were duplicated. Further, there was the difficulty brought on by the vastly differing sizes of the administrative regions and populations served. In 1974 an integrated, three-tiered structure was imposed to establish centralized control and decentralized execution. At the top was the Department of Health and Social Security. At the next level were 14 regional health authorities. The local level was governed by approximately 100 area health authorities. In 1982 the lowest level was abolished in favor of 192 smaller units called district health authorities. This move was intended to allow for more local decision making.

The NHS has revived the importance of the general practitioner (GP). The GP is the first health care contact for most people. Ninety-seven percent of all residents are registered with a GP. The GP refers difficult cases to specialists or hospitals. Many GPs also take paying patients, however, and some GPs are very reluctant to take many NHS patients. In fact, private practice is growing. This trend threatens to undermine the whole idea of the NHS and to produce two levels of care.

The goal of service equality in the United Kingdom has not met with complete success. Regional imbalances in the amount and quality of care exist in spite of attempts to redress them. Services for the mentally ill, mentally handicapped, and the elderly do not receive their fair share of support. Another problem is the lengthy wait, weeks or even months, for medical services like major operations.

The role of power groups in the NHS has been investigated by Eyles and Woods (1983). They found that professional, especially medical, opinion is an integral part of the planning and management of the service at all administrative levels and in all areas of its activity. Responsiveness to the opinions of groups representing the public is far more limited. These groups consist of lay members of the regional and district health authorities and community health councils (CHCs), the last being the most important counterbalance to professional groups. The representatives of the public have limited time to devote to health care issues, have few resources to investigate situations that arise, and are external to the management of the NHS. In some areas the CHCs have been effective; in other places they have been submissive; overall, their influence is declining.

Australia

The struggle between advocates of private and advocates of public provision of health care is well illustrated in Australia. National political parties and medical and lay groups are split between two different viewpoints of man and society, including how politics, the economy, and the institutions of society should be run.

Before 1941 attempts to introduce compulsory health insurance in Australia were unsuccessful. From 1941 to 1949 an unsuccessful effort was made to introduce a system like the United Kingdom's NHS. The health care system in place from 1950 to 1969 was heavily influenced by the ideas of Sir Earle Page. His Voluntary Health Plan provided for government subsidies for health costs, but individuals received payments only if they were voluntary subscribers to a private health insurance fund. Page's scheme was designed to help those who helped themselves. This plan implies that individuals and their rights are paramount and tries to minimize the role of and costs to the national government.

By the 1970s dissatisfaction with this system had grown. It was criticized as inefficient and not comprehensive enough. The view of the Labour government, in power from 1970 to 1975, was radically different from the Page idea. Labour advocated heavy government involvement in health care provision and the payment of benefits to all those who were in need. The state, the government felt, was responsible for the welfare of all its citizens. The Labour government put its ideas into effect in July 1975 under the Medibank scheme. Funds came from collective taxation, and everyone was entitled to basic medical and hospital care without a financial means test.

Medibank was established after much political infighting, and opposition to it remained strong. In December 1975 a Liberal government returned to power. Today the system differs little from the Voluntary Health Plan. Primary and specialist care are mostly in private hands. Some hospital services are largely state-funded, however, and about 30% of the 28,000 doctors are paid from public funds. In comparison to most countries, Australia pays heavily for health care—9% of its gross domestic product.

A few other features of the Australian system are worthy of note. The six states and the Northern Territory are quite autonomous in their health policies. There are underserved areas both within cities and in the vast hinterland. This latter area has been served by the Flying Doctor Service since 1928. The aborigines have suffered from lack of health care. Recently, attempts have been made to help them, partly by training aborigines to administer health care.

United States

The health care delivery system in the United States comes closest to the free enterprise idea. Although there has been a long-term trend toward government intervention in health care provision, the United States relies mainly on the private sector. This means that most providers seek profits; those who can pay have a wide choice of care and generally receive high-quality treatment. Many doctors and patients have tried to keep government intervention to a minimum. However, the federal government pays a large share of medical bills. Most people are covered for health care by a variety of public and private insurance schemes.

The result of this political and economic philosophy has been fragmentation, a collection of systems and subsystems that are based on local and institutional initiatives. There have been attempts at national, state, and local planning, but for the most part each segment (e.g., Hospital Services, Community Health Services, Emergency Medical Services, Pharmaceutical Supplies) has gone its own way. As in other sectors of life, in the United States a wide range of special interest groups has had an impact on health care policy. These groups include major industrial interests; organized labor; federal, state, and local health planning agencies; hospital organizations; and professional physician groups. Attempts to bring order out of this chaos have met with limited success.

A brief history of the United States system will help explain the current situation. Health care in early colonial days followed European traditions of private treatment for those who could pay, and church-run charity hospitals for those who could not. William Penn established the first almshouse in 1713 in Philadelphia. Benjamin Franklin is credited with setting up the first hospital in 1751 in the same city. Other cities followed suit. During the 18th and 19th centuries, hospital growth was slow, but by the end of the period many hospitals had established medical schools. Germ theory and new medical knowledge gave impetus to construction of many health facilities in the first half of this century. Voluntary (community organization, church, fraternal, and so forth), short-term hospitals predominated during this free-wheeling era, when there was no overall planning. Meanwhile the Public Health Service was increasing its size and scope, and some local governments were assuming responsibility for the indigent.

Several federal legislative acts have attempted to organize the United States health care delivery system. The Hill-Burton Act of 1946 was the first of these. The main idea of Hill-Burton was to provide government support for hospital construction in needy areas. The states had responsibility for needs assessment and planning. Much of the data that had to be collected was spatial. Ideal hospital-bed-to-population ratios or standards were set, and the next 20 years saw a better distribution of hospital facilities. Hill-Burton also led to the establishment of many health planning agencies, mostly voluntary. In many instances friction arose between planning agencies, hospitals, and medical associations.

New planning legislation was passed in the 1960s. Medicaid was established to provide medical care to the poor who received government payments and to others unable to pay medical costs. Medicare made similar provision for the elderly. Both programs are administered by the states, with financial support from the federal government. Both pay for outpatient treatment, physician fees, hospitalization, surgery, and equipment such as walkers and pacemakers. Regional Medical Programs (RMPs) were also begun in the mid-1960s. These programs were intended to address degenerative diseases, mainly heart disease, stroke, and cancer. The geographic boundaries of RMPs were not clearly defined, so territorial conflicts resulted. Also, the establishment in 1966 of Comprehensive Health Planning Agencies (CHPAs) led to interprogram conflicts. By the mid-1970s RMPs were taken over by CHPAs.

Another landmark in United States health care legislation came in 1974, with passage of the National Health Planning and Resources Development

Act. This act was an attempt to update and revise existing programs and to foster cooperation among the national, state, and local health care planning agencies. The new program called for each state to create clearly defined Health Systems Agencies (HSAs). HSA populations were to be between 500,000 and 3 million. HSA boundaries were to coincide with other administrative units as much as possible. The HSA became the basic unit for determining health care needs, administering federal funds for approved programs, and reviewing and evaluating programs.

The 1974 act has by no means provided a smooth road to overall planning. Although consumers are supposed to be a majority on state planning boards, many feel that providers have too strong a voice in planning decisions. Providers, on the other hand, often resent government interference and complain about the bureaucratization of health care delivery, the costs of maintaining HSAs, and the long review process.

Health care costs have escalated dramatically in the United States over the last several decades. From 1950 to 1981 the amount spent on health care rose from $12.7 billion to $286.6 billion, or from 4.4% of the gross national product (GNP) to 9.8% of the GNP (Public Health Service, 1982). There were several reasons for this. The growth in health expenditures exceeded that of most other goods and services. An aging population meant more illness and therefore a higher demand for treatment. Desegregation of society led to better health treatment for minority groups. Most people also had better access to good health care.

Costs have risen differentially within the medical system. Whereas practitioner fees have kept pace with rising costs, hospital room costs have escalated above this level of growth. Reasons for this include new medical technologies, procedures, and techniques; more hospital personnel; higher employee wages; and more services demanded by patients.

The current health care delivery system in the United States shows very little spatial rationalization, despite the intention of the 1974 act to establish a kind of spatial–functional system of HSAs. There are spatial imbalances in resources, accessibility, and utilization, however these may be measured. Planning has had varied success in different regions of the country, partly because power groups with conflicting interests combat each other.

In a study of urban areas from the late 19th to the early 20th century, Knox, Bohland, and Shumsky (1983) argue that scientific advances, medical technology, and the influence of doctors were less important in determining access to care than economic, political, social, and spatial changes in the structure of cities. For example, physician home visits or a short journey to a family doctor gave way to a longer journey to hospitals or specialists because services were more spatially concentrated. On the positive side, quality services have expanded to cover people and areas that were poorly served before.

Three recent developments in the United States health care system should be mentioned. The concept of a health maintenance organization (HMO), although it has been in existence since World War II, has gained more acceptance as a health care alternative in recent years. The basic idea is that health care is prepaid. Subscribers to the HMO pay a set fee in advance for certain types of services and then receive whatever treatments they require. Two of the more well-known HMOs are the Kaiser Foundation Health Plan and the Health Insurance Plan (HIP) of Greater New York. Like Britain's NHS, access to specialty physicians in an HMO is through a "primary" physician. This is a major change in health care for many people and may affect future mixes of specialties.

Advantages of the HMO system include physician group practice, which gives doctors flexibility and a chance to consult with colleagues; quicker patient access to physicians; and an emphasis on prevention. Disadvantages are that patients are often preselected, which means that groups like the unemployed or seriously ill may not be able to participate; and that the traditional right to choose a physician is surrendered. Another type of prepaid plan has been gaining rapid acceptance: the preferred-provider organization (PPO). PPOs share most of the advantages of HMOs but offer enrollees a wider range of choices in participating hospitals and physicians. It is estimated that 30 million people in the United States belong to either an HMO or a PPO.

The second recent development concerns the number of physicians available to the public. In 1980, the Graduate Medical Education National Advisory Committee (GMENAC) published a detailed report on medical personnel for the Department of Health and Human Services (DHSS). It stated that there would be 70,000 more physicians than required in 1990 and that most specialties would have a surplus. The report made several specific recommendations on how to reduce the over-supply of domestic and foreign-trained physicians and other medical personnel (Graduate Medical Education National Advisory Committee: Summary Report, 1980).

The third development is an alteration in the way Medicare payments are made to hospitals (40% of all hospital revenues). Payments were based on hospital reports of what treatments cost; now uniform payments to all hospitals are based on predetermined rates for each of 468 diagnosis-related groups (DRGs). This change, which began a phased transition in 1983, was a response to findings that there were wide differences among hospitals in treatment charges, types of treatment, length of stay for similar problems, and use of ancillary resources.

DRGs are obviously an attempt to cut down on costs. They are intended to foster better and more efficient hospital management and improved cooperation between hospitals and physicians. However, many questions as to their efficacy remain. Will hospitals now concentrate on pushing people through the system as quickly as possible? Shorter hospital stays may save money but may also lower the quality of care. Will case mixes change because hospitals attempt to provide the services that will produce the most revenue for them? What are the spatial implications, such as regional equity in service provision?

CONCLUSION

This small sample of health care profiles suggests the tremendous variety in the ways that nations attempt to deal with illness and health. Cultural attitudes and beliefs, types and severity of illness, government policies, economic resources, the history of traditional and biomedical health services, and the roles of various groups within different societies all play a part. Each type of system has its advantages and disadvantages.

REFERENCES

Ackerknecht, E. H. (1955). *A short history of medicine.* New York: Ronald Press.

Agren, H. (1974). Patterns of tradition and modernization in contemporary Chinese medicine. In A. Kleinman, P. Kunstadter, E. R. Alexander, & J. L. Gale (Eds.), *Medicine in Chinese cultures* (pp. 37–59). Washington, DC: U. S. Department of Health, Education, and Welfare.

Basham, A. L. (1976). The practice of medicine in ancient and medieval India. In C. Leslie (Ed.), *Asian medical systems: A comparative study* (pp. 18–43). Berkeley, CA: University of California Press.

Bibeau, G. (1981). Current and future issues for medical social scientists in less developed countries. *Social Science and Medicine, 15A,* 357–370.

Bryant, J. (1969). *Health and the developing world.* Ithaca, NY: Cornell University Press.

Burgel, J. C. (1976). Secular and religious features of medieval Arabic medicine. In C. Leslie (Ed.), *Asian medical systems: A comparative study* (pp. 63–81). Berkeley, CA: University of California Press.

Carter, J. R. (1984). *Computer mapping: Progress in the '80s.* Washington, DC: Association of American Geographers.

Crozier, R. (1974). Medicine and modernization in China: An historical overview. In A. Kleinman, P. Kunstadter, E. R. Alexander, & J. L. Gale (Eds.), *Medicine in Chinese cultures* (pp. 21–35). Washington, DC: U. S. Department of Health Education, and Welfare.

Dunn, F. L. (1976). Traditional Asian medicine and cosmopolitan medicine as adaptive systems. In C. Leslie (Ed.), *Asian medical systems: A comparative study* (pp. 133–158). Berkeley, CA: University of California Press.

Eliade, M. (1976). *Shamanism: Archaic techniques of ecstasy* (W. R. Trask, Trans.). Bollingen Series 76. Princeton: Princeton University Press.

Eyles, J., & Woods, K. J. (1983). *The social geography of medicine and health.* London: Croom Helm.

Gesler, W. M. (1984). *Health care in developing countries.* Washington, DC: Association of American Geographers.

Graduate Medical Education National Advisory Committee: Summary report (Vol. 1). (1980). U. S. Department of Health and Human Services Publication No. (HRA) 81–651. Washington, DC: U. S. Government Printing Office.

Hippocrates. (1939). Aphorisms. In F. Adams (Trans.), *The Genuine Works of Hippocrates.* Baltimore: Williams & Wilkins.

Janzen, J. M. (1978). The comparative study of medical systems as changing social systems. *Social Science and Medicine, 12B,* 121–129.

Joseph, A. E., & Phillips, D. R. (1984). *Accessibility and utilization: Geographical perspectives on health care delivery.* New York: Harper & Row.

King, M. (Ed.). (1966). *Medical care in developing countries.* Nairobi: Oxford University Press.

Knox, P. L., Bohland, J., & Shumsky, N. L. (1983). The urban transition and the evolution of the medical care delivery system in America. *Social Science and Medicine, 17,* 37–43.

Monmonier, M. S. (1982). *Computer-assisted cartography: Principles and prospects.* Englewood Cliffs, NJ: Prentice-Hall.

Polgar, S. (1962). Health and human behavior: Areas of interest common to the social and medical sciences. *Current Anthropology, 3,* 159–179.

Primary health care. (1978). Report of the International Conference on Primary Health Care, Alma-Ata, USSR, 6–12 September. Geneva: World Health Organization.

Public Health Service. (1982). *Health, United States, 1982.* Hyattsville, MD: U. S. Department of Health and Human Services.

Pyle, G. F. (1979). *Applied medical geography.* Washington, DC: V. H. Winston.

Shannon, G. W., & Dever, G. E. A. (1974). *Health care delivery: Spatial perspectives.* New York: McGraw-Hill.

Veith, I. (Trans.). (1972). *The yellow emperor's classic of internal medicine.* Berkeley, CA: University of California Press.

Vladeck, B. C. (1984). Medicare hospital payment by diagnosis-related groups. *Annals of Internal Medicine, 100,* 576–591.

Wood, C. S. (1979). *Human sickness and health: A biocultural view.* Palo Alto, CA: Mayfield.

Further Reading

Elling, R. H. (1981). Political economy, cultural hegemony, and mixes of traditional and modern medicine. *Social Science and Medicine, 15A,* 89–99.

Fabrega, J., Jr. (1980). *Disease and social behavior: An interdisciplinary perspective.* Cambridge: Cambridge University Press.

Henderson, G., & Primeaux, M. (1981). *Transcultural health care.* Menlo Park, CA: Addison-Wesley.

Kleinman, A. (1978). Concepts and a model for the comparison of medical systems as cultural systems. *Social Science and Medicine, 12,* 85–93.

Mechanic, D. (1962). The concept of illness behavior. *Journal of Chronic Disease, 15,* 189–194.

Vignette 9-1

COMPUTER CARTOGRAPHY

Among the most dramatic changes affecting the way geographers go about their work during the past 15 years has been the growth of computer-assisted cartography. The proliferation of increasingly accessible computers has enabled researchers to examine ever larger sets of data, using sophisticated statistical analysis techniques marketed in "packaged," and usually easily available, software programs. The result of this explosion in research capacity is that we can ask research questions, examine sets of data, and find answers previously beyond our capacity to carry out the complex analysis. The impact of the high-speed computer and its massive memory on cartography has been less well recognized but, for that subdiscipline, equally important.

The ability to create maps by computer is based on the ability to generate and store for repeated use computer files of geographic coordinates of boundary patterns that are the base for mapping. Boundaries can be represented as a series of discrete line segments, with their end points defined by pairs of x- and y-coordinates. A series of these line segments joined together creates an enclosed space, called a polygon; that may be a census tract, state, or whatever.

Perhaps the best known mapping procedure in the United States is the Dual Independent Map Encoding (DIME) file, developed as a listing of state and county boundaries by the Department of Transportation and expanded and popularized by the Census Bureau. DIME files consist of line segments identified numerically by beginning and end points. Thus, the segment has direction and a left and right side. For a city block, defined by the four streets that form its sides, there will be four DIME-file line segments (Vignette Figure 9-1). These segments are joined together in a chain defined by the four nodes of their intersections. These are in fact the street intersections at each corner of the block. All DIME files are composed of these three basic elements—line, node, and enclosed area.

Using the example of the block in the figure, the computer identifies the block by the series of lines, or chain links, from nodes 18 to 19, 19 to 30, 30 to 29, and 29 to 18. The Census DIME file includes beginning and ending addresses on each side of each line segment. In this chain block 43 is always on the right side of each chain link.

Geographers can use DIME files to assign data to appropriate areal units for analysis. For example, the address ranges may be used with patient addresses, to assign patient residents to a specific census tract so that they can be analyzed with the socioeconomic and demographic data available at the tract level. A complete assignment of all patients would enable the calculation of standard rates. Many other pieces of data can be similarly assigned to a gradually emerging data set.

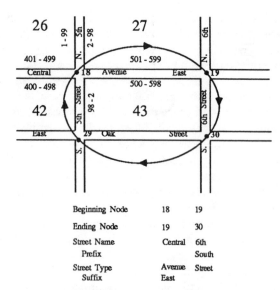

Vignette Figure 9-1. DIME file street segments and chains. The file is read as a series of block chains beginning at the upper left and running clockwise. From *Computer Mapping: Progress in the '80s* (p. 18) by J. R. Carter, 1984, Washington, DC: Association of American Geographers. Copyright 1984 by Association of American Geographers. Reprinted by permission.

The Census Bureau has made available in computer-readable form the more recent censuses of population and housing, agriculture, and all other general censuses, as well as the wealth of material in the periodically updated *City and County Data Book.* All of these can be used with the DIME file. Statistics Canada has produced a comparable mapping program called CART-LIB, available for Canadian areas as small as census tracts for the country's census metropolitan areas.

The Central Intelligence Agency created two widely available boundary files called World Data Bank I and World Data Bank II. Both map the world and all of its countries and colonies, the first at the scale of 1:12,000,000 and the second (generally more appropriate for a continent or smaller area) at 1:3,000,000. An increasing variety of global data collected by such agencies as the World Health Organization and the World Bank is available in computer-readable form and thus directly usable with the World Data Bank files.

There are also a number of grid or fixed geographic coordinate programs available. One divides the globe into a pattern of grid units each measuring 1 degree of latitude by 1 degree of longitude. It has been used to map things such as vegetation and land use. Other programs, available at less than a global scale, provide data grids of smaller size. If no appropriate mapping file is available for a study area, geographic boundary coordinates of data point locations may be converted into machine-readable form with a device called a digitizer. This is expensive machinery that requires training to use, but computer mapping courses are common in geographic curriculums. Digitizers are usually available in universities and government planning agencies.

Maps may generally be produced in "hard copy," that is, on paper, through computer mapping in one of two forms. They may be created on a standard printer. Individual characters on the printer can be used to symbolize boundaries or data categories, and they can be overprinted to create intensities of shading from almost blank to black. Such maps are inexpensive and have the advantage of being produced on nearly universally available printers. However, they are of coarse quality because of the large individual character size that printers use. Alternatively, maps may be produced on computer-driven plotters that generate line maps. Plotter maps are generally visually superior, offer a potentially far greater variety of data symbols for data categories, usually have much finer geographic detail, and are more expensive.

Two of the most common of the many mapping programs available are SYMAP and SAS-Graph. SYMAP, created at the Laboratory for Computer Graphics and Spatial Analysis at Harvard University in the mid-1960s, was the first widely used computer mapping program. It is still available and popular. SYMAP can utilize either point or areal data. Among the many other computer programs created by the Harvard lab, probably the most widely used are CALFORM, a plotter-based choropleth program, and SYMVU, which provides three-dimensional, plotter-generated maps.

The newer SAS-Graph, a product of the Statistical Analysis Systems (SAS) Institute, was created as a map and graph component of a comprehensive package of statistical programs. Unlike SYMAP, SAS-Graph has many coordinate boundary files built into the system and is part of a system of comprehensive and flexible file management. It can be used to create a variety of plotter-based choropleth maps, including three-dimensional figures. It does not interpolate well, however, and is inappropriate at present for generating surface or contour maps.

Several of the basic computer mapping programs, notably SYMAP, have been used for two decades. Indeed, programs such as SYMAP and SAS-Graph are likely to be heavily used on large mainframe computers for years. Active development in the near future will be in material for small microcomputers, either through the adaptation of existing programs to the smaller micros or the development of new programs especially designed for them.

10

Health Care Resources

One of the main concerns of medical geographers has been the assessment of health care systems to discern spatial patterns among health care resources. Several approaches have been used to assess existing resource distributions and to plan for more equitable ones. Some researchers have devised ways to map imbalances. Others have looked for reasons for inequalities among areas, including why physicians and facilities are located where they are. One geographic concept that has potential for examining the spatial distribution of health care resources is central place theory. Models of the location of facilities and their human resources are used to optimize physical accessibility and economic efficiency. Regionalization helps to organize health care resources with a view to equity. This chapter first addresses the gathering of data on health care resources and proceeds to deal with the geographic approaches to resource distribution.

RESOURCE INVENTORIES

Health care resources include practitioners like physicians, nurses, and native doctors; facilities like hospitals, clinics, and the homes of indigenous healers; materials like antibiotics and medicinal herbs; equipment that ranges from body scanners to tongue depressors; institutions like the American Medical Association (AMA) and herbalist associations; and financial support. The first two resources, personnel and facilities, are emphasized in most geographic studies, perhaps because they are more easily assessed. The other types of resources should not be forgotten, however, because they influence the quality, costs, and administration of care.

In Western countries the kind, quality, and quantity of health care resources is information available most often in records open to the public. For example, one can find lists of physicians and their specialities in government and medical association publications and telephone directories. Hospitals and other facilities publish financial statements and can supply information on the equipment and medicines they use. Access to patient records is less open because of the necessity for confidentiality. Investigators may have to rely on interviews to collect information on patient attitudes or the role of community groups in making decisions about health care financing.

In developing countries, researchers have to rely more on field surveys to obtain the same kinds of information. It is harder to define who is a health care practitioner; traditional healers may play other roles in society and practice medicine only part time. Also, it is hard to categorize native doctors because they often have several specialities. Even information on biomedical systems is not as reliable as in developed countries. In many developing areas the distribution of pharmaceutical drugs is not controlled, and much open marketing and street selling of drugs goes on, which would be very difficult to trace.

THE SPATIAL DISTRIBUTION OF RESOURCES

Why do some areas have fewer or more resources than other areas? Spatial analysis is the geographical approach to understanding inequalities in income distribution, food supplies, or health care service. The complexity of health care makes demonstrating its unequal distribution a challenge. For example, Figure 10-1 shows physician-to-population ratios for 1982 for the 50 states of the United States. There is a wide range of ratios. Statistical tests could be performed to show that there are significant differences in ratios among various regions of the country, but they would mask many other issues of health care delivery.

What constitutes an equitable distribution of resources is not well defined. The recent report of the Graduate Medical Education National Advisory Committee (GMENAC) states flatly that "valid criteria for designating geographic areas as adequately served or underserved have not been developed" (*Graduate Medical Education National Advisory Committee: Summary report*, 1980, p. 99). The measurement of resource-to-population ratios, like the ratios in Figure 10-1, is a common geographic approach to this issue. A more sophisticated ratio, the locational quotient, is sometimes used. (The location quotient for a spatial unit such as a county or state is calculated by dividing the proportion of all resources a unit possesses by the proportion of the total population in that unit.)

Not all planners or social scientists consider measures of spatial maldistribution to be of primary importance. Many economists would argue that market forces should determine where resources are located and that open competition is the best assurance of economic efficiency. Others (e.g., Hemenway, 1982) might say that physicians, if in limited supply, should be located where they would prolong the most lives. This might mean that areas where the people were relatively healthy would have more doctors, depriving less healthy populations.

Inequalities might exist at one scale, but not at another (see Vignette 7-1). What level of unit—township, county, health systems agency, or state—should

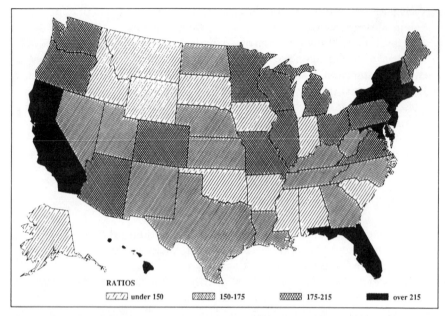

Figure 10-1. Physician-to-population ratios in the United States, 1982. Data from *Current Population Reports* (Series P-52, No. 944, p.) by the U. S. Census Bureau, 1984, Washington, DC: U. S. Government Printing Office; and *Physician Characteristics and Distribution in the United States* (1982 ed., pp. 50–51) by M. A. Eiler, 1983. Copyright 1983 by American Medical Association, Chicago.

be used to evaluate resource distributions? Should the basic unit be a political unit like a county or a functional unit like a hospital catchment area? The type of resource must also be considered; the distribution of brain surgeons is more likely to approach equality on the state level than on the township level.

An important issue of equitable service distribution is the contrast between *need* for services and *demand* for them. The economic efficiency criterion emphasizes demand, which is based on use of resources and thus is biased toward those who have social and financial access to care. Demand can be determined from patient records. A health system, however, must meet needs. Needs assessment involves costly and time-consuming procedures such as administering health status questionnaires; demographic, epidemiologic, and social indicator surveys; and interviews with key personnel within communities.

Needs analysis discloses the *inverse care law*, which operates at different geographic scales. The law states in part that "the availability of good medical care tends to vary inversely with the needs of the population served" (Hart, 1971, p. 412).

For planning purposes, quantitative measures of resource availability are essential, as are measures of the quality of the personnel, facilities, and other resources in an area. Quantity alone is inadequate to judge equality of resource distribution. For example, the United States has one of the world's highest ratios of health care resources to population, but there are a few countries with fewer resources and lower infant mortality rates. One reason for a situation like this is that medical care can itself be a detriment to health. This is termed *iatrogenic* disease, caused by the treatment itself. Hospitals, for example, may produce infections that spread rapidly, especially in crowded conditions. In a poor country the establishment of more health clinics in a rural area might look good in a ministry of health report, but if a clinic's staff is poorly trained, beds are dirty, and drug supplies are low, more harm than good may have been done. Quality involves continuing medical education, keeping facilities open at convenient hours, good staff attitudes toward all patients, short waiting times for services, and many other factors.

GEOGRAPHIC STUDY OF RESOURCE DISTRIBUTION

Health care resource distribution has been studied at scales from the national to the neighborhood. Macrostudies have distinguished resource items based on administrative units from those based on attributes within areas, such as rich versus poor areas or urban versus rural areas. Determining physician-to-population ratios by state in the United States (Figure 10-1) is an example of a macroscale investigation. Figure 10-2 shows inequalities in work load (the ratio of population served to general-use inpatient beds) in hospital "spheres of influence" throughout the country of Malawi in Africa. Three hospitals were found to have inadequate facilities when compared to the national norm. A macroscale study of resource imbalance in England compared the number of people on general practitioners' lists of patients. Although England makes a serious attempt to balance physician-to-population ratios geographically, list sizes vary greatly over the eight major regions of the country.

Administrative unit studies may mask differences within regions. Resource inequality is consistently found *within* such areas as cities or countries. In most countries urban areas have more biomedical resources than do rural areas. There has been a trend in the United States over the past century toward greater urban-rural physician imbalances. This trend may be lessening or reversing because the pressure of a greater supply of physicians is prompting more doctors to establish practices in smaller towns. Given developed transportation routes and modes, urban facilities may be more accessible to rural populations than many rural facilities. In India, for example, most villagers can get to the market town but may not be able to cross a river to another part of their rural district.

There are other within-region imbalances. Wennberg and Gittelsohn (1982) reported that they found much variation in surgery rates among 193

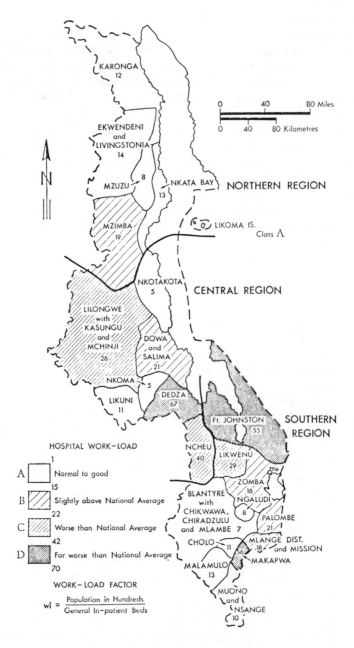

Figure 10-2. Hospital areas and work-load factors in Malawi. From "The Distribution of Population and Medical Facilities in Malawi" by N. D. McGlashan, 1972, in N. D. McGlashan (Ed.), *Medical Geography: Techniques and Field Studies* (p. 92), London: Methuen. Copyright 1972 by Methuen. Reprinted by permission.

medical service areas in New England. The overall surgery rate varied as much as two to one from unit to unit. The total rate in an area was strongly associated with the number of surgeons and hospital beds per capita, both of which also varied widely. The key factor was local medical practice.

Microscale studies have usually focused on urban areas. Their results might not generalize to entire regions, but they can provide information for a more detailed analysis. As one would expect, the more deprived urban areas in terms of income and other measures of socioeconomic status are medically underserved. Ethnicity may be an important intraurban factor. It is not unusual to find large differences between mostly white and mostly black census tracts in numbers of physicians available. Even where maps of ethnicity and resources seem to indicate equality, the health care provided is not equally accessible to all ethnic groups. For example, minority groups may be living in the zone of transition of a city close to a downtown university medical center but will rarely go there for treatment (a zone of transition being an area which once housed the wealthy, but became populated by low-income groups after the wealthy moved out of the city toward the suburbs).

Central Place Theory

Models for the existing distribution of goods and services are often based on central place theory. Basically, this theory focuses on hierarchies of places that sell goods and services, and thresholds and ranges of goods and services at different levels in the hierarchy. High-order centers market a wide range of goods and services; low-order places market few goods and services. High-order items (like a Rolls Royce) will be found in only a few large places, while low-order items (like a loaf of bread) will be found in many very small places, as well as in high-order places. Low-order goods and services require a lower population threshold to be profitable. They also require a smaller territory or range than high-order items. Places of each order are distributed across space in a regular pattern, a set of nested hexagons. The theory as Walter Christaller (1933) first formulated it had many simplifying assumptions, such as an isotropic plane and an evenly distributed population with equal purchasing power.

Central place theory can be linked to the functional and spatial organization of an idealized health care delivery system. The levels of the hierarchy range from individual practitioners or paramedics in offices or homes up to large medical centers providing the whole range of health care services. Each level of service has a certain population size or threshold and area or range. Sweden and the Soviet Union have this kind of system (Chapter 9). An ideal *spatial* organization is harder to realize in practice than is a *functional* hierarchy.

In Malaysia in the 1960s, the national health care delivery system was explicitly structured on a central place theory framework (Figure 10-3). Three

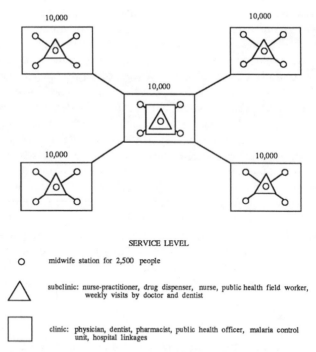

SERVICE LEVEL

○   midwife station for 2,500 people

△   subclinic: nurse-practitioner, drug dispenser, nurse, public health field worker,
        weekly visits by doctor and dentist

▢   clinic: physician, dentist, pharmacist, public health officer, malaria control
        unit, hospital linkages

Figure 10-3. Central place health care: the ideal distribution is illustrated by the health care delivery structure of the Federation of Malaysia.

levels of care were nested within service areas intended eventually to serve 50,000 people. These service goals have not been achieved, but there has been great progress in training personnel, constructing the infrastructure, and reaching the entire population. At the lowest level, government-trained midwives are stationed in villages and are expected to care for women in a total population of 2,500. Five midwives are associated with a subclinic in more central locations. At the subclinic are a midwife and nurses who can treat simple infections, do blood smears for malaria, treat skin problems and intestinal worms, give vaccinations, and perform screening tests for patients to see a doctor who attends once a week. A dispenser provides a range of pharmaceuticals. A public health field-worker chlorinates wells against cholera, educates people about garbage burial, and monitors mosquito-breeding conditions. Each of five subclinics look to a main clinic, at the most central location, for the full range of medical and public health services, linkage to one local hospital, and referral to the major hospitals in the capital.

As the Malaysian system was implemented in the 1970s, there were difficulties in stretching it over sparsely populated, forested rural areas, as well as the more densely populated, more developed west coast. It was often difficult for villagers in low-density areas to travel to the subclinic, and so they delayed seeking care. Once they had found transportation, they often preferred to travel to the main clinic to see a physician rather than to the subclinic to see a nurse. The middle-level services tended to be underutilized. Efforts have been made to upgrade the skills of the midwives so that they can assist with primary care at the subclinic level.

Applying central place theory must take into account many factors that distort the ideal. These factors include physical barriers, transport networks, and income distribution. The result is a system that usually retains its functional nature, but whose spatial configuration rarely resembles nested hexagons.

The pattern of admissions for hospitals in the cities and towns of the Snake River Valley in southern Idaho illustrates the real-world impact of central place and the urban hierarchy on the spacial distribution of health care (Figure 10-4). There is a clear, direct relationship between city size and number of patients admitted. Boise is the valley's largest city, followed by Pocatello, Idaho Falls, and Twin Falls. Burley and Rexberg are the smallest

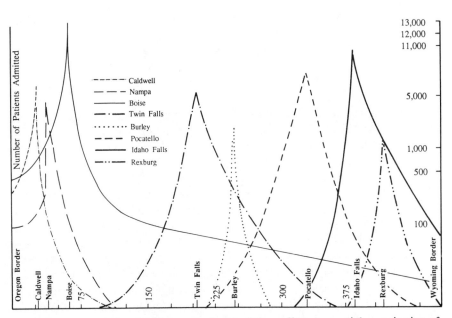

Figure 10-4. Central place health care: real central place effects on spatial organization of health services resemble those in Idaho. From *Hospital Service Areas in Idaho: A Geographic Approach to Spatial Efficiency* (p. 273) by J. M. Meade, 1971, an unpublished doctoral dissertation, University of North Carolina at Chapel Hill, Chapel Hill, NC.

towns with hospitals. Contrary to the central place ideal, the service areas of hospitals of similar size (e.g., Twin Falls and Pocatello) overlap to some degree. The service cones of the smallest communities are steep and extend only a few miles, indicating a relatively large friction of distance. People residing in Burley will most likely use the Burley hospital, but those just a few miles both east and west are more likely to travel to Twin Falls. Boise, by comparison, with its more sophisticated hospital facilities and physician specialties, extends its service to people in all other communities in the valley. Twin Falls's service area is somewhat larger than that of more populous Pocatello because it serves a comparatively sparsely populated area with no nearby competitors. The east slopes for both Caldwell and Napa are very steep, probably the result of competition from nearby Boise, which lies to the east.

Some geographers (e.g., Okafor, 1982) contend that central place theory should not be used as an ideal at all. They criticize the theory's claim that market forces should dictate thresholds and ranges for each level of service and argue that social welfare issues should override issues of economic efficiency. Thus, for the good of a population's health, a government might have to establish facilities with lower thresholds and smaller ranges than central place theory justifies. Under a social welfare scheme, however, providing levels of service in overlapping spatial units is valid and, in fact, is used in several countries. The number of levels and their corresponding thresholds and ranges will vary according to the type of system. Perhaps central place theory should be used as an explanatory rather than as a prescriptive model.

Reasons for Personnel and Facility Location

Given the existence of inequalities in health care resource distribution, no matter how one chooses to measure inequality, the next logical step is to try to devise means to correct imbalances. This requires some knowledge of what caused the inequalities; this causation is complex because of the influence of many interrelated factors.

Two types of analysis have been carried out, macroscale and microscale. Macrolevel analysis concentrates on demographic, socioeconomic, and environmental correlates of location decisions. One of the most consistent findings is that larger populations attract more resources. In both Western and non-Western countries, more facilities and higher personnel-to-population ratios are associated with higher levels of urbanization or population concentration. This finding makes sense, particularly in a capitalist economy where health care delivery requires a sufficiently large market demand to be viable. As the threshold and range of a specialty increases, the specialty will be tied more closely to population size: that is, rarer specialties tend to be found in more populous places. Urban places also provide the professional, cultural, and recreational facilities that many physicians and other medical personnel desire. Some studies have found that the income of populations in different places was important (Rinlinger & Steele, 1963) to health care distribution, but other investigations have concluded that income's significance was limited (Joseph & Phillips, 1984). Age structure can be quite important; for example, doctors might be attracted to older populations that have relatively high morbidity levels. Other significant population characteristics studied include social or ethnic composition of a population, educational level, and population density.

Macroscale studies, because they deal with aggregates, should be used to describe general patterns rather than to explain resource distributions. This is a situation that encourages ecological fallacy (see Vignette 7-2). Also, some variables associated with resource locations may be correlated among themselves, and it is often difficult to know which variables are the most important ones. Furthermore, the relative importance of these variables may change over time. In developing countries the data for macroscale analysis are often simply not available.

Microscale studies are the preferred approach to analyzing resource location factors; at the microlevel the actual mechanisms behind resource distribution decisions are clearer. The microscale study may complement results obtained at the macrolevel. A microscale study might focus on practitioner motives for selecting a location. Doctors like to be fairly close to hospitals whose facilities they can use and colleagues with whom they can discuss cases. Group practices are now common; they allow professional consultation and flexibility of leisure time. An important consideration may be where office space is available; medical offices often require special features such as heavy wiring for technical equipment, so that possible sites are limited. Thus doctors tend to practice in professional buildings or in malls rather than in residential areas.

Western-trained doctors in developing countries may have additional reasons for locating where they do. Some doctors use their influence with ministries of health to avoid assignments in remote or otherwise undesirable places. Although there has been little investigation of the locational decisions of traditional healers, kinship connections and inherited practices probably affect where they locate. Ramesh and Hyma (1981) mapped out the locations of all registered professional traditional healers in Madras City, India. These practitioners of the ayurvedic, siddha, and unani systems tended to concentrate in the older, higher density residential areas of the city and also in the older, commercial and manufacturing sectors. New residential areas had fewer of these healers. Techatraisak (1985) found the recent expansion of Chinese practitioners to newer suburbs of Bangkok to be evidence of the continuing vitality and importance of the Chinese medical system in Thailand.

The wishes of individual practitioners and health care planners are often blocked because of various social, economic, and political constraints, discussed in Chapter 9. To take a simple example, a hospital may be located on a particular site because the land was cheap or was a gift; original capital outlay concerns would override possible long-term inefficiencies (perhaps in transporting supplies to the hospital) and patient and staff accessibility.

The role of the government in providing care operates to control health care allocation. When Europeans controlled colonies, agriculturally productive areas and the more productive groups of people had the greatest share of resources. A series of maps showing biomedical facility location over time in many Third World countries would show a diffusion from the coast inland as the European colonists advanced. The maps would also show concentrations of facilities in places near iron mines and rubber plantations. The owners of plantations and mines wanted to keep their labor forces healthy. In China, however, although paramedics serve their own townships, fully trained doctors tend to gather in major urban centers, despite periods of strong government opposition to such behavior.

Power groups other than governments are also involved. Professional medical organizations set standards of admission, which control practitioner supply; this in turn affects overall distribution. Professional organizations set the tone for the expected life-style of their members. Another constraint involves the mix of specialties. A study of Phoenix, Arizona (Gober & Gordon, 1980), showed that specialists tend to cluster around hospitals more than primary care physicians, so that in countries with relatively high proportions of specialists patients must travel farther for care.

Another set of constraints, common particularly in Third World countries, has been pointed out by Navarro (1974). He feels that the maldistribution of health care resources in Latin American countries is due to the same factors that help keep these areas underdeveloped: the cultural, technological, and economic dependency of the countries and economic and political control of resources by local elites and foreign interests. Unless these conditions change, he contends, resource imbalances will continue to exist.

Location/Allocation Modeling

Just as there are many ways of measuring health resource imbalances, so there are many criteria for locating facilities to try to redress these imbalances. Calvo and Marks (1973) suggest three, based on the three sectors of society involved in locational decisions, health care consumers, facility operators (management, practitioners, and staff), and the community (residents and businesses). Each sector is interested in different aspects of the location decision. Consumers would like to have good quality care, short travel times, and low fees; operators desire accessibility to support services, operational freedom, and a guarantee to financial success; and the community is interested in the economic, environmental, and socio-political impacts on the area. It is impossible to satisfy all these wishes; what is done in practice is to select a few, often surrogate, variables and try to model them.

Location/allocation modeling is used to locate a set of facilities and allocate groups of people to the facilities, so that selected criteria will be optimized. Model populations are taken as fixed at a particular time, but adjustments can be made if populations shift. Three factors may vary: the number of facilities, their capacity or size, and their location. Solutions depend on whether facilities are constrained to certain places or can be located anywhere and whether populations in certain spatial units will be assigned to a single facility or can be split up among several facilities. Most solutions must be worked out using computer programs. For simple situations there may be unique mathematical solutions, but there are no general mathematical solutions to more complex situations. Vignette 10-1 shows how to locate a single facility to minimize travel distance from five populations.

The techniques of location/allocation analysis are useful in a variety of settings. Probably the most well-known example comes from Sweden. The Swedish government asked geographer Godlund to locate two new specialist facilities (Godlund, 1961). The location of six centers was already established; Godlund's problem was to decide in which of five possible places the two new hospitals should go. Godlund used aggregate travel time as his criterion. He mapped isochrons (lines of equal travel time) around the six fixed centers and 10 trial combinations of the other centers. By calculating total travel time from his maps, he selected the two best locations.

Much of the seminal work in the area of hospital location and patient allocation models comes out of the Chicago Regional Hospital Study of the mid-1960s. Morrill and Earickson (1969) used data collected during this study to build a *revealed demand* model, which assigned patients to hospitals in a manner that minimized travel distance. Their model considered the characteristics and spatial distribution of patients and hospitals, hospital hierarchies based on their scale of services, and the role of physicians in the patient–hospital relationship. Pyle (1971) extended this model by bringing in the patient morbidity factor, focusing on need for care. He collected data for 1960–1967 on heart disease, stroke, and cancer (which account for at least 75% of urban deaths) and mapped out disease patterns for Chicago. Then he found associations between disease rates and demographic variables and used these to forecast 1980 disease patterns. After analyzing facilities currently available for treating these chronic diseases, he developed a model to allocate hospitals on the basis of forecast morbidity.

Location/allocation models attempt to remedy the usual lack of consideration for location in planning decisions. For example, planners in Third World countries often do not consider locations based on accessibility and cost efficiency but respond instead to requests from individual villages. Modeling can lead to economic efficiency and, potentially, overcome resource imbalances. The models' best use is as an aid to planning, and they should not be applied too rigidly. They can demonstrate ideal location decisions, thereby acting as a screening device for various planning options. The study by Morrill and Earickson of the locational efficiency of Chicago hospitals illustrates this last point. They found that relocating hospital beds could substantially decrease patient travel but pointed out that the same object could be achieved by relaxing existing constraints of income and race.

One difficulty with location/allocation models is deciding which criteria are best. Should efficiency be optimized in terms of mean travel distance, or equity in terms of standard deviations of travel distances? Does one concentrate on consumer, operator, or community? What compromises can be made among all these considerations? Vignette 10-2 discusses one compromise strategy. In addition to these questions, it is not clear that maximal accessibility necessarily has a positive impact on health. People closer to facilities may overuse them, and health care services may create ill health.

Another disadvantage of these models is that they usually leave out nonspatial aspects of health care delivery, in particular, population characteristics and health status. Location/allocation models generally assume that the populations assigned to facilities have unvarying needs, demands, and degrees of mobility. Although the models are often complicated enough without trying to incorporate these other concerns, it is imperative that the effort be made.

The limits of location/allocation modeling have been demonstrated by research byMohan (1983) in northeastern England. A conflict arose between two levels of the health care planning hierarchy over the possible construction of a new hospital. Durham Area Health Authority (AHA) wanted a new facility constructed in the eastern part of their jurisdiction, but the plan was blocked by the higher level Northern Regional Health Authority (RHA) on grounds of inefficiency. Durham AHA asked Mohan to study the situation, and, based on aggregate travel distance, he recommended constructing a new hospital in Peterlee. This was what Durham AHA wanted to hear. The final decision favored Northern RHA, but Peterlee did receive a community hospital. Mohan observed that the final solution was not decided on technical grounds and that academics need to become wise in the ways of planners and politicians.

Some attempts have been made to locate facilities within a wider social, cultural, and political context. A good example is the use Bosanac and Hall (1981) made of small-area data to help select feasible sites for primary care clinics in West Virginia. They produced detailed tables and maps of population characteristics and medical data for a variety of spatial units (statewide traffic assignment model zones, minor civil divisions, counties, and the entire state) and used these as a planning guide. Geographic information systems (GIS) are important in linking data to geographic locations, conducting spatial analyses, and producing maps.

Another approach to locating facilities, suggested by Pyle (1974), is based on mapping potentials (see Vignette 8-2). If several clinics were to be located in a health service area and one wanted to ensure their fullest use, an *income potential* map could be developed, and the clinics could be located where income was highest. Alternatively, clinics could be located where *disease potential* was greatest.

Regionalization

Health care regionalization is used to divide up an area in such a way that there will be adequate services within the area and some degree of equality of service among subareas. Regionalization schemes may recognize that health and social conditions vary among subareas and therefore attempt to satisfy each subarea's needs. Some schemes try to establish a hierarchy of services within each subarea. Regionalization might appear to be an easy task; however, at least three basic decisions have to be made. The geographic regions may follow the convenient boundaries of administrative units or may be delineated by function. The second decision, which overlaps with the first, is whether to base regions on existing patterns of patient-to-resource flows or to somehow "rationalize" or alter flows to conform to some scheme based on efficiency or equality. Rationalization is based on one or more of the following: patient flows, facility location, or personnel location. Choosing rationalization leads to a further decision, whether to decentralize or concentrate services. If the aim is to increase resource accessibility, one might decentralize. If cost efficiency is important, or if the services are of relatively high functional order, resources might be concentrated.

Regionalization decisions and the emphasis placed on regionalization depend on the type of medical system a country has. Welfare state and socialist state systems use regionalization as part of their overall approach. All spatial-functional schemes, which follow central place theory, are rational regionalizations. They have not always been successful.

In the United States there has been less concern about regionalization than in other developed countries. The free-enterprise system often leads to competition among resources and regional imbalances that are very difficult to rationalize. However, at least 10 attempts have been made to regionalize health care delivery in the United States. Several of these schemes resulted from the legislative acts that were discussed in Chapter 9. Some of the legislation was explicit about the types of units to be used in health care delivery plans, while other programs were vague about boundaries or left such decisions up to the states. The resulting regionalizations tended to be nodal, functional, and based on administrative regions. The regions were made up of two types of units: central or core units with relatively large populations and health care resources, and hinterland or peripheral units with less population and resources.

Regionalization can be attempted for many different kinds of health care resources. The United States and other countries have given much attention to delineating hospital service areas. Vignette 10-3 shows one way to determine a hospital's service area. Different strategies may have to be applied in rural areas where single hospitals are fairly far apart and in urban areas where there is substantial overlap among hospital catchment areas.

Burn centers are medical services of very high order that have been regionalized in the United States. Praiss, Feller, and James (1980) recommended the following criteria for these centers: (1) only 10% of all acute general hospitals should be directly involved, (2) there should be a hierarchy of three levels of care based on burn severity and intensity of care required, and (3) burn care should be organized within a comprehensive regionalized emergency medical service. The last recommendation illustrates a concept that is an essential part of regional schemes: integration of various institutions to satisfy the particular needs and structures of each region.

Emergency medical services (EMS) were created in developed countries in recent decades in response to a demand for on-site treatment where a serious problem occurs. They are good examples of hierarchical regionalization. The Medic-I program run by the Seattle Fire Department has received international recognition for its effectiveness (Mayer, 1979). The system is three-tiered. At the highest level are four Medic-I units staffed by paramedics who have received 1,000 hours of intensive theoretical and clinical training. They can carry out procedures such as cardiac defibrillation and administering intravenous medication. Unit personnel are in constant radio communication with emergency and specialist physicians. At the second level are nine aid units staffed by emergency medical technicians (EMTs) who have had 81 hours of training. Fire company personnel administer aid at the third level of the system.

As with location/allocation modeling, the best-laid regional plans often run into practical difficulties. Kunitz and Sorensen (1979) provided an example of this in their report on the effects of regional planning on a rural hospital in upstate New York. They traced the history of the attempt to build a hospital near Syracuse through changes in regionalization legislation and implementation of this legislation. Several different types of conflict arose over a period of many years among government administrative agencies, regions, local and area-wide concerns, social classes, and different political groups. Government regulation, which was intended by the dominant local industrial interests to contain costs, resulted in competition for the limited number of beds mandated by the regulatory agencies. The consumers involved in the struggle turned out to be a very small group of prominent people in the area.

CONCLUSION

Analysis of spatial inequalities in health care resources requires sophisticated mapping techniques and consideration of equality criteria at different scales. The concept of central place theory has shed some light on the organization of facilities in terms of size and specialty hierarchies. Both micro- and macrolevel studies have investigated the reasons why medical personnel and facilities locate where they do. Besides these assessments of existing health care resource distributions, researchers have contributed important insights and practical guidelines to the impovement of health care delivery. One way is location/ allocation modeling and another is regionalization. Both of these methods include theoretical models and practical applications.

REFERENCES

Abler, R. Adams, J. S., & Gould, P. (1971)., *Spatial organization: The geographer's view of the world.* Englewood Cliffs, NJ: Prentice-Hall.

American Medical Association. (1983). *Physician characteristics and distribution in the U. S.* (1982 ed.). Chicago, IL: Author.

Bosanac, E. M., & Hall, D. S. (1981). A small area profile system: Its use in primary care resource development. *Social Science and Medicine, 15D,* 313–319.

Calvo, A. B., & Marks, D. H. (1973). Location of health care facilities: An analytical approach. *Socio-economic Planning Sciences, 7,* 407–422.

Christaller, W. (1933). Die zentralen orte in Süddeutschland. Jena: Gustav Fischer Verlag.

Elesh, D., & Schollaert, P. T. (1972). Race and urban medicine: Factors affecting the distribution of physicians in Chicago. *Journal of Health and Social Behavior, 13,* 236–250.

Eyles, J., & Woods, K. J. (1983). *The social geography of medicine and health.* London: Croom Helm.

Florin, J. W. (1980). Health service regionalization in the United States. In M. S. Meade (Ed.), *Conceptual and methodological issues in medical geography* (pp. 282–298). Chapel Hill, NC: University of North Carolina, Department of Geography.

Fox, R. T., & Fox, D. H. (1974). The use of central place theory for the location of maternal and infant care projects. *American Journal of Public Health, 64,* 898–903.

Gober, P., & Gordon, R. J. (1980). Intraurban physician location: A case study of Phoenix. *Social Science and Medicine, 14D,* 407–417.

Godlund, S. (1961). *Population, regional hospitals, transportation facilities, and regions: Planning the location of regional hospitals in Sweden.* Lund Studies in Geography Series B: Human Geography No. 21. Lund, Sweden: Department of Geography, Royal University of Lund.

Graduate Medical Education National Advisory Committee: Summary Report (Vol. 1). (1980). U. S. Department of Health and Human Services Publication No. (HRA) 81–651. Washington, DC: U. S. Government Printing Office.

Hart, J. T. (1971) The inverse care law. *Lancet, 1,* 405–412.

Hemenway, D. (1982). *The optimal location of doctors. New England Journal of Medicine, 306,* 397–401.

Joseph, A. E., & Phillips, D. R. (1984). *Accessibility and utilization: Geographical persepectives on health care delivery.* New York: Harper & Row.

Kunitz, S. J., & Sorensen, A. A. (1979). The effects of regional planning on a rural hospital: A case study. *Social Science and Medicine, 13D,* 1–11.

Mayer, J. D. (1979). Seattle's paramedic program: Geographical distribution, response times, and mortality. *Social Science and Medicine. 13D,* 45–51.

Mayer, J. D. (1981). Urban health care planning in a spatial context: A critical analysis. In W. F. J. Lierop & P. Nijkamp (Eds.), *Locational developments and urban planning* (pp. 323–343; NATO Advanced Behavioral and Social Sciences Series, No. 5). Sitjhoof and Noordhoff: Kluwer Academic.

McGlashan, N. D. (1972). The distribution of population and medical facilities in Malawi. In N. D. McGlashan (Ed.), *Medical geography: Techniques and field studies* (pp. 89–95). London: Methuen.

Meade, J. M. (1971). *Hospital service areas in Idaho: A geographic approach to spatial efficiency.* Unpublished doctoral dissertation, University of North Carolina, Chapel Hill, NC.

Mohan, J. (1983). Location-allocation models, social science and health service planning: An example from northeast England. *Social Science and Medicine, 17,* 493–499.

Morrill, R. L., & Earickson, R. (1968). Variation in the character and use of Chicago area hospitals. *Health Services Research, 3,* 224–238.

Morrill, R. L., & Earickson, R. (1969). Locational efficiency of Chicago hospitals: An experimental model. *Health Research, 4,* 128–141.

Navarro. V. (1974). The underdevelopment of health or the health of underdevelopment: An analysis of the distribution of human health resources in Latin America. *International Journal of Health Services, 4,* 5–27.

Okafor, S. I. (1982). *The case of medical facilities in Nigeria. Social Science and Medicine, 16,* 1971–1977.

Praiss, I. L., Feller, I., & James, M. H. (1980). The planning and organization of a regionalized burn care system. *Medical Care, 18,* 202–210.

Pyle, G. F. (1971). *Heart disease, cancer, and stroke in Chicago: A geographical analysis with facilities, plans for 1980* (Research Paper No. 134). Chicago: University of Chicago, Department of Geography.

Pyle. G. F. (1974).The geography of health care. In J. M. Hunter (Ed.), *The geography of health and disease* (pp. 154–184). Chapel Hill, NC: University of North Carolina, Department of Geography.

Ramesh, A., & Hyma, B. (1981). Traditional Indian medicine in practice in an Indian metropolitan city. *Social Science and Medicine, 15D* 69–81.

Rimlinger, G. V., & Steele, H. B. (1963). An economic interpretation of the spatial distribution of physicians in the United States. *Southern Economic Journal, 30,* 1–12.

Rushton, G. (1975). *Planning primary health services for rural Iowa: An interim report* (Technical Report No. 39). Iowa City, IA: University of Iowa, Center for Locational Analysis, Institute of Urban and Regional Research.

Schwartz, W. B., Newhouse, J. P., Bennett, B. W., & Williams, A. P. (1980). The changing geographic distribution of board-certified physicians. *New England Journal of Medicine, 303,* 1032–1038.

Shannon, G. W., & Dever, G. E. A. (1974). *Health care delivery: Spatial perspectives.* New York: McGraw-Hill.

Shonick, W. (1976). *Elements of planning for area-wide personal health services.* St. Louis: C. V. Mosby.

Techatraisak, B. (1985). *Traditional medical practitioners in Bangkok: A geographic analysis.* Unpublished doctoral dissertation, University of North Carolina, Chapel Hill, NC.

U. S. Census Bureau. (1984). *Current population reports* (Series P-52, No. 944). Washington, DC: U. S. Government Printing Office.

Wennberg, J., & Gittelsohn, A. (1982). Variations in medical care among small areas. *Scientific American, 245,* 120–133.

Further Reading

Bashshur, R. L., Shannon, G. W., & Metzner, C. A. (1970). The application of three-dimensional analogue models to the distribution of medical care facilities. *Medical Care, 8,* 395–407.

Dear, M., Taylor, S. M., & Hall, G. B. (1980). External effects of mental health facilities. *Annals of the Association of American Geographers, 70,* 342–352.

Gesler, W. M. (1984). *Health care in developing countries*. Resource Publications In Geography. Washington, DC: Association of American Geographers.

Illich, I. (1976). *Medical nemesis: The expropriation of health*. New York: Pantheon.

Joseph, A. E., & Phillips, D. R. (1984). *Accessibility and utilization: Geographical perspectives on health care delivery*. New York: Harper & Row.

Lankford, P. M. (1974). Physician location factors and public policy. *Economic Geography, 50*, 244–255.

Meade, M. S. (Ed.). (1980). *Conceptual and methodological issues in medical geography*. Chapel Hill, NC: University of North Carolina, Department of Geography.

Monroe, C. B. (1980). A simulation model for planning emergency response systems. *Social Science and Medicine, 14D*, 71–77.

Pyle, G. F. (1979). *Applied medical geography*. Washington, DC: V. H. Winston.

Scott, A. J. (1970). Location-allocation systems: A review. *Geographical Analysis, 2*, 95–119.

Shannon, G. W. (1977). Space, time and illness behavior. *Social Science and Medicine, 11*, 683–689.

Smith, C. J. (1983). Locating alcoholism treatment facilities. Economic Geography, *59*, 368–385.

Vignette 10-1

LOCATING A SINGLE HEALTH CARE FACILITY

How do we locate a hospital among five populations so that the criterion of minimum travel distance will be met? The areas occupied by the five populations are shown in Vignette Figure 10-1.

To begin with, we make a simplifying assumption that each population can be represented by a single point, such as the population centroid or the main town in each of the five areas. Laying a coordinate system over the map, the five populations are represented by the coordinates (1,7), (2,5), (3,7), (5,5), and (6,2).

Locating the facility follows a step-by-step process called an algorithm: we make an educated guess and then try to improve upon it. The final result should be the location of a point whose total distance from all five population points is as small as possible. This is a unique point, but there is no single-step procedure to find it.

The first step is the educated guess; we guess that the point we require is the mean center of the five population points. We call this the first iteration

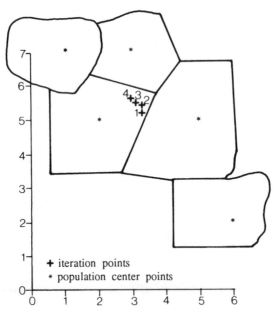

Vignette Figure 10-1. Locating a single facility to minimize total travel distance for five populations.

point, the first approximation to the final point. This is analogous to the mean of a set of data, like the mean age of a population, but here we are dealing with two dimensions. The mean center of a set of points is given by (x_c, y_c), where x_c is the average of the x-coordinates of the points and y_c is the average of the y-coordinates: $x_c = \Sigma x_i/n$ and $y_c = \Sigma y_i/n$, where n is the number of points. In our example, $(x_c, y_c) = (3.40, 5.20)$. The mean center does not satisfy the minimum distance requirement. What is sought is analogous to the median of a set of data, and since the mean and median are often fairly close, the mean is a good first try.

The second step of the algorithm is to find the distance, $d(p_i, p_c)$ of each population point from the mean center, or first iteration point. This requires the use of the two-dimensional distance formula, based on the Pythagorean theorem:

$$d(p_i, p_c) = \sqrt{(x_i - x_c)^2 + (y_i - y_c)^2}.$$

Distances for all five population points are shown in Vignette Table 10-1.

Step 3 consists of calculating three ratios for each population point: $1/d(p_i, p_c)$, $x_i/d(p_i, p_c)$, and $y_i/d(p_i, p_c)$. These are also shown in the table.

In step 4 we calculate a second iteration point that will be a little closer to the final point desired. The coordinates of this point are given by

$$x_c = \frac{\Sigma \dfrac{x_i}{d(p_i, p_c)}}{\Sigma \dfrac{1}{d(p_i, p_c)}}$$

and

$$y_c = \frac{\Sigma \dfrac{y_i}{d(p_i, p_c)}}{\Sigma \dfrac{1}{d(p_i, p_c)}}.$$

The second iteration point in our example is $(3.24, 5.42)$.

Vignette Table 10-1. Locating a Single Facility

Coordinates		$d(p_i, p_c)$	$\dfrac{1}{d(p_i, p_c)}$	$\dfrac{x_i}{d(p_i, p_c)}$	$\dfrac{y_i}{d(p_i, p_c)}$
x	y				
1	7	3.00	0.33	0.33	2.33
2	5	1.41	0.71	1.41	3.54
3	7	1.84	0.54	1.63	3.80
5	5	1.61	0.62	3.10	3.10
6	2	4.12	0.24	1.46	0.48
Totals 17	26	11.98	2.44	7.93	13.25

Algorithm steps 2, 3, and 4 are repeated, using the second iteration point, to find a third iteration point, and so on. The third and fourth iteration points for this example would be (3.12, 5.51) and (3.04, 5.56). All four iteration points are plotted on the map. They tend to drift to the northwest. Also, the distance between each new point and the preceding one tends to decrease.

We do not know when the final location is reached, but we select an arbitrarily small number, say 0.01, and stop the algorithm when the distance between two successive iteration points is less than 0.01. This example, programmed on a microcomputer, stopped at the eighth iteration point, using the 0.01 criterion.

Vignette 10-2

STRATEGY COMPROMISES IN LOCATION/ALLOCATION MODELING

Suppose we wish to locate a series of health care facilities in a state. One approach would be to site the facilities in a regular spatial pattern. They might be located in towns that were spread approximately 30 miles apart. If these equal areas had varying population density, the facilities would serve populations of different sizes. Their capacity or the types of services they provided could be varied accordingly. This is strategy A in Vignette Figure 10-2. It emphasizes an important geographic concern, accessibility, as it tries to insure that everyone is within a certain distance from a facility.

An alternative strategy would be facilities of identical capacity, which performed the same levels of service or medical specialties. All facilities would provide surgery, outpatient care, and emergency services. This plan might lower the costs of constructing facilities, by building them all on a similar scale. To serve the state population most efficiently, more facilities would be located in more densely populated areas, resulting in uneven spacing of uniformly sized facilities. Vignette Figure 10-2 shows this as strategy B.

Both strategies cannot be satisfied at the same time; strategy C represents a series of compromises between A and B. Interested parties would have to come to an agreement about the final compromise strategy to use. The trade-offs are geographic accessibility against economic efficiency. The geographer can offer rational alternatives but the final decisions are up to health care planners and consumers.

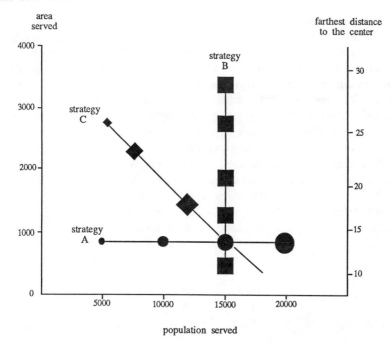

Vignette Figure 10-2. Strategy compromises in location/allocation modeling. Three strategies for locating health care facilities are based on accessibility, capacity, and a compromise solution. From *Planning Primary Health Services for Rural Iowa: An Interim Report* (p. 24) by G. Rushton, 1975, Iowa City, IA: University of Iowa, Center for Locational Analysis, Institute of Urban and Regional Research. Copyright 1975 by Gerard Rushton. Reprinted by permission.

Vignette 10-3

ESTIMATING A HOSPITAL'S SERVICE AREA: THE MARKET PENETRATION MODEL

There are two ways to find out what territory or catchment area a hospital serves and where most of the hospital's patients live. A survey in the community around the hospital would show how many people attended the hospital from certain areas but would be costly and time-consuming. An alternative is to use patient records at the hospital and apply the market penetration model.

To use this model, one needs to know the patient population in spatial units surrounding the hospital for a time period, such as a year, and the total population in these same units. If the patient population is restricted, perhaps to maternity cases only, then the total population should be reduced to include only potential patients or people at risk. For each spatial unit calculate the ratio $m_i = p_i/c_i$, where m_i is the percentage of patients in each unit, p_i is the patients attending the hospital from unit i, and c_i is the population of each unit.

Vignette Figure 10-3 shows a situation in which 12 census tracts are served by a hospital. The percentage of tract populations (m_i) who attended the hospital in the last year are displayed on the map. A choropleth map could be constructed using four data classes, 0.0–9.9%, 10.0–19.9%, 20.0–29.9%, and 30.0% and above. For the isopleth map in Vignette Figure 10-3, each m_i value was assigned to the centroid of its census tract and isolines were drawn for 10%, 20%, and 30% usage by the population.

The general pattern displays at least three typical facility market penetration qualities. Patient use of or attendance at the hospital tends to be greater, closer to the hospital. In other words, the distance decay phenomenon is at work here (see Vignette 6-1). Also, distance decay is not uniform in all directions. Various factors, such as illness patterns, income, and transport availability, distort a perfectly uniform decline in attendance. Thirdly, in the northwest corner is an "island" of fairly high use which may be caused by the factors above.

What are the boundaries of the service area? Some people might travel hundreds of miles to attend the hospital. In practice we would set a limit to the percentage that would include a particular unit in the effective service area; for

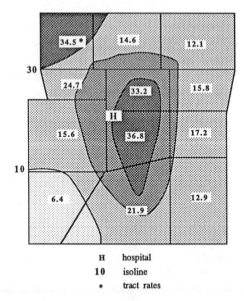

H   hospital
10   isoline
•   tract rates

Vignette Figure 10-3. Estimating a hospital's service area: the market penetration model. The isopleth map depicts different ratios of hospital visits per population in a hospital service area.

example, at least 5% of a unit's population had to attend the hospital in the past year.

If an area has several hospitals, market penetration maps for each of them would probably show, especially in urban areas, that in some places there is overlap. This might indicate wasteful service redundancies. Some areas might appear to be outside the effective penetration of all the hospitals. Future plans could attempt to provide services to these underserved places. Market penetration models are also useful for examining the service areas of specific items of equipment such as a CAT scan or a premature-baby support unit.

11

Accessibility and Utilization

The most important link or interaction in any health care delivery system is that between consumer and provider. An optimal distribution of health care resources is possible only if this relationship is understood. Many factors are involved in the relationship, most of them nonspatial.

Two aspects of the consumer–provider interaction have been intensively examined, accessibility and utilization. Accessibility can be thought of as a *potential* link. Health care planners attempt to provide accessibility by placing patient and practitioner in the best possible position in relationship to each other. Of course this position may not lead to utilization of resources, which can be thought of as an *actual* link. Both accessibility and utilization are influenced by the same factors, and their patterns are often quite similar.

The following discussion assesses the part distance plays in accessibility and utilization. A variety of nonspatial factors are discussed. The interaction of distance and other factors is examined. Utilization processes and models are discussed. At the close of the chapter two current topics in medical geography—changes in medical systems and integration of traditional and modern medicine—are briefly described.

FACTORS IN PATIENT-PRACTITIONER CONTACTS

The Role of Distance

Studies have often determined that physical proximity is an important factor in accessibility and utilization of health care resources. Closeness to a particular doctor or facility is one of the main reasons for using that resource. The importance of distance seems obvious, but, unfortunately, it has often been overlooked in planning decisions. Vignette 11-1 illustrates what happens to the distance people have to travel to hospitals when the hospital nearest them closes.

The provider–consumer link weakens as distance increases, following a distance decay curve (see Vignette 6-1). Distance decay derives from the gravity model, which states that the attractional force between two objects is directly proportional to their masses and inversely proportional to the square of the distance between them. When distance decay is applied to health care resources, it usually measures the interactions between a single facility or practitioner and people at varying distances from this resource.

Distance decay is often studied in connection with the use of mental health facilities. In the middle of the 19th century, Edward Jarvis, who was studying the occurrence of lunacy in the United States, stated his famous law: "The people in the vicinity of lunatic hospitals send more patients to them than those at a greater distance" (Jarvis, 1851/1852, p. 344). Several studies have confirmed the law but have found that the distance patients travel to mental institutions is affected by other important factors, such as type of mental problem.

The actual form that the distance decay function should take has been widely debated. There are several possibilities. One of the simplest equations that has been applied is $f = k/d^b$, where f is the frequency of patient-resource contact, d is distance, and k and b are parameters that must be determined for any particular situation. There is no formula that fits all interactions, which means that the best type of equation can only be determined empirically.

Of particular interest is the "friction of distance," or how rapidly interaction decreases as distance increases. If the equation above were used, friction could be measured by the size of the distance exponent, b. If b is relatively large, then there is a rapid drop-off of contact. Figure 11-1 shows the friction of distance for the use of two types of facilities in Harlem Hospital, New York City: clinics and the emergency room. The slope of the distance decay lines represents the b parameter. For clinic use the slope is -7.1 and for emergency room use it is -3.3, indicating that people are willing to travel farther to use the emergency room than to use clinics. The friction of distance is greater for clinic use.

Distance decay, or friction of distance, is useful in determining central place hierarchies and functional regionalization (Chapter 10). If the friction of distance is high for a certain level of health care service, then this service should be decentralized and locally accessible. This usually applies to low-order services such as first aid. High-order services, like the treatment of rare diseases or heart transplants, are not as sensitive to distance. People are willing to travel farther for these services, so resources can be centralized. Friction of distance can establish threshold distances for levels of service. For example, a distance decay curve shows that people in low-income countries will normally walk up to 3 kilometers (1.86 miles) to a primary health care clinic. A low-level clinic beyond this threshold has limited usefulness.

There are many ways to measure distance. *Map distance* from a patient's residence to a health care resource is commonly used because it is relatively easy to determine. Other distance measures may be more meaningful in certain

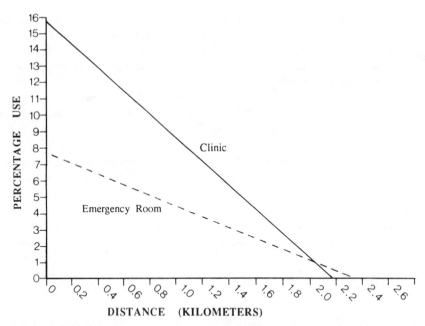

Figure 11-1. Distance decay curves. Harlem Hospital clinic and emergency room use shown as distance decay lines. From 1967–1970 one in 25 households was surveyed about their hospital use during the past year. From "Studying Spatial Patterns of Illness and Hospital Use in a Central Harlem Health District" by W. M. Gesler and J. Cromartie, 1985. *Journal of Geography, 84*, p. 215. Copyright 1985 by National Council for Geographic Education. Reprinted by permission.

situations, however. *Road distance* takes into account the actual or supposed route taken from home to practitioner. This measure can be weighted by road quality. In societies where time is often more important than distance, the time it takes to reach a facility, *time distance*, may be the best measure. The degree of patient *mobility*, which involves the type of transportation available, is also implicated in distance measures. Harder to determine accurately, because it is subjective, is *perceived distance*, a patient's idea of how far away health care resources are. Some important distance measures that do not involve distance in the geographic sense are *socio-cultural distances*. *Social distance* is the gap between consumer and provider in terms such as social status or illness beliefs. *Economic distance* is the ability to pay for services. These socio-cultural types of distance measures can be used in accessibility and utilization studies through such techniques as multidimensional scaling.

Distance is not the overriding factor in all patient–resource situations, and sometimes it is of very little importance. Where there are few health resources, as in much of rural North America, people might travel a great distance for treatment. In some developing areas facilities located within a few miles of most of the population may be rarely used because their quality is so poor. Geographers should be aware of the danger of a narrow focus on distance; it can mask other factors. Most social scientists are aware of this and consider distance in relation to other variables.

Nonspatial Factors

In Chapter 9 we made the point that analysis of health care delivery systems requires a multidisciplinary approach. Cultural, social, economic, and political constraints on health care–seeking behavior require a similar approach to understanding client–provider interactions.

Some population characteristics that are important to accessibility and utilization are age, sex, social class, and ethnicity. Older people, for example, typically use health care resources more often than other age groups. Several studies have shown that women in the United States use facilities such as hospitals more than men do. People having lower social status or belonging to minority groups often have relatively less accessibility to health care. What people believe about the causes of illness and the appropriate treatment of illness also affects the kinds of help they will seek. The type and severity of illness are considerations, too. Various ethnic groups may use different criteria to legitimize sick-role behavior.

Economic constraints on health care access may display unexpected complexity. Pyle and Lauer (1975) investigated disease rates and hospital service use in Akron and Summit County, Ohio. They found a consistent progression from high to low mortality, from heart disease, strokes, and cancer, that was associated with low to high income groups. However, a consistent relationship was not found between income and use of hospitals. The poorest people, who received Medicaid payments, demanded services at higher levels than the marginally poor, who could not claim these benefits.

Provision of care involves the level and quality of care, staff attitudes toward patients, and the possible religious affiliation of hospitals. In addition health-seeking behavior has a political and cultural context. The system of government and the actions of power-wielding groups are important. For example, most people living in a welfare state will have access to a minimum level of health care but little access to top-quality care. Therapy managing groups, mentioned in Chapter 9, exemplify the power-wielding group.

Distance as It Interacts with Other Factors

Distance can be combined with nonspatial variables to aid in understanding patient–practitioner contacts. Distance is distorted by political, cultural, and economic considerations; it may be a surrogate for other variables or a mask for the importance of other variables. Although people of higher income live closer to a hospital than people of lower income, a higher rate of hospital use by the higher income people may be a function of ability to pay rather than distance. The most likely case is that distance and income are both important; the problem is to assess how these two variables interact.

Most research has looked at distance and at least two other factors. Weiss and Greenlick (1970) investigated the behavior of members of a prepaid group practice, the Kaiser Foundation Health Plan, in the Portland, Oregon, standard metropolitan statistical area. All the patients they studied were treated in a uniform way at three clinics and a hospital. Subjects were classified as either working or middle class, based on the occupation of the household head. Distance was measured along the best route from home to facility. People approached the medical care system in four ways, by telephone, previously scheduled appointment, walk-in, and emergency room use. Use of these four approaches varied by distance and social class. For example, scheduled appointments by middle-class patients dropped substantially at distances over 15 miles from a facility. Overall, social class was more important than distance as an influence on the likelihood of contact.

Travel distance was analyzed by Morrill, Earickson, & Rees et al. (1970) in their study of hospital use in Chicago. One important factor that is often overlooked is that physicians, and not patients, usually choose the hospital a patient will use. Patients often travel beyond their closest facility because their doctor is affiliated with, or closer to, a different hospital. Doctors' offices were closer to hospitals than patients' homes were to either doctors or hospitals. A patient might go to a more distant facility for one of two sociocultural reasons. They traveled farther because of attractive features of certain hospitals; for example, the religious affiliation of the hospital was especially important to Jews and Catholics. Some people were denied access to certain hospitals because of admission or referral practices; this was most often true for blacks and lower income people.

Other Geographic Considerations

Residential location affects aspects of health care delivery other than overall utilization rates. Lasker (1981) found that rural versus urban residence could be an important factor in illness behavior among subjects in the Ivory Coast. When both traditional and biomedical healers were available, villagers used native healers first, and urban people used biomedical practitioners first.

People living in certain sections of urban areas have more difficulty than others in reaching health care. Students of the social geography of the city have identified areas of deprivation where the residents do not receive their fair share of services, including health care. Residential relocation affects the relative location of people and practitioners. In his study of patient behavior in West Glamorgan, Wales, Phillips (1979) discovered that some people traveled quite far to GPs. This was due to historical inertia. These people went to certain GPs before moving away and liked their GPs well enough to make the long trip back to consult them.

Another geographic consideration is the link between disease and its treatment. Girt (1973) asked a sample of adults in rural Newfoundland several questions about their attitudes toward disease and health care. He found that people who lived farther away from health care resources were more aware of the development of disease but less likely to consult a physician for treatment. The balance point between awareness and consultation varied by type of disease.

Utilization Models

Faced with the wide variety of spatial and nonspatial variables that are associated with patient–practitioner contacts to varying degrees in various situations, how can one make any useful generalizations? Models of health care utilization behavior assess the relative importance of different factors and try to arrange these factors in ways that demonstrate causal relationships.

Different disciplinary perspectives have different approaches to utilization. The *sociocultural* approach emphasizes factors such as family structure, religion, economic status, health beliefs, and friendship networks. *Sociodemographic* studies deal with population characteristics like age, sex, education, occupation, ethnicity, and health status. Knowledge, beliefs, and attitudes about disease and health care are the focus of the *social–psychological* approach. Those who take the *organizational* approach believe that utilization is mainly determined by the structure of the health services system; they look at things such as government policies and payment plans. The *social systems* approach (and the approach of this text) attempts to fuse the other approaches by considering health care as a system with various components and interrelationships

A second type of utilization model might focus on important variables. One could include *enabling* factors like income, health insurance status, and education of the head of household. Factors *predisposing* to utilization include attitudes toward health care, having a regular source of care, and knowledge of treatment possibilities. Some *accessibility* factors are distance of people from services, waiting time at facilities, and availability of services. Still other factors have to do with *perceived health level* and can be measured as sick days, disability days, or restricted activity days. Finally, there are individual and area-wide *exogenous* variables like age, sex, family size, and location.

CURRENT TOPICS

Health care delivery is a rapidly developing area of research for medical geographers. Two areas where investigation has begun, but which would benefit from further study are the process of change in a medical system and the integration of traditional health care and biomedicine.

Change in Medical Systems

Disease and health care delivery are embedded in the social fabric of a group of people; therefore a cultural system will elicit change in its medical systems. Because disease patterns change, technologies are improved, and perceptions change, health care systems are bound to change as well. Examples of this are fluctuations in public and private support of health care in Australia, the changing roles of groups in Zaire's medical system, and changes in accessibility to doctors in urban areas of the United States (see Chapter 9).

The relationships among biomedicine and other health systems in the United States have shifted over the last century. Medical pluralism in the United States is represented by the primary, or establishment, biomedical system, plus alternatives such as holistic health care, lay midwifery, chiropractic, Christian Science, homeopathy, root doctors, and *espiritismos.* These alternatives are based upon paradigms that differ from those of the dominant medical system within a society. Some of these alternatives cross ethnic and socioeconomic boundaries, while others are confined to certain groups and localities.

The relative importance of the biomedical and alternative subsystems in the United States has shifted over space and time in response to cultural change. The temporal shift can be partially understood in terms of the model in Figure 11-2, which shows the progress of biomedicine since about 1850 to the present. The three stages of the model correspond to the phases of a dialectical process: thesis, antithesis, and synthesis. At the beginning of Stage 1 humoral ideas dominated the practice of medicine. Following discoveries in germ theory toward the end of the 19th century, modern biomedical practice began to emerge and gain ground. At the same time, some traditional practices, like lay midwifery, declined in importance. In Stage 2 some successful innovations (for example, certain surgical procedures such as anesthesia, skin grafting, and artery clamps) proliferated, biomedicine expanded rapidly, and doctors began to enjoy a high status socially and financially. However, around the middle of this century the establishment system began to be criticized for neglecting the whole person; for its overemphasis on reductionist, biologistic thinking; for a slavish devotion to high-technology cures; and for accelerating costs. Some people began to turn to other systems of health care. During this stage, for example, holistic health care (HHC) started to be adopted by significant

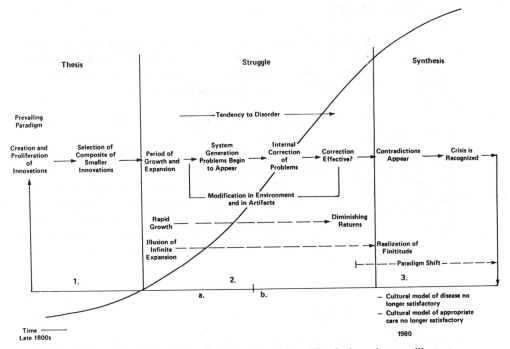

Figure 11-2. A framework for medical care transition. The S-shaped curve illustrates change in medical paradigms. From *Holistic Health: A Changing Paradigm for Cultural Geography* (p. 135) by D. E. Stribling, 1983, an unpublished doctoral dissertation, University of North Carolina at Chapel Hill, Chapel Hill, NC. Reprinted by permission.

numbers of people. Stage 2 witnessed clashes among medical subsystems for control of resources and patients. Attacks by the American Medical Association (AMA) on chiropractors is one example of this struggle. In Stage 3, treatment innovations began to gain ground and old ideas were revived; each of the new subsystems began its own evolutionary cycle. The possibility for new syntheses arose at this stage. Thus nurse midwives have emerged as a synthesis of lay midwifery and biomedical nursing, and many doctors have begun to practice in holistic settings.

The temporal shifts in health care subsystems have their spatial parallels. The modern system of biomedicine did not diffuse uniformly over space. There were areas of both acceptance and rejection, areas where it was more or less profitable to practice biomedicine. The beginnings of HHC were in Virginia and Florida in the 19th century. In the 20th century the system spread slowly at first. By 1950 it had reached New York, Maine, Wisconsin, Illinois, North Carolina, and Louisiana. In the 1950s organized groups appeared in California, Oregon, Missouri, and Pennsylvania; in the 1960s Kansas, Colorado, Arkansas, and Maryland became involved; and in the 1970s there was a proliferation of HHC establishments throughout the country. The diffusion of HHC has hierarchical aspects; it tends to begin in larger cities first and spreads to smaller places. There is also a distinct element of resurfacing or readoption of older folk methodologies in some relatively remote areas of the United States.

Chiropractic, another important alternative to biomedicine, began around the turn of the century in the Midwest. Records from 1965 show that chiropractor-to-population ratios were highest in the North Central, northern Mountain, and West Coast states. Between 1965 and 1978 chiropractic made some gains in several states but showed losses in others. No clear spatial pattern of change is evident, however. There is some evidence to suggest that relatively more chiropractors are found in states where there are low doctor-to-population ratios; perhaps they "fill in" for doctors in these states. Chiropractors tend to compete with osteopaths (doctors who deal with musculoskeletal problems) over the same territory.

Integrating Traditional and Modern Medicine

In most countries, traditional (professional and nonprofessional) and biomedicine coexist. In most instances there is little cooperation between these two systems. Some Western-trained personnel feel that there is no place for traditional medicine at all. However, traditional healers are accessible to most of the world's population, and it will be a very long time before the same can be said of Western practitioners. Although integration of the two systems would be a good idea in wealthy societies, the benefits of integration would be even greater in low-income countries.

Several policy options are open to those in charge of health delivery systems that have traditional and modern components. One is to make traditional medicine illegal, as the Ivory Coast has tried to do; this is not a realistic approach. Some places, like Hong Kong and Singapore, have informally recognized traditional medicine, but healers have no legal status, and the government is only concerned that they obey the laws of medical practice. A third approach has been to pass simple legislation to license traditional healers, as Nigeria and Ghana have done. However, the license is no guarantee of good quality traditional practice, and there are few attempts at integration under this option. A fourth tactic is to gradually increase cooperation between modern and traditional practitioners. This approach is supported by the World Health Organization (WHO) and, if done intelligently, offers the best solution. It takes advantage of medical pluralism and is ecologically sound, providing system diversity and maturity.

An essential part of the integration process is a rational examination of the strengths and weaknesses of different medical systems. Western medicine, especially following World War II, has brought mortality and morbidity rates down dramatically around the world. Immunizations, antibiotic injections, and various drugs have been effective against many infectious diseases. Western medicine can also boast a systematic body of scientific knowledge, great advances in surgical techniques, and the effective use of high technology.

On the negative side, biomedicine's emphasis on cures and costly technology is not suitable to areas where prevention would solve far more health problems and where people are very poor. Most doctors trained in biomedicine, whether in their own countries or abroad, are not trained to deal with local health problems. They know very little about the cultural, political, and economic environments in which disease is experienced and help is sought. Indigenous doctors trained in Western medicine often are reluctant to serve outside of cities. Some leave their countries for more lucrative jobs in industrialized countries, where they can use the technology they have studied. In addition, the elite (which includes doctors) in many poor countries control ministries of health and perpetuate the hegemony of biomedicine; building a prestigious teaching hospital may take precedence over providing a minimum level of health care for all the people.

The main positive quality of traditional medicine is that it is part of the culture of the people it serves. Thus traditional healers can convey social and psychological benefits through sympathy for a patient's beliefs and feelings. Traditional medicine is holistic; that is, it treats body and mind and attempts to integrate the person, society, and physical environment. Some of the drugs developed by traditional healers over many centuries are very effective, and specialists like bonesetters and traditional birth attendants may be very effective.

Indigenous healers can be criticized for several possible shortcomings. Many of their herbs may be ineffective, and cures are often based on trial and error. Ignorance of proper drug dosage can be dangerous. Witchcraft and sorcery practices are potentially harmful. Western medicine is quite expensive, but indigenous healers have likewise been known to have their eye on the marketplace. Both systems attract quacks.

Successful integration of modern and traditional practice is most likely if it follows the goals of primary health care (Chapter 9). These goals include an emphasis on self-reliance and decision making at the local level, the use of paramedical personnel for lower levels of care, appropriate technology, and geographic, financial, cultural, and functional accessibility to prevention and treatment.

The following steps have been suggested as necessary to achieve integration (Good, 1979, pp. 150–151): (1) Systematically evaluate the knowledge and skills of practitioners of both systems; (2) Identify and train traditional healers as health aids for each basic spatial unit, defined by community social boundaries; (3) Identify and train traditional birth attendants; (4) Identify and use selected traditional healers as psychiatric aids; (5) Supply communities with small stocks of drugs; and (6) Establish a simple, flexible, referral system. Figure 11-3 diagrams a proposed cooperative health care system.

There is evidence that traditional healers are becoming somewhat accepted by biomedically trained people in some societies. Two examples will serve to show what can be done in small ways. In the United States, the Internal Revenue Service accepts payments to Navaho medicine men as legitimate medical expenses. In Zambia a traditional healer has made powdered milk part of his pharmacopoeia; he dissolves his own herbal remedies in the milk, and thus his patients receive the benefits of two medical systems. If integration is to succeed, however, it must do so on a large scale. China has gone farther along the route of health care integration than any other nation.

CONCLUSION

The work of medical geographers and others using geographic techniques has made a clear contribution to our understanding of accessibility and utilization, which are measures of the vital links among health care providers and consumers. The concept of distance decay, although its exact form may be difficult to determine, has played an important role. However, medical geographers have become aware of the limitations of an overemphasis on distance measures alone and have proceeded to the far more important task of determining how distance interacts with nonspatial factors in utilization and accessibility. Other geographic contributions in this area have included the concept of the spatial location of patients and providers and attempts to encompass the factor of illness in research. Finally, medical geographers have become involved in up-to-date issues such as how medical systems change over time and space and how the integration of traditional and modern medicine might be effected to improve health care delivery in many countries of the world.

REFERENCES

Anderson, J. G. (1973). Health services utilization: Framework and review. *Health Services Research, 8*, 184–199.

Annis, S. (1981). Physical access to utilization of health services in rural Guatemala. *Social Science and Medicine, 15D*, 515–523.

Basu, R. (1982). Use of emergency room facilities in a rural area: A spatial analysis. *Social Science and Medicine, 16*, 75–84.

Efird, C. (1985). *The changing geography of lay midwifery practice.* Unpublished doctoral dissertation, University of North Carolina, Chapel Hill, NC.

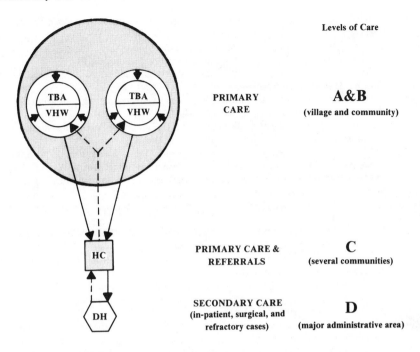

Levels of Care

PRIMARY
CARE

A&B
(village and community)

PRIMARY CARE &
REFERRALS

C
(several communities)

SECONDARY CARE
(in-patient, surgical, and
refractory cases)

D
(major administrative area)

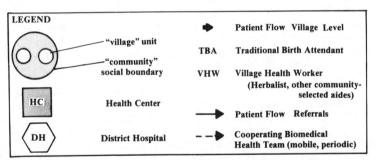

LEGEND

"village" unit

"community"
social boundary

Health Center

District Hospital

Patient Flow Village Level

TBA   Traditional Birth Attendant

VHW   Village Health Worker
(Herbalist, other community-
selected aides)

Patient Flow   Referrals

Cooperating Biomedical
Health Team (mobile, periodic)

Figure 11-3. The structure of a proposed cooperative health care system that integrates traditional and biomedical practices. From *Ethnomedical Systems in Africa* (p. 313) by C. M. Good, 1987, New York: Guilford Press. Copyright 1987 by The Guilford Press. Reprinted by permission.

Fuller, G. (1974). On the spatial diffusion of fertility decline: The distance-to-clinic variable in a Chilean community. *Economic Geography, 50,* 324–332.

Gesler, W. M. (1984). *The geography of chiropractic in the United States.* Paper presented to the Seventh Annual Conference on Applied Geography, Tallahassee, FL.

Gesler, W. M., & Cromartie, J. (1985). Patterns of illness and hospital use in Central Harlem Hospital District. *Journal of Geography, 84,* 211–216.

Girt, J. L. (1973). Distance to general medical practice and its effect on revealed ill-health in a rural environment. *Canadian Geographer, 17,* 154–166.

Good, C. M. (1979). The interface of dual systems of health care in the developing world: Toward health policy initiatives in Africa. *Social Science and Medicine, 13D,* 141–154.

Good, C. M. (1984, April). *Structure of proposed cooperative health care system.* Paper presented at the Annual Meeting of the Association of American Geographers, Washington, DC.

Gross, P. F. (1972). Urban health disorders, spatial analysis, and the economics of health facility location. *International Journal of Health Sciences, 2,* 63–84.

Jarvis, E. (1851/1852). On the supposed increase of insanity. *The American Journal of Insanity, 8,* pp. 333–364.

Joseph, A. E., & Phillips, D. R. (1984). *Accessibility and utilization: Geographical perspectives on health care delivery.* New York: Harper & Row.

Lasker, J. M. (1984). Choosing among therapies: Illness behavior in the Ivory Coast. *Social Science and Medicine, 15A,* 157–168.

McLafferty, S. (1984, April). *Analyzing the impact of hospital closure on geographical accessibility to hospital services.* Paper presented at the Annual Meeting of the Association of American Geographers, Washington, DC.

Morrill, R. L., Earickson, R., & Rees, P. (1970). Factors influencing distances traveled to hospitals. *Economic Geography, 46,* 161–171.

Phillips, D. R. (1979). Spatial variations in attendance at general practitioner services. *Social Science and Medicine, 13D,* 169–181.

Press, I. (1978). Urban folk medicine: An overview. *American Anthropologist, 80,* 71–84.

Pyle, G. F., & Lauer, B. M. (1975). Comparing spatial configurations: Hospital service areas and disease rates. *Economic Geography, 51,* 50–68.

Shannon, G. W., Bashshur, R. L., & Metzner, C. A. (1969). The concept of distance as a factor in accessibility and utilization of health care. *Medical Care Review, 26,* 143–161.

Shannon, G. W., Bashshur, R. L., & Spurlock, C. W. (1978). The search for medical care: An exploration of urban black behavior. *International Journal of Health Services, 8,* 519–530.

Stock, R. F. (1983). Distance and the utilization of health facilities in rural Nigeria. *Social Science and Medicine, 17,* 563–570.

Stribling, D. E. (1983). *Holistic health: A changing paradigm for cultural geography.* Unpublished doctoral dissertation, University of North Carolina, Chapel Hill, NC.

Weiss, J. E., & Greenlick, M. R. (1970). Determinants of medical care utilization: The effect of social class and distance on contacts with the medical care system. *Medical Care, 8,* 456–462.

Further Reading

Cohen, M. A., & Lee, H. L. (1985). The determinants of spatial distribution of hospital utilization in a region. *Medical Care, 23,* 1003–1014.

Dubos, R. (1959). *Mirage of health.* New York: Harper & Row.

Eckholm, E. (1977). *Picture of health.* New York: W.W. Norton.

Eyles, J., & Woods, K. J. (1983). *The social geography of medicine and health.* London: Croom Helm.

Gesler, W. M. (1984). *Health care in developing countries*. Resource Publications in Geography. Washington, DC: Association of American Geographers.

Good, C. M. (1977). Traditional medicine: An agenda for medical geography. *Social Science and Medicine, 11*, 705–713.

Good, C. M. (1980). Ethnomedical systems in Africa and the LDCs: Key issues for the geographer. In M. S. Meade (Ed.), *Conceptual and methodological issues in medical geography* (pp. 93–116). Chapel Hill, NC: University of North Carolina, Department of Geography.

Joseph, A. E., & Boeckh, J. L. (1981). Locational variation in mental health care utilization dependent upon diagnosis: A Canadian example. *Social Science and Medicine, 15D*, 395–404.

Luft, H. S. (1985). Regionalization of medical care. *American Journal of Public Health, 75*, 125–126.

McLafferty, S. (1982). Urban structure and geographical access to public services. *Annals of the Association of American Geographers, 72*, 347–354.

Meade, M. S. (Ed.) (1980). *Conceptual and methodological issues in medical geography*. Chapel Hill, NC: University of North Carolina, Department of Geography.

Person, P. H., Jr. (1962). Geographic variation in first admission rates to a state mental hospital. *Public Health Reports, 77*, 719–731.

Prothero, R. M. (1972). Problems of public health among pastoralists: A case study from Africa. In N. D. McGlashan (Ed.), *Medical geography: Techniques and field studies* (pp. 105–118). London: Methuen.

Pyle, G. F. (1979). *Applied medical geography*. Washington, DC: V. H. Winston.

Shannon, G. W., & Dever, G. E. A. (1974). *Health care delivery: Spatial perspectives*. New York: McGraw-Hill.

Shannon, G. W., & Spurlock, C. W. (1976). Urban ecological containers, environmental risk cells, and the use of medical services. *Economic Geography, 52*, 171–180.

Shannon, G. W., Spurlock, C. W., Gladin, S. T., & Skinner, J. L. (1975). A method for evaluating the geographic accessibility of health services. *The Professional Geographer, 27*, 30–36.

Vignette 11-1

HOSPITAL CLOSURES AND PATIENT TRAVEL DISTANCE

There are times when health officials decide to close a hospital, often for economic reasons. One of the effects of a closure is that some people will have to travel farther to attend a hospital. Without knowing exactly which people used which hospital in an area, we can roughly estimate how travel distance will be affected.

The tables and map below were provided by a medical geographer, Sara McLafferty, who investigated an actual hospital closure (McLafferty, 1984). Sydenham Hospital, located in Central Harlem, New York City, was closed in 1980. The hospital was an important source of care to local residents, many of whom were poor, had little access to physicians, and had difficulty finding transportation.

Central Harlem people reside in 20 census tracts and were served by four other hospitals (Vignette Figure 11-1). The figures in Vignette Table 11-1a show census tract populations and how far each of the five hospitals was from

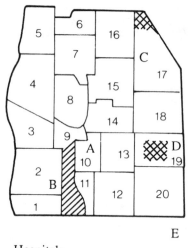

Hospitals

A Sydenham

B Saint Lukes

C Harlem

D North General

E Mount Sinai

Park

Barrier

Vignette Figure 11-1. The study area in Central Harlem. Location of 20 tracts and five hospitals. From *Analyzing the Impact of Hospital Closure on Geographical Accessibility to Hospital Services* by S. McLafferty, 1984, presented to the Annual Meeting of the Association of American Geographers, Washington, DC. Reprinted by permission.

Vignette Table 11-1a. Travel Distances (in Miles) to Hospitals in Central Harlem

Tract	1980 Population	A Sydenham	B Saint Lukes	C Harlem	D North General	E Mount Sinai
1	9,758	1.56	.45	2.68	2.28	2.00
2	12,133	.93	.43	2.30	1.90	2.34
3	11,609	.65	.95	1.93	1.78	2.90
4	17,816	1.10	1.45	1.38	2.22	3.35
5	17,472	1.65	1.95	1.48	2.78	3.90
6	5,160	1.33	1.93	1.15	2.45	3.58
7	2,655	1.03	1.68	.85	2.15	3.28
8	4,053	.53	1.33	1.15	1.65	2.78
9	3,850	.20	.65	1.58	1.33	2.45
10	3,839	.25	.95[a]	1.58	1.18	2.00
11	3,582	.70	1.00[a]	1.93	1.53	1.55
12	10,954	1.00	1.38[a]	1.68	1.25	1.25
13	6,171	.55	1.30[a]	1.18	.78	1.70
14	7,612	.48	1.33	.90	1.05	2.18
15	11,442	.93	1.78	.45	1.40	2.53
16	14,453	1.45	2.30	.53	1.83	2.95
17	15,362	1.65	2.50	.23	1.08	2.20
18	7,003	1.13	1.98	.75	.55	1.68
19	4,739	1.08	1.78[a]	1.18	.33	1.25
20	15,583	1.45	1.40[a]	1.63	.73	.80

[a]Distance computed around barrier.

All distances calculated according to the Manhattan metric, from centroid of tract to hospital.

Note. From *Analyzing the Impact of Hospital Closure on Geographical Accessibility to Hospital Services* (p. 6) by S. McLafferty, 1984, presented to the Annual Meeting of the Association of American Geographers, Washington, DC. Reprinted by permission.

the center of each tract. The assumptions (probably not entirely correct) are that Sydenham's service area consisted of all tracts whose nearest hospital was Sydenham and that, following closure, each of these tracts was served by the closest alternative hospital.

The first step is to select the tracts whose nearest hospital is Sydenham; these are shown in the first column of Vignette Table 11-1b. The populations in these tracts had to travel farther to receive health care after Sydenham's closure. The main question addressed here is, how much farther? Table 11-1b lists the populations and distance to Sydenham of the tracts in question. Column 4 calculates people-miles, or the total travel distance from Sydenham's service area to the hospital, 51,530.7 people-miles. Dividing this figure by the total service-area population, 69,486, gives an average figure of 0.74 miles that each person had to travel to Sydenham. With the hospital closed, people in the nine tracts had to travel to the next nearest hospital; column 5 records these

Vignette Table 11-1b. Calculating Extra Travel Distances following a Hospital Closure

Tract	Population (P)	Distance to Sydenham (D1)	Total Travel Distance (P × D1)	Distance to 2nd Nearest (D2)	Total Travel Distance (P × D2)
3	11,609	0.65	7,545.9	0.95	11,028.6
4	17,816	1.10	19,597.6	1.38	24,586.1
8	4,053	0.53	2,148.1	1.15	4,660.95
9	3,850	0.20	770.0	0.65	2,502.5
10	3,839	0.25	959.8	0.95	3,647.05
11	3,582	0.70	2,507.4	1.00	3,582.0
12	10,954	1.00.	10,954.0	1.25	13,692.5
13	6,171	0.55	3,394.1	0.78	4,813.4
14	7,612	0.48	3,653.8	0.90	6,850.8
Totals	69,486		51,530.7		75,363.9

Note. From *Analyzing the Impact of Hospital Closure on Geographical Accessibility to Hospital Services* (p. 7) by S. McLafferty, 1984, presented to the Annual Meeting of the Association of American Geographers, Washington, DC. Reprinted by permission.

distances. The new travel total is 75,363.9 people-miles. Dividing this figure by the same service-area population, 69,486, produces a new average travel distance of 1.08 miles. This represents a 45.95% increase in travel distance for this population because of Syndenham's closure.

Travel distance is not the only consideration in facility closures. Planners would also be concerned about, among other things, possible overcrowding in the hospitals still in use and how much money the closure would save the city.

12

Conclusion

Geographical research has two concerns: integrating all the phenomena at a point in space in order to understand the nature of place; and understanding and explaining the distribution of varied phenomena over space. This textbook describes the systems, such as biology, ecology, economics, politics, culture, meteorology, demographics, and medicine, whose interactions form the status of human health. Health-related phenomena have a certain distribution, move in certain directions at varying speeds, and affect people's perceptions about their communities and surrounding environments. Medical geographers who examine these spatial processes draw upon the concepts and techniques of all the subdisciplines of geography.

The medical geography of health care freely incorporates the concerns and findings from other disciplines about the social, economic, political, and cultural behavior of individuals and systems. In turn it contributes its geographical perspective to the emergence of a social science of health care. Similarly, epidemiological design and methodology, parasitology, entomology, microbiology, and the anthropology of medical belief, along with many other disciplinary insights, are synthesized and used in explaining the spatial distribution of disease occurrence. Medical geographers are becoming better trained in the cognate fields relevant to their specialization, as economic, political, and climatological geographers have before them.

We have discussed problems that arise in available medical statistics and in obtaining microarea data. More important is awareness that ignorance of basic and relevant biology can cause research hypotheses even about spatial form to be deficient and that ignorance of basic sociocultural processes can result in simplistic genetic or environmental explanations. Spatial perspective, knowledge about differences of scale and spatial autocorrelation, and general familiarity with physical and social sciences can result in fresh geographical insights, new hypotheses, and sounder planning and policy.

Most medical geographers recognize their medical limitations, and are eager to collaborate with other health professionals and scientists. The limitations of methodology and the limits to medical contribution have been critiqued by Mayer (1983), Stimson (1983), and others. For example, changes in types of human behavior change the structure of various spatial systems:

however, deducing process, to which behavioral change may contribute, from spatial structure is a risky enterprise, because more than one change can generate the same structure. Yet reasoning from spatial form to process is a basic tool of geographical method.

PROSPECTIVE MEDICAL GEOGRAPHY

The health care delivery system in the United States is in a state of great flux and uncertainty. Increasing numbers of people are choosing nontraditional forms of care. Competition between insurance companies and prepaid health care plans is growing in intensity. Privatization of the health care industry is progressing. Concern for the increasing cost of health care has led to the emergence of many cost-containment proposals. Will the concern for cost result in a sacrifice of quality, or of the goal of universal access to health care? New patterns of living, increased pollution and crowding, an aging population, and the continued social and geographic mobility of the population contribute to new patterns of ill health and shifts in demand for health care. Geographic integration and research from a spatial perspective can help us understand and form future patterns. As Brownlea (1981) has pointed out, this pattern of change and uncertainty represents a great opportunity for medical geographers to play a more active role in policy formation and in the maintenance (or creation) of a healthful environment.

It has been the fashion to split medical geography into two parts, one focusing on the geography of health and disease, the other on health care delivery. The research paradigms, place within geography, and cognate fields of the two were considered very different. As we hope this text illustrates, the literature of medical geography is far too complex and interrelated to be dichotomous. There is a rich diversity to the research in medical geography. Indeed, some would say that diversity is too great. They suggest that a concentration on a smaller number of problems might lead to a more rapid development of significant answers. Yet, the variety of effort has promoted hybrid vigor. Much of the enthusiasm of the field derives from the excitement of the mixing of many new ideas. Individual medical geographers may feel most comfortable with some segment of the subdiscipline, but a single medical geography exists.

It can be argued that the individual research efforts, in all their diversity, are too narrow and limiting. Often the focus is upon a point of etiology or delivery of a specific intervention. The broader health questions of the role of social institutions, the consequences of environmental management or misuse, and the impact of sociocultural roles and perception are too frequently ignored. No understanding of place can be complete without including and addressing them.

FINIS

Medical geography barely existed 25 years ago in North America. A few individuals were producing sometimes excellent work. They were laying the groundwork for the next quarter century, and much of their work has still not been superseded. Their work was limited in volume and scope, however, mostly by the small pool of active researchers with whom they could interact. Medical geographers existed, but a well-defined subdiscipline of medical geography did not.

Much has changed since then. The number of active medical geographers has doubled and redoubled. A generation who specialized at both the master's and doctoral levels in the field is now productive. Medical geography is maturing and becoming more introspective and critical. The result is an emerging body of literature sufficient in both volume and substance to create a vibrant medical geography. The existence of that body of research effort means many things. Geographers have been introduced to new sources of health-related data. Research techniques have been applied in many innovative ways. Most important, the intellectual ties that link together different projects, whether they be data sources, techniques, specific hypotheses, or a more general paradigm, help create the building blocks of learning and understanding. Medical geography has reached the point where the student can expect to draw upon the work of other medical geographers before initiating a project. It also means that a text like this, drawing from the work of dozens of individuals and scores of publications, can be written.

REFERENCES

Brownlea, A. A. (1981). From public health to political epidemiology. *Social Science and Medicine, 15D*, 57–68.

Mayer, J. D. (1983). The role of spatial analysis and geographic data in the detection of disease causation. *Social Science and Medicine, 17*, 1213–1221.

Stimson, R. J. (1983). Research design and methodological problems in the geography of health. In N. D. McGlashan & J. R. Blunden (Eds.), *Geographical aspects of health* (pp. 321–334). London: Academic Press.

Index